MOLECULAR METHODS FOR VIRUS DETECTION

MOLECULAR
METHODS FOR
VIRUS DETECTION

Edited by

Danny L. Wiedbrauk
Daniel H. Farkas

William Beaumont Hospital
Royal Oak, Michigan

ACADEMIC PRESS

San Diego   New York   Boston   London
Sydney   Tokyo   Toronto

Front cover illustration represents an electron micrograph of vesicular stomatitis virus (family Rhabdoviridae). Illustration by Mary Haddad.

This book is printed on acid-free paper. ∞

Academic Press, Inc.
A Division of Harcourt Brace & Company
525 B Street, Suite 1900, San Diego, California 92101-4495

United Kingdom Edition published by
Academic Press Limited
24-28 Oval Road, London NW1 7DX

Library of Congress Cataloging-in-Publication Data

Molecular methods for virus detection / edited by Danny L. Wiedbrauk,
    Daniel H. Farkas.
         p.   cm.
    Includes index.
    ISBN 0-12-748920-7 (case)
     1. Virus diseases--Molecular diagnosis.   2. Polymerase chain
reaction.   3. Nucleic acid probes.   I. Wiedbrauk, Danny L.
II. Farkas, Daniel H.
    RC114.5.M656   1994
    616.9' 250756-dc20   94-28603
                                                                       CIP

Printed and bound in the United Kingdom
Transferred to Digital Printing, 2011

Contents

3. Nucleic Acid Blotting Techniques for Virus Detection   39

Daniel L. Stoler and Nelson L. Michael

4. *In Situ* Hybridization   75

Jeanne Carr

Contributors

Numbers in parentheses indicate the pages on which the authors' contributions begin.

Mark S. Berninger (351), Life Technologies, Inc., Gaithersburg, Maryland 20844

Irena Bronstein (147), Tropix, Inc., Bedford, Massachusetts 01730

John D. Burczak (315),[1] Probes Diagnostic Business Unit, Abbott Laboratories, Abbott Park, Illinois 60064

Eric Buxton (193), The Immune Response Corporation, Carlsbad, California 92008

Jeanne Carr (75), Thomas F. Puckett Laboratory, Hattiesburg, Mississippi 39402

Max A. Chernesky (219), McMaster University Regional Virology and Chlamydiology Laboratory, St. Joseph's Hospital, Hamilton, Ontario, Canada L8N 4A6

Shanfun Ching (315), Probes Diagnostic Business Unit, Abbott Laboratories, Abbott Park, Illinois 60064

Anne Daigle (193), The Immune Response Corporation, Carlsbad, California 92008

M. S. Dey (329), Becton Dickinson Research Center, Research Triangle Park, North Carolina 27709

J. A. Down (329), Becton Dickinson Research Center, Research Triangle Park, North Carolina 27709

Ann M. Drevon (1), Molecular Probe Laboratory, Department of Clinical Pathology, William Beaumont Hospital, Royal Oak, Michigan 48073

[1] *Current address:* Department of Molecular Diagnostics, SmithKline Beecham Pharmaceuticals, King of Prussia, Pennsylvania 19406

Chris Duffy (193), The Immune Response Corporation, Carlsbad, California 92008

Eoin Fahy (287), Applied Genetics, San Diego, California 92121

Francois Ferre (193), The Immune Response Corporation, Carlsbad, California 92008

Soumitra S. Ghosh (287), Applied Genetics, San Diego, California 92121

Thomas R. Gingeras (287), Affymetrix, Inc., Santa Clara, California 95051

Richard L. Hodinka (103), Departments of Pathology and Pediatrics, Children's Hospital of Philadelphia; and School of Medicine, University of Pennsylvania, Philadelphia, Pennsylvania 19104

D. R. Howard (329), Becton Dickinson Research Center, Research Triangle Park, North Carolina 27709

Hsiang-Yun Hu (315), Probes Diagnostic Business Unit, Abbott Laboratories, Abbott Park, Illinois 60064

S. R. Jurgensen (329), Becton Dickinson Research Center, Research Triangle Park, North Carolina 27709

W. E. Keating (329), Becton Dickinson Research Center, Research Triangle Park, North Carolina 27709

Helen H. Lee (315), Probes Diagnostic Business Unit, Abbott Laboratories, Abbott Park, Illinois 60064

M. C. Little (329), Becton Dickinson Research Center, Research Triangle Park, North Carolina 27709

James B. Mahony (219), McMaster University Regional Virology and Chlamydiology Laboratory, St. Joseph's Hospital, Hamilton, Ontario, Canada L8N 4A6

Lawrence T. Malek (261), Cangene Corporation, Mississauga, Ontario, Canada L4V 1T4

Annie Marchese (193), The Immune Response Corporation, Carlsbad, California 92008

Bruce J. McCreedy (175), Roche Biomedical Laboratories, Research Triangle Park, North Carolina 27709

Nelson L. Michael (39), Department of Retroviral Research, Walter Reed Army Institute of Research, Rockville, Maryland 20850

J. G. Nadeau (329), Becton Dickinson Research Center, Research Triangle Park, NC 27709

V. R. Neece (329), Becton Dickinson Research Center, Research Triangle Park, North Carolina 27709

Gerard J. Nuovo (237), Department of Pathology, State University of New York at Stony Brook, Stony Brook, New York 11794

C. M. Nycz (329), Becton Dickinson Research Center, Research Triangle Park, North Carolina 27709

Corinne E. M. Olesen (147), Tropix, Inc., Bedford, Massachusetts 01730

Patrick Pezzoli (193), The Immune Response Corporation, Carlsbad, California 92008

Ayoub Rashtchian (351), Life Technologies, Inc., Gaithersburg, Maryland 20844

David M. Schuster (351), Life Technologies, Inc., Gaithersburg, Maryland 20844

Roy Sooknanan (261), Cangene Corporation, Mississauga, Ontario, Canada L4V 1T4

C. A. Spargo (329), Becton Dickinson Research Center, Research Triangle Park, North Carolina 27709

Jay Stoerker (25), Applied Technology Genetics Corporation, Malvern, Pennsylvania 19355

Daniel L. Stoler (39), Department of Molecular and Cellular Biology, Roswell Park Cancer Institute, Buffalo, New York 14263

Mickey S. Urdea (131), Nucleic Acid Systems, Chiron Corporation, Emeryville, California 94608

Bob van Gemen (261), Organon Teknika, 5281RM Boxtel, The Netherlands

G. T. Walker (329), Becton Dickinson Research Center, Research Triangle Park, North Carolina 27709

A. H. Walters (329), Becton Dickinson Research Center, Research Triangle Park, North Carolina 27709

Danny L. Wiedbrauk (1,25), Departments of Clinical Pathology and Pediatrics, William Beaumont Hospital, Royal Oak, Michigan 48073

Judith C. Wilbur (131), Nucleic Acid Systems, Chiron Corporation, Emeryville, California 94608; and Department of Laboratory Medicine, University of California, San Francisco, San Francisco, California 94143

P. Zwadyk, Jr. (329), Veterans Administration Hospital, Durham, North Carolina 27705

Patrick Perzelt (193), The Immune Response Corporation, Carlsbad, California 92008

Ayoub Rashtchian (351), Life Technologies, Inc., Gaithersburg, Maryland 20844

David M. Schuster (351), Life Technologies, Inc., Gaithersburg, Maryland 20844

Roy Sooknanan (261), Cangene Corporation, Mississauga, Ontario, Canada L4V 1T4

C. A. Spargo (329), Becton Dickinson Research Center, Research Triangle Park, North Carolina 27709

Jay Stoerker (25), Applied Technology Genetics Corporation, Malvern, Pennsylvania 19355

Daniel L. Stoler (39), Department of Molecular and Cellular Biology, Roswell Park Cancer Institute, Buffalo, New York 14263

Mickey S. Urdea (131), Nucleic Acid Systems, Chiron Corporation, Emeryville, California 94608

Bob van Gemen (261), Organon Teknika, 5281RM Boxtel, The Netherlands

G. T. Walker (329), Becton Dickinson Research Center, Research Triangle Park, North Carolina 27709

A. H. Walters (329), Becton Dickinson Research Center, Research Triangle Park, North Carolina 27709

Danny L. Wiedbrauk (1,25), Departments of Clinical Pathology and Pediatrics, William Beaumont Hospital, Royal Oak, Michigan 48073

Judith C. Wilber (131), Nucleic Acid Systems, Chiron Corporation, Emeryville, California 94608, and Department of Laboratory Medicine, University of California, San Francisco, San Francisco, California 94143

P. Zwadyk, Jr. (329), Veterans Administration Hospital, Durham, North Carolina 27705

Preface

Nucleic acid-based virus detection methods were first described in the 1970s. Since that time, molecular biologists and commercial diagnostics companies have labored to introduce these methods into the clinical laboratory. Southern, dot, and slot blots (Chapter 3) were among the first applications of the new DNA hybridization methods to be introduced. *In situ* hybridization (Chapter 4) was developed soon thereafter in response to morphologists' desires to identify individual virus-infected cells. Despite these successes, most of the early diagnostic procedures were not widely adopted in the clinical laboratory because they were usually more expensive, more labor intensive, and had longer turnaround times than existing methods. In addition, many of the early molecular diagnostic procedures used radioisotopic probes that had a number of disadvantages in the clinical laboratory. Recent improvements in signal amplification (Chapter 5), nucleic acid amplification (Chapters 8–16), and nonisotopic detection methods (Chapter 6) have significantly improved the sensitivity and specificity of molecular diagnostic procedures. More importantly, these improvements have made molecular diagnostic procedures available to a wider variety of clinical laboratories. Emerging clinical areas such as antiviral susceptibility testing (Chapter 5) have also relied on molecular diagnostic methodologies.

Of all the recent technological developments, the polymerase chain reaction (PCR) and nonisotopic detection methods have had the greatest impact on the clinical virology laboratory. This book includes four chapters (Chapters 8–11) on PCR methods (Chapter 8) and applications such as quantitative PCR (Chapter 9), multiplex PCR (Chapter 10), and PCR *in situ* hybridization (Chapter 11). Written by experts in the field, these chapters provide detailed procedures for detecting viral (or chlamydial) nucleic acids in the clinical laboratory. The two chapters on nonisotopic detection (Chapters 6 and 7) also contain procedures for identifying viral nucleic acids in clinical specimens. These nonisotopic detection systems have significantly improved turnaround times, reduced initial start-up costs, and eliminated many of the regulatory hurdles facing laboratories that perform nucleic acid-based tests.

One unique feature of this book is the presence of clinical detection protocols for seven emerging diagnostic methodologies. Written by the inventors or principal developers of the technologies, the chapters on branched chain signal amplification (Chapter 6), chemiluminescence (Chapter 7), nucleic acid sequence-based amplification (NASBA; Chapter 12), self-sustained sequence replication (3SR; Chapter 13), ligase chain reaction (LCR; Chapter 14), strand displacement amplification (SDA; Chapter 15), and ligase-activated transcription (LAT; Chapter 16) all contain diagnostic procedures suitable for the clinical laboratory. In contrast with the other chapters, the SDA chapter contains a protocol for detecting *Mycobacterium tuberculosis,* which is obviously not a virus. This protocol was included because it is representative of an entire class of emerging assays and because viral applications will soon be developed.

In this book, we have attempted to bring together a wide variety of methods that have been, or soon will be, used in the clinical virology laboratory. Chapter 1 provides an introduction to these molecular diagnostic methods while Chapter 2 describes the important elements of a quality assurance program for the molecular virology laboratory. Procedures that ensure the accuracy, precision, and reproducibility of the test results are important parts of every molecular virology program. However, the major difference between an academic research laboratory and the clinical diagnostic laboratory is the formality and pervasiveness of the quality control process. Another unique feature of this book is the incorporation of quality assurance protocols within each methodological chapter. This type of quality assurance information is often assumed or overlooked in other publications.

Our purpose in compiling this book was to assemble a representative, but by no means exhaustive, group of procedures for using molecular diagnostic procedures to detect viral infections. Several recent publications have concentrated on PCR methodologies. We did not intend to duplicate these efforts, but rather, we have gone beyond PCR to provide clinically relevant procedures for many of the newer diagnostic methodologies. This compilation of diverse protocols illustrates the methodological diversity available to laboratories that employ molecular methods for virus detection.

Danny L. Wiedbrauk
Daniel H. Farkas

Nucleic Acid Detection Methods

Danny L. Wiedbrauk
*Departments of Clinical Pathology
and Pediatrics
William Beaumont Hospital
Royal Oak, Michigan 48073*

Ann M. Drevon
*Molecular Probe Laboratory
Department of Clinical Pathology
William Beaumont Hospital
Royal Oak, Michigan 48073*

I. Introduction
II. Specimen Processing
III. Target Amplification
    A. Polymerase Chain Reaction
    B. Nucleic Acid Sequence-Based Amplification and Self-Sustained Sequence Replication
    C. Ligation-Activated Transcription
    D. Strand-Displacement Amplification
IV. Probe Amplification
    A. Ligase Chain Reaction
    B. Q-Beta Replicase System
    C. Cycling Probe Technology
V. Detection Systems
    A. Enzyme Immunoassay-Based Detection
    B. Immunochromatography (One-Step) Assays
    C. Sequence-Based Detection
    D. Signal Amplification
    E. Chemiluminescence Detection
VI. Potential Applications
VII. Difficulties and Disadvantages
VIII. Conclusions
References

I. INTRODUCTION

Viral infections are the most common cause of human disease, and are responsible for at least 60% of the illnesses that prompt patients to visit a physician (Ray, 1979). Despite the large number of viral illnesses that affect humans, diagnostic virology has only recently entered the mainstream of

1

clinical medicine. This increased use of viral diagnostics is due to the availability of an increasing number of antiviral drugs and associated rapid viral diagnostic procedures. The pace of the virology laboratory changed significantly in the 1980s with the introduction of fluorescent antibody methods, shell vial procedures, and enzyme immunoassays for viral antigen detection. The major shaping force of the 1990s promises to be the availability of molecular methods for virus detection.

Molecular diagnostic procedures have been available since the 1970s, when researchers first began using cloned DNA probes to detect viral nucleic acids. Proponents of the new molecular diagnostic methods predicted that nucleic acid tests would rapidly replace traditional virus detection methods (Kulski and Norval, 1985; Tenover, 1988). Despite these optimistic predictions, molecular diagnostic procedures were not widely adopted by clinical virology laboratories because these tests used radioisotopic detection systems, were more expensive and more labor intensive, and had unacceptably long turnaround times compared with traditional antibody-based detection methods. The first nonisotopic detection systems were exquisitely specific but were not as sensitive as traditional antibody–antigen methods (Lowe, 1986). Clearly, some type of amplification methodology was required. To date, two general amplification techniques (signal amplification and target amplification) have been used to improve the sensitivity of viral nucleic acid assays. As the name implies, signal amplification methods are designed to increase the signal-generating capability of the system without altering the number of target molecules. In contrast, target amplification procedures generate more viral nucleic acids, thereby allowing the user to employ less sensitive (and generally less expensive) signal detection methods.

Signal and target amplification systems have reduced or eliminated the need for radioisotopes, reduced the turnaround times, and simplified testing protocols. Several of these methods are discussed in this chapter.

II. SPECIMEN PROCESSING

Extraction of nucleic acids from biological materials is one of the most important steps performed in the molecular virology laboratory. High quality nucleic acids are required for most applications, but they are especially important when nucleic acid amplification and sequencing are contemplated. The purposes of specimen processing steps are (1) to make the nucleic acids available for amplification, hybridization, or detection; (2) to concentrate the nucleic acids; and (3) to remove any inhibitory substances that might be present in the specimen.

Molecular diagnostic methods can be inhibited by (1) chelating divalent cations such as Mg^{2+}, (2) degradation of the target and/or primer nucleic acids, or (3) inactivation of the enzymes used in these procedures. Several reports have demonstrated that heme (Mercier *et al.*, 1990; Ruano *et al.*, 1992), heparin (Beutler *et al.*, 1990; Holodniy *et al.*, 1991), phenol (Katcher and Schwartz, 1994), polyamines (Ahokas and Erkkila, 1993), plant polysaccharides (Demeke and Adams, 1992), and urine (Kahn *et al.*, 1991) can inhibit polymerase chain reactions (PCR). Unfortunately, little is known about the substances that inhibit other enzymes used in molecular diagnostic procedures. Specimen processing procedures must therefore remove these substances without degrading the nucleic acid target or adding any other inhibitory substances.

The most conventional methods for extracting nucleic acids from clinical specimens involve proteinase K digestion(s) followed by multiple phenol and chloroform:isoamyl alcohol (24:1) extractions. The resulting nucleic acids are precipitated in the presence of salts and cold ethanol. The DNA pellet is washed with cold 70% ethanol to remove any contaminants, dried, and dissolved in a suitable buffer system for the ensuing procedures. Although these procedures have proven useful for extracting genomic DNA from tissues, they are often too lengthy and laborious for routine use in a molecular virology laboratory that only tests spinal fluids. In addition, multiple chloroform extraction of specimens containing low copy numbers of viral DNAs can produce false negative reactions because of sample loss. The use of copious amounts of phenol in the laboratory is often undesirable because of the caustic and poisonous nature of these chemicals. The nonorganic salting-out procedures of Miller and Polesky (1988) and Buffone and Darlington (1985) provide alternatives to organic chemical extractions. In general, extraction procedures must be tailored to the individual specimen type and to the suspect agent. Viruses present in high concentrations in highly inhibitory substances (e.g., rotavirus in fecal specimens) can be extracted extensively before performing nucleic acid testing. Viruses that are present in low copy numbers [e.g., herpes simplex virus (HSV) in vitreous or spinal fluids] should be handled as little as possible to prevent nucleic acid loss.

Specimen storage time and temperatures can have a significant impact on nucleic acid recovery and the efficiency of subsequent diagnostic procedures. Lysis of red blood cells can influence PCR reactions by inhibition of *Taq* DNA polymerase by heme. Heme can also bind to and damage DNA at the elevated temperatures used in PCR reactions (Winberg, 1991). In addition, lysis of granulocytes releases proteases and nucleases that degrade viral particles and nucleic acids. Cuypers *et al.* (1992) reported a 1000- to 10,000-fold reduction in hepatitis C virus (HCV) RNA concentrations when whole blood and serum were stored at room temperature. However, degradation was even faster when whole blood was stored at 4°C, presumably because of increased granulocyte lysis at 4°C relative to room temperature. The recommended method for storing specimens for HCV testing is allowing the blood to clot, removing the serum, and storing the serum at 4°C or −20°C.

The best DNA yields and diagnostic results are achieved when the nucleic acids are extracted from fresh specimens. Once the nucleic acids are purified and precipitated in ethanol, they are stable at −20°C for years. If shipping whole blood is absolutely necessary [e.g., for human immunodeficiency virus (HIV) testing], specimens should be sent on wet ice and the nucleic acids should be extracted as soon as the samples are received in the reference laboratory (Cushwa and Medrano, 1993). Minimizing exposure of blood samples to temperatures ≥23°C is also important because of decreased DNA yields from specimens stored at these temperatures (Cushwa and Medrano, 1993).

III. TARGET AMPLIFICATION

Since the development of PCR in 1983, target (nucleic acid) amplification methods have been used by an increasing number of laboratories. In the clinical virology laboratory, target amplification methods are generally used to detect viruses that are difficult to grow in culture (e.g., human parvovirus, rubella virus, and caliciviruses) and viruses that are present in low numbers in clinical specimens (e.g., HIV and HSV in cases of suspected HSV encephalitis). Although target amplification procedures are often more complicated to use than signal amplification procedures, the commercial availability of high quality reagents, primers, and controls has made these procedures much more accessible. Test kits that use target amplification methodologies have further improved the ease of use. However, the extreme sensitivity of target amplification procedures has advantages and disadvantages. Target amplification procedures require impeccable laboratory technique because even minuscule quantities of contaminating nucleic acids can produce false positive results.

A. Polymerase Chain Reaction

PCR (Chapter 8) is an elegantly simple method for making multiple copies of a DNA sequence. Developed by researchers at the Cetus Corporation (Saiki *et al.*, 1985; Mullis and Faloona, 1987), PCR uses a thermostable DNA polymerase to produce a 2-fold amplification of target genetic material with each temperature cycle. The PCR procedure uses two oligonucleotide primers that are complementary to the nucleic acid sequences flanking the target area (Fig. 1).

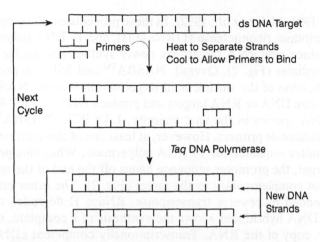

Figure 1 Polymerase chain reaction. The double-stranded DNA (ds DNA) target is heated to separate the strands. As the solution cools, the two oligonucleotide primers bind to the target DNA, and the thermostable (*Taq*) DNA polymerase extends the primers according to the nucleotide sequence of the target strand. Reprinted with permission from Wiedbrauk (1992). Copyright © 1992 by the American Society of Clinical Pathologists.

In the PCR procedure, the oligonucleotide primers are added to the reaction mixture containing the target sample, the thermostable DNA polymerase, a defined solution of salts, and excess amounts of each of the four deoxyribonucleotide triphosphates. The mixture is heated to approximately 95°C to separate the DNA strands. As the temperature is lowered to about 60°C, the primers bind to the target nucleic acids; at 72°C, the DNA polymerase extends the primers according to the sequence of the target DNA. When the reaction mixture is again heated to the strand-separation temperature, the extended primers and the original nucleic acids serve as templates for another round of DNA replication.

PCR is fast and relatively simple to perform, and can produce a 10^5- to 10^6-fold increase in target sequence in 25–30 cycles. Because PCR was the first target amplification technology to be widely used, numerous viral primer sets for this technique are commercially available.

B. Nucleic Acid Sequence-Based Amplification and Self-Sustained Sequence Replication

Nucleic acid sequence-based amplification (NASBA™; Chapter 12) and self-sustained sequence replication (3SR; Chapter 13) reactions are very similar; both these procedures are isothermal reactions that are patterned after the events that occur during retroviral transcription (Guatelli *et al.*, 1990; Fahy *et al.*, 1991; Malek *et al.*, 1992). In these procedures, the activities of reverse transcriptase, ribonuclease H (RNase H), and T7 RNA polymerase combine to produce new RNA targets via newly synthesized double-stranded DNA intermediates (Fig. 2). Overall, NASBA™ and 3SR can produce a 10^7-fold amplification of the nucleic acid target within 60 min. NASBA™ and 3SR can utilize DNA or RNA targets and produce DNA and RNA products, with the RNA species in the vast majority. Like PCR, NASBA™ and 3SR use oligonucleotide primers. However, at least one of these primers also contains a promoter sequence for T7 RNA polyermase. When this primer anneals to the target, the promoter sequence hangs off the end of the template because it is not complementary to the target (Fig. 2). The other end of the primer is extended by reverse transcriptase. RNase H degrades the RNA in the RNA:DNA hybrid and allows the synthesis of a complete, double-stranded cDNA copy of the RNA. Transcriptionally competent cDNAs are used by T7 RNA polymerase to produce 50–1000 antisense RNA copies of the original target. These antisense transcripts are converted to T7 promoter-containing double-stranded cDNA copies and used as transcription templates. This process continues in a self-sustained, cyclical fashion under isothermal conditions until components in the reaction become limited or inactivated. In this procedure, each DNA template generates not one but many RNA copies, and transcription takes place continuously without thermocycling.

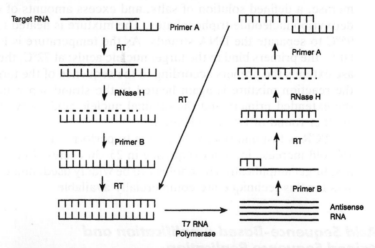

Figure 2 Self-sustained sequence replication. Primer A, which contains the promoter sequence for the T7 RNA polymerase, binds to the target RNA strand. Reverse transcriptase (RT) extends the primer according to the sequence of the target strand; then ribonuclease H (RNase H) degrades the RNA portion of the DNA:RNA hybrid molecule. After Primer B binds to the complementary DNA, RT extends the primer to make a complete, transcription-competent, double-stranded DNA intermediate. Reprinted with permission from Wiedbrauk (1992). Copyright © 1992 by the American Society of Clinical Pathologists.

C. Ligation-Activated Transcription

Ligation-activated transcription (LAT; Chapter 16) is an isothermal target amplification system that is based on the simultaneous action of four enzymes: DNA ligase, reverse transcriptase, T7 RNA polymerase, and RNase H (Rashtchian *et al.*, 1987). In this procedure, the target DNA is ligated to a partially double-stranded hairpin primer that contains the T7 RNA polymerase promoter. The single-stranded portion of the primer is complementary to the 3′ end of the target DNA. After hybridization of the primer to the target DNA, and subsequent ligation, the T7 RNA polymerase produces multiple copies of complementary RNA. The second primer binds to the RNA and reverse transcriptase produces a DNA copy of the RNA. The RNA portion of the DNA:RNA hybrid is degraded by RNase H and the free DNA strand is available to ligate to the original primer and serve as a template to produce more RNA copies. The LAT process continues in a self-sustained manner until the components become limiting. RNA can be amplified in this system by starting with the reverse transcriptase step. LAT amplification is rapid and very powerful because each ligated DNA strand can produce 50–1000 RNA copies. An overall amplification of 10^7- to 10^8-fold can usually be achieved in 3 hr.

D. Strand-Displacement Amplification

The strand displacement amplification (SDA; Chapter 15) method was developed by Walker *et al.* (1992a,b). In this procedure, a primer containing a *Hinc*II restriction site (5′-GTTGAC) binds to a complementary target nucleic acid and the primer and target are extended by an exonuclease deficient (exo⁻) Klenow fragment of DNA polymerase in the presence of dGTP, dCTP, dUTP, and a derivatized dATP that contains an alpha-thiol group (dATPαS). The resulting DNA synthesis generates a double-stranded *Hinc*II recognition site, one strand of which contains phosphorothioate linkages located 5′ to each dA residue (5′-GUC$_s$A$_s$AC) (Chapter 15). *Hinc*II nicks the recognition site between the T and the G in the sequence 5′-GTT ↓ GAC without cutting the complementary thiolated strand (5′-GUC$_s$A$_s$AC). The exo⁻ Klenow fragment initiates another round of DNA synthesis at the nick. However, the DNA located downstream of the nick is not degraded because the exo⁻ Klenow fragment lacks 5′ exonucleolytic activity. Instead, the downstream DNA fragment is displaced as the new DNA molecule is synthesized. The displacement step regenerates the *Hinc*II site, so the nicking and strand displacement steps cycle continuously until the reaction components become limited. This process can produce a new DNA target every 3 min (Chapter 15). This linear amplification scheme becomes exponential when sense and antisense primers containing *Hinc*II sites are used. This procedure doubles the number of target sequences every 3 min until the reaction components become rate limiting. With the exception of an initial boiling step to denature the nucleic acids, all SDA reactions are isothermal and are carried out at 41°C for 2 hr.

IV. PROBE AMPLIFICATION

A. Ligase Chain Reaction

Like the target amplification procedures, the ligase chain reaction (LCR) also uses target-specific oligonucleotides that are complementary to specific sequences on the target DNA (Barany, 1991a,b). However, the oligonucleotides used in LCR do not flank the target nucleic acid sequence; instead, they completely cover the target sequence (Fig. 3). One set of oligonucleotides is complementary to the left half of the target and the second set of oligonucleotides matches the right half. In the LCR procedure, high concentrations of four oligonucleotides and a thermostable DNA ligase are added to the test sample. The target DNA is denatured by heating to 94°C; as the reaction mixture cools to 65°C, the oligonucleotides anneal to the target sequences. Because both oligomers are situated adjacent to each other, DNA ligase interprets the gap between the ends of these oligonucleotides as a nick in need of repair and covalently links the oligonucleotides. The oligomers remain joined during subsequent target denaturation cycles and serve as templates for the hybridization and ligation of other oligonucleotides. LCR methods are extremely sensitive but the specificity of this procedure is relatively low because of false positive reactions caused by target-independent ligation. Gap-junction LCR methods (Chapter 14) have corrected this problem. In gap-junction LCR, the oligonucleotides anneal to the target so that a single nucleotide gap is formed between the primers. A thermostable DNA polymerase and a single deoxynucleotide triphosphate (dNTP) are included in the reaction mix to fill the gap. Target-independent ligation products cannot serve as ligation templates and are not amplified. PCR-like extension of the 5′ primer does not occur because only one dNTP is present. Once the gap has been filled, the thermostable DNA ligase joins the two oligonucleotides and the cycle continues as described.

After 10–30 cycles, the presence of the target sequence can be determined electrophoretically by the appearance of oligonucleotides that are twice their original size. In the Abbott system (Abbott Diagnostics, Abbott Park, IL), one side of the oligomeric pair is biotinylated whereas the other oligomer has a fluorescent label (Fig. 3). After amplification, the reaction mixture is

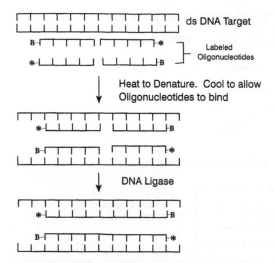

Figure 3 Ligase chain reaction. The double-stranded target DNA (dsDNA) is heated to separate the strands; as the solution cools, the labeled oligonucleotides bind to the target strands. The oligonucleotides are synthesized so they will lie next to each other on the target DNA. The thermostable DNA ligase covalently mends the nick between the two oligonucleotides. Biotin (B) and a fluorescent or enzyme label (*) can be used to capture and detect full-length oligonucleotides. Reprinted with permission from Wiedbrauk (1992). Copyright © 1992 by the American Society of Clinical Pathologists.

mixed with streptavidin-coated microparticles. The microparticles are captured in the Abbott IMx® and the free primers are removed by washing. The presence of full-length oligonucleotides results in a fluorescent signal in the instrument.

B. Q-Beta Replicase System

Gene-Trak's Q-Beta replicase (QBR; Fig. 4) system is based on the activity of an RNA-dependent RNA polymerase from the bacteriophage Qβ (Q-Beta replicase) that acts on and replicates a single-stranded Q-Beta RNA sequence called MDV-1 (Lomeli *et al.*, 1989; Klinger and Pritchard, 1990). In this procedure, an RNA sequence that is complementary to a particular target is inserted into the Q-Beta RNA strand so it does not interfere with the action of the replicase. The solution containing the target nucleic acid is heated in the presence of the RNA probe to separate any double-stranded structures; then the mixture is cooled to allow the probe to hybridize with the target sequences. After the unbound probe is removed, the Q-Beta replicase and excess ribonucleotide triphosphates are added to the amplification mixture. The immobilized RNA probe serves as a template for probe amplifi-

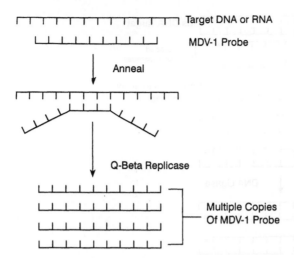

Figure 4 Q-Beta replicase system. The target DNA or RNA is heated to remove any double-stranded structures. The MDV-1 probe containing sequences complementary to the target sequence is added. The Q-Beta replicase is added after any unbound probe is removed. The resulting reaction rapidly produces multiple copies of the MDV-1 probe, which can be detected electrophoretically. Reprinted with permission from Wiedbrauk (1992). Copyright © 1992 by the American Society of Clinical Pathologists.

cation. The Q-Beta replicase system is extremely rapid and the probe can be amplified a millionfold in 10–15 min. Once the initial heating step is completed, the remainder of the reaction is isothermal and thermocyclers are not required. The main challenges of this procedure have been generation of stable RNA probes, the efficient removal of unbound probes, and the elimination of background signals.

C. Cycling Probe Technology

The cycling probe technology (CPT; Duck *et al.*, 1990) is an extremely rapid probe amplification method. Developed by ID Biomedical Corporation (Vancouver, Canada), the CPT system is an isothermal linear probe amplification system that utilizes a chimeric DNA–RNA–DNA probe (Fig. 5). In the CPT procedure, the internal RNA portion of the probe is complementary to, and hybridizes with, the target DNA. RNase H, which is present in the reaction mix, degrades the RNA of the DNA:RNA hybrid and the noncomplementary DNA portions of the probe dissociate from the target. Another intact probe segment binds to the target DNA and repeats the cycle. Detection of the probe fragments by gel electrophoresis, chemiluminescence, or fluorescence indicates that the test is positive. Cycling probe technology is

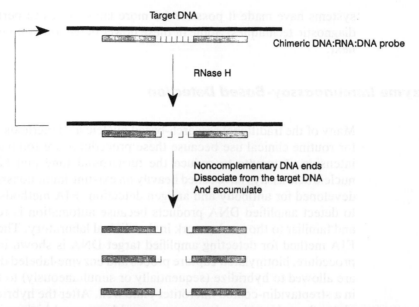

Figure 5 Cycling probe technology. This procedure utilizes a chimeric DNA:RNA:DNA hybrid that is constructed so that only the RNA portion is complementary to the target DNA sequence. After the chimeric probe binds to the DNA target, RNase H degrades the RNA portion of the DNA:RNA hybrid. The noncomplementary DNA portions of the probe dissociate from the target and accumulate in solution. The target DNA strand is then available to bind another chimeric probe.

especially interesting because probe fragments are not amplifiable and the product carryover problems associated with other amplification methods are minimized.

V. DETECTION SYSTEMS

Nucleic acid detection systems have been used since researchers first began using cloned DNA probes in the 1970s. Early clinical diagnostic procedures depended almost exclusively on the Southern blot (Southern, 1975) and on the related dot and slot bot methods (Chapter 3). Although still an important research and confirmatory testing tool, the Southern blot is being replaced by enzyme immunoassay (EIA), chemiluminescence, and other signal amplification systems. In addition, target and probe amplification systems have progressed to such an extent that relatively insensitive agarose gel electrophoresis methods can be used to detect the presence of amplified target nucleic acids. These advances and improvements in nonisotopic detection systems have made it possible for more laboratories to perform molecular diagnostic techniques. Some of these newer detection systems are discussed next.

A. Enzyme Immunoassay-Based Detection

Many of the traditional research-based nucleic acid methods are not suitable for routine clinical use because these procedures are too lengthy and labor intensive. Attempts to reduce the turnaround time and labor content of nucleic acid assays have relied heavily on existing immunoassay technologies developed for antibody and antigen detection. EIA methods are often used to detect amplified DNA products because automation is readily available and familiar to those who work in the clinical laboratory. The most common EIA method for detecting amplified target DNA is shown in Fig. 6. In this procedure, biotinylated capture probes and enzyme-labeled detection probes are allowed to hybridize (sequentially or simultaneously) to the target DNA in a streptavidin-coated microtiter plate well. After the hybridization is complete, the unbound probes are removed by several high stringency washes using a standard microtiter plate washer. An appropriate chromogen/substrate is added to the wells and the absorbance of the solution is measured using a standard microtiter plate spectrophotometer. In this procedure, the concentration of target DNA is usually proportional to the final absorbance of the chromogen/substrate solution.

EIA detection systems are sensitive, relatively fast, and easy to perform. EIA systems that use nucleic acid probes are very specific; primer dimers and other nonspecific amplification products are not detected. Several procedural

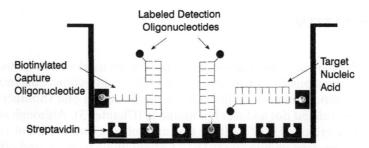

Figure 6 Enzyme immunoassay (EIA) based detection. Biotinylated capture oligonucleotides and enzyme-labeled detector oligonucleotides are allowed to hybridize with the target DNA in a streptavidin-coated microtiter plate well. After several stringency washes, an appropriate chromogen/substrate is added and the absorbance of the solution is measured with a microtiter plate reader.

modifications have been described for different solid supports (Chapter 13) or detection systems (e.g., fluorescence or chemiluminescence).

Another interesting modification of the EIA procedure has been introduced by Digene Diagnostics, Inc. (Silver Springs, MD) for use with their RNA probe systems. In this system, DNA amplification occurs using one 5'-biotinylated primer and one normal primer. The amplified target DNA is denatured and allowed to hybridize to an unlabeled RNA probe. The hybridization mix is transferred to a streptavidin-coated microtiter plate and the DNA:RNA hybrids are captured. The unbound materials are removed by washing and the DNA:RNA hybrids are detected using a unique enzyme-labeled monoclonal antibody that is specific for DNA:RNA hybrids. The chromogen/substrate is added to the well after several washes. A colorimetric signal is developed and read on a conventional microtiter plate reader. This procedure is very sensitive because multiple enzyme-labeled antibodies bind to each captured hybrid. This assay can detect as few as 10 target copies; the resulting signal is proportional to the number of copies of the original target nucleic acid present in the specimen (Bukh *et al.,* 1992).

B. Immunochromatography (One-Step) Assays

The immunochromatography, or one-step, assay for the detection of amplified DNA has been modified from traditional immunoassay methods that are used for over-the-counter pregnancy tests. Although these 5-min immunochromatography assays are not as sensitive as 3-hr EIA or radioimmunoassay (RIA) tests, they have more than enough sensitivity to detect amplified DNA products. The immunochromatography assay utilizes digoxigenin-labeled primers that are extended during nucleic acid amplification. After amplification, the nucleic acids are denatured and allowed to hybridize with biotin-labeled sense and antisense capture probes. The hybridization cocktail is then added to a cassette containing a membrane strip with an absorbent pad at the far end (Figs. 7 and 8). The hybridization cocktail flows across the membrane toward the absorbent pad. Several bands of reagents are printed on the membrane, perpendicular to the direction of the fluid flow. Streptavidin-coated colored latex or colloidal gold particles are printed in the band nearest the sample addition point. As the hybridization cocktail flows over the colored particles, the particles are rehydrated and mixed with sample. The biotinylated capture probes bind to the microparticles (Fig. 7). The fluid then flows over a band containing immobilized digoxigenin antibodies. Digoxigenin-labeled DNA is immobilized at this band. If the hybridization mix contains the appropriate DNA, the colored microparticles accumulate at the anti-digoxigenin band and at the biotin band. If the appropriate DNA is absent, the microparticles accumulate only at the biotin band. In a modifi-

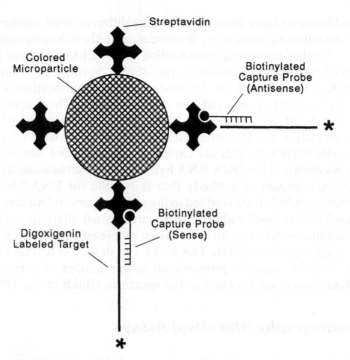

Figure 7 Immunochromatography—first step. The reaction mixture containing digoxigenin-labeled amplicons (target nucleic acids) is allowed to hybridize to biotinylated sense and antisense capture probes. The reaction mixture is added to the detection cassette. As the hybridization mixture flows over the band containing the streptavidin-labeled colored microparticles, the particles are rehydrated and mixed with the sample. The capture probes bind to the microparticles to produce the complexes shown.

cation of this procedure, the bands can be arranged perpendicularly to each other in a cross formation where the biotin band is in the horizontal plane and the anti-digoxigenin band is in the vertical plane. In this way, negative specimens will produce a minus (−) pattern and positive specimens will produce a plus (+) pattern. Immunochromatography is an extremely fast (<5 min) homogeneous immunoassay system that can provide qualitative results for any number of different nucleic acid amplifications. One caveat for this system is that colloidal gold must be used with formamide-based hybridization mixtures; latex cannot be used because formamide dissolves latex microparticles.

C. Sequence-Based Detection

The term "sequence-based detection" (SBD) is somewhat of a misnomer because all hybridization reactions are sequence based. In this chapter,

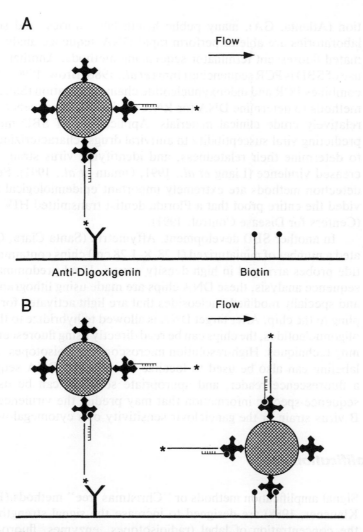

Figure 8 Immunochromatography—detection steps. As the hybridization mixture flows over the band containing immobilized antibodies to digoxigenin (anti-digoxigenin), the colored microparticles will accumulate at the anti-digoxigenin band if digoxigenin-labeled DNA is present (A). The band containing immobilized biotin serves as a positive control band because it will capture microparticles with or without the digoxigenin-labeled nucleic acids (B). The presence of two colored bands indicates the presence of the specific nucleic acid target whereas the presence of a single band indicates the absence of the target. Tests with no bands are invalid.

however, SBD refers to the identification of viruses and their phenotypes (e.g., susceptibility to antiviral drugs, virulence, etc.) based on DNA sequence analysis (Leitner *et al.*, 1993). SBD has benefitted greatly from methods developed for the Human Genome Project; currently, some specialized laboratories including those at the Centers for Disease Control and Prevention (Atlanta, GA), many public health laboratories, and some reference laboratories are able to perform rapid DNA sequence analyses using automated fluorescent terminator sequencing methods. Another advance in the use of SBD is PCR sequencing (Innis *et al.*, 1988; Brow, 1990). This procedure combines PCR and dideoxynucleotide chain termination (Sanger *et al.*, 1977) methods to determine DNA (or RNA via RT-PCR) sequences directly from relatively crude clinical materials. Applications of SBD methods include predicting viral susceptibility to antiviral drugs, characterizing virus isolates to determine their relatedness, and identifying virus strains that have increased virulence (Liang *et al.*, 1991; Omata *et al.*, 1991). Sequence-based detection methods are extremely important epidemiological tools and provided the entire proof that a Florida dentist transmitted HIV to his patients (Centers for Disease Control, 1991).

In another SBD development, Affymetrix (Santa Clara, CA) has generated a number of miniaturized (1.28 × 1.28 cm) chips containing oligonucleotide probes arranged in high density arrays. Used predominantly for DNA sequence analysis, these DNA chips are made using lithographic techniques and specially modified nucleosides that are light activated for chemical coupling to the chip. After target DNA is allowed to hybridize to the immobilized oligonucleotides, the chips can be read directly using fluorescence microscanning techniques. High-resolution microscopy, radioisotopes, and enzymatic labeling can also be used to localize the hybrids. DNA sequencing chips, a fluorescence reader, and appropriate software can be used to provide sequence-specific information that may predict the virulence of a hepatitis B virus strain or the ganciclovir sensitivity of a cytomegalovirus isolate.

D. Signal Amplification

Signal amplification methods or "Christmas tree" methods (Fahrlander and Klausner, 1988) are designed to increase the signal strength by increasing the concentration of label (radioisotopes, enzymes, fluorochromes, etc.) attached to the target nucleic acid. Numerous Christmas tree approaches have been used to detect viral nucleic acids (Fig. 9), the simplest of which involves attaching multiple labels to each probe (Fig. 9A). However, early protocols using multiple enzyme labels produced unpredictable results, presumably due to steric hindrance problems. These problems were eliminated when spacer elements were inserted between the enzyme labels. The use of spacer elements and multiple enzyme labels allowed each probe to generate a greater signal than probes containing a single label. A different Christmas tree method has been used by Orion Diagnostica (Espoo, Finland). This method employs several short probes that are complementary to different regions of the target (Fig. 9B). Multiple probe systems have significantly improved the signal strength over single probe systems.

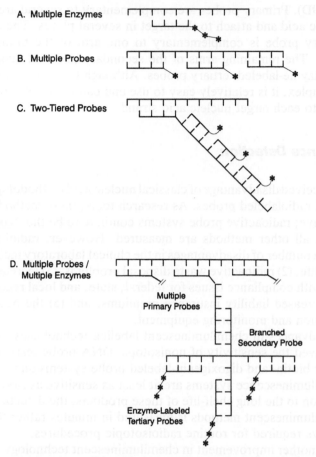

Figure 9 Signal amplification methods. Reprinted with permission from Wiedbrauk (1992). Copyright © 1992 by the American Society of Clinical Pathologists.

ImClone Systems (New York, NY) developed a two-tiered probe system consisting of an unlabeled primary probe and a series of enzyme-labeled secondary probes (Fig. 9C). The primary probes are longer than usual and sequences on only one end of the probe are complementary to the target nucleic acids. The remainder of the probe is used to bind target-independent secondary probes. This two-tiered method can produce a significant improvement in signal strength, especially when multiple enzymes are attached to the secondary probes.

Chiron Corporation (Emeryville, CA) produces a multiple probe/multiple enzyme system (Urdea *et al.*, 1987,1990) that is one of the most powerful signal amplification systems described to date (Chapter 6). This intricate network of oligonucleotide fragments consists of a series of primary probes, a novel branched secondary probe, and short enzyme-labeled tertiary probes

(Fig. 9D). Primary probes are complementary to several areas on the target nucleic acid and attach to the target in several places. The distal end of the primary probe is complementary to one arm of the branched secondary probe. The remaining arms of the secondary probe are complementary to the enzyme-labeled tertiary probes. Although this multitiered probe system is complex, it is relatively easy to use and can attach 60–300 enzyme molecules to each target nucleic acid strand.

E. Chemiluminescence Detection

A perceived disadvantage of classical nucleic acid methodologies involves the use of radiolabeled probes. As research tools, these methods are extremely sensitive; radioactive probe systems continue to be the "gold standard" by which all other methods are measured. However, radioisotopic methods have a number of disadvantages in the clinical laboratory including (1) limited shelf-life, (2) radioactive waste disposal problems, (3) increased costs associated with compliance issues for federal, state, and local regulatory agencies, (4) increased liability insurance premiums, and (5) the need for expensive detection and monitoring equipment.

Advances in chemiluminescent labeling technologies have significantly improved the sensitivity of nonisotopic DNA probe tests (Chapter 7). The newer biotin- and digoxigenin-labeled probe systems and enzyme-activated chemiluminescence systems are at least as sensitive as isotopic methods. In addition to the long shelf-life of these products, the detection time for these chemiluminescent methods is measured in minutes rather than in the hours or days required for routine radioisotopic procedures.

Another improvement in chemiluminescent technology is the hybridization protection assay produced by Gen-Probe (San Diego, CA). This assay uses highly chemiluminescent acridium ester labels that are covalently attached to oligonucleotide probes via an pH-sensitive ether bond (Arnold et al., 1989). Once the esters are hydrolyzed, the label becomes permanently nonluminescent. However, probes that are bound to target nucleic acids are protected from hydrolysis and retain their chemiluminescence. The amount of chemiluminescence produced in the hybridization protection assay is proportional to the amount of probe–target hybrid formed.

Of all the technological developments in nucleic acid detection, the availability of practical chemiluminescence detection systems has had the greatest impact on the clinical laboratory. Chemiluminescence detection systems have significantly lowered the emotional hurdles ("We can't do that here") facing laboratories that want to begin offering nucleic acid-based testing. In addition, chemiluminescence methods have significantly improved turnaround times, reduced initial start-up costs, and eliminated many of the regulatory hurdles facing laboratories that offer nucleic acid-based tests.

VI. POTENTIAL APPLICATIONS

In his 1988 review, Tenover stated that the goal of DNA probe technology was to eliminate the need for routine viral, bacterial, and fungal cultures (Tenover, 1988). Although this goal could eventually be reached, the principal advantage of molecular diagnostic methods in current clinical virology laboratories is in the detection of nonculturable agents such as human papilloma virus, human parvovirus, astroviruses, caliciviruses, hepatitis B virus, and hepatitis C virus. Molecular methods are also valuable for detecting viruses that are difficult to culture, including enteric adenoviruses, some coxsackie A viruses, and hantavirus. Indeed, PCR methods played a significant role in confirming the presence of hantavirus in the fatal respiratory disease outbreak in the Four Corners region of the southwestern United States (Centers for Disease Control, 1993), and rapid sequencing methods helped establish that the suspect agent was a new hantavirus. Molecular diversity studies helped establish that this hantavirus had been in the southwestern United States for a long time and provided important epidemiological evidence that the United States was not facing an epidemic caused by a single virus that was spreading across the country.

Molecular diagnostic methods are especially useful when trying to detect viruses that are dangerous to culture, such as HIV. PCR is the method of choice for detecting HIV infections in neonates born to HIV-infected mothers. Molecular methods are also useful when trying to determine the HIV status of patients with unusual antibody reactivities (e.g., HIV antibody positive with only the p24 band present on the Western blot). We have also used molecular methods to test for HIV in needles that children found at the beach or in a parking lot.

DNA amplification methods can assist laboratories in detecting viruses that are present in low numbers, for example, HIV in antibody-negative patients or cytomegalovirus in transplanted organs. We have used molecular diagnostic methods to detect HSV in culture- and antibody-negative cerebrospinal fluids from patients with biopsy-proven HSV encephalitis. Molecular diagnostic methods are also important when a tiny volume of specimen is available (e.g., forensic samples or intra-ocular fluid specimens). For example, we routinely perform five PCR tests (HSV, cytomegalovirus, varicella-zoster virus, Epstein–Barr virus, and human herpesvirus 6) on a single 100-μl intra-ocular fluid specimen. This specimen volume is barely sufficient for a single culture procedure.

Molecular diagnostic methods allow the laboratory to predict antiviral drug susceptibilities (Chapter 5) and to detect infections when viable virus cannot be obtained (e.g., latent viral infections or viruses that are present in immune complexes or environmental samples). Molecular diagnostic methods may also be used to differentiate antigenically similar viruses such as adenoviruses types 40 and 41 and to detect viral genotypes that are associated with human cancers (e.g., human papilloma virus). Molecular epidemiological techniques have been used to identify point sources for hospital- and community-based virus outbreaks, and have been used to predict viral virulence (Liang *et al.*, 1991; Omata *et al.*, 1991).

Overall, the potential applications for molecular diagnostic procedures appear to be limitless. The most critical near-term application of molecular diagnostic methods is the detection of fastidious viruses that grow poorly or not at all in cell cultures. In the not so distant future, molecular virology procedures will become more widespread as more and more antiviral drugs are released by the Food and Drug Administration. In the long run, Tenover's predictions may prove to be correct.

VII. DIFFICULTIES AND DISADVANTAGES

Although researchers have been using molecular diagnostic methods to detect viral nucleic acids in clinical specimens for nearly 30 years, the transition from the research laboratory to the clinical laboratory has been slow and painful. Early nucleic acid hybridization tests were more expensive and more labor intensive, and had unacceptably long turnaround times compared with existing antibody methods. In addition, these tests often used radiolabeled probes, which were disadvantageous in the clinical virology laboratory because of limited shelf-life and radioactive waste disposal problems. The cost of additional equipment and the increased space needed to separate pre- from postamplification procedures have also hindered the introduction of molecular methods into some clinical laboratories.

Although diagnostics manufacturers have provided exquisitely sensitive and specific solutions to these problems, many laboratories still do not use nucleic acid tests because high reagent costs and low reimbursement rates make them unprofitable. Clearly, the reagent and labor costs for molecular diagnostic methods must be lowered or these tests may never become widely accepted in the clinical laboratory.

Another disadvantage of nucleic acid detection is that these tests cannot detect unsuspected agents. Current organism-specific nucleic acid detection methods assume that the physician knows exactly which virus is causing the disease. This assumption is not true in most hospitals; in our laboratory, HSV is isolated from 30% of all varicella-zoster virus cultures. Dual infections are also a problem for nucleic acid methods because dual infections will not be detected unless the laboratory is specifically instructed to look for both viruses. In our laboratory, about 2% of all positive respiratory specimens contain more than one virus and significantly more specimens contain both bacterial and viral agents. In the future, the most useful molecular diagnostic tests will be those that can simultaneously test for more than one agent (see Chapter 11). Nucleic acid detection methods also have difficulty detecting new viruses that come into the community; exclusive use of molecular methods would overlook these infections. One could argue that diagnosis of the hantavirus outbreak in the Four Corners region of the southwestern United States (Centers for Disease Control, 1993) refutes this observation. However, the first laboratory evidence of hantavirus infection came from the serology laboratory, not the molecular biology laboratory (Le Guenno, 1993). Once the suspect agent was established, PCR methods were able to quickly confirm the diagnosis and determine that the infectious agent was a new hantavirus.

"Turf wars" associated with molecular diagnostic procedures have already become a reality in many departments (Diamandis, 1993; Farkas, 1994), demonstrated by increasing tensions between technology-oriented laboratories and discipline-oriented laboratories. This struggle between centralized testing advocates and decentralized testing advocates is not new. The same struggle occurred when RIA and EIA methods were introduced 20 years ago. Proponents of centralized testing point to the relative scarcity of trained personnel and the increased costs associated with equipment duplication as justification for centralized testing in their laboratory. However, the limited personnel in these laboratories and the lack of support by established discipline-oriented laboratories mean that the development and implementation of molecular methods will be slow. Decentralized molecular biology testing in discipline-oriented laboratories such as those focused on microbiology and virology has a number of advantages, because these laboratories have long-standing experience with infectious agents and the physicians who order tests for them. Microbiology and virology laboratories are methodologically diverse and as such, they are better equipped to handle indeterminate test results and coordinated quality control programs because they have both the infectious agents and the "gold-standard" methods. Finally, discipline-oriented laboratories are more likely to encourage the development and implementation of molecular diagnostic techniques if the testing volume stays within their laboratory rather than going to another laboratory.

An alternative implementation model is being employed at William Beaumont Hospital's Molecular Probe Laboratory (Royal Oak, Michigan). The Molecular Probe Laboratory serves as a core facility for the development and validation of new test methods that are transferred to the appropriate discipline-oriented laboratories. In addition to performing clinical testing, the Molecular Probe Laboratory provides a service to the other laboratories in the department by training technicians and assisting with assay troubleshooting.

VIII. CONCLUSIONS

Despite the relatively slow start, the potential of molecular diagnostics remains undiminished. Molecular methods can be used to detect viruses that are difficult to cultivate in cell culture and, with the help of the amplification methods described in the chapters in this book, molecular methods allow laboratories to detect viruses that are present in low numbers in clinical specimens. However, molecular diagnostic methods must be simpler, faster, and less expensive before their full potential is realized. Once molecular probe assays become more efficient and cost effective, they will certainly live up to the expectations of the last two decades.

ACKNOWLEDGMENTS

The work described in this chapter was supported by Grant RI-93-04 from the William Beaumont Hospital Research Institute, Royal Oak, Michigan.

REFERENCES

Ahokas, H., and Erkkila, M. J. (1993). Interference of PCR amplification by the polyamines, spermine and spermidine. *PCR Meth. Appl.* **3,** 65–68.

Arnold, L. J., Hammond, P. W., Wiese, W. A., and Nelson, N. C. (1989). Assay formats involving acridinium-ester-labeled probes. *Clin. Chem.* **35,** 1588–1594.

Barany, F. (1991a). Genetic disease detection and DNA amplification using cloned thermostable ligase. *Proc. Natl. Acad. Sci. USA* **88,** 189–193.

Barany, F. (1991b). The ligase chain reaction in a PCR world. *PCR Meth. Appl.* **1,** 5–16.

Beutler, E., Gelbart, T., and Kuhl, W. (1990). Interference of heparin with the polymerase chain reaction. *BioTechniques* **9,** 166.

Brow, M. A. D. (1990). Sequencing with *Taq* polymerase. *In* "PCR Protocols: A Guide to Methods and Applications" (M. A. Innis, D. H. Gelfand, J. J. Sninsky, and T. J. White, eds.), pp. 189–196. Academic Press, San Diego.

Buffone, G., and Darlington, G. (1985). Isolation of DNA from biological specimens without extraction with phenol. *Clin. Chem.* **31,** 164–165.

Bukh, J., Purcell, R. H., and Miller, R. H. (1992). Importance of primer selection for the detection of hepatitis C virus RNA with the polymerase chain reaction assay. *Proc. Natl. Acad. Sci. USA* **89,** 187–191.

Centers for Disease Control (1991). Update: Transmission of HIV infection during an invasive dental procedure—Florida. *Morbid. Mortal. Wkly. Rep.* **40,** 21–33.

Centers for Disease Control (1993). Outbreak of acute illness—Southwestern United States. *Morbid. Mortal. Wkly. Rep.* **42,** 421–423.

Cushwa, W. T., and Medrano, J. F. (1993). Effects of blood storage time and temperature on DNA yield and quality. *BioTechniques* **14,** 204–207.

Cuypers, H. T. M., Bresters, D., Winkel, N., Reesink, H. W., Weiner, A. J., Houghton, M., van der Poel, C. L., and Lelie, P. N. (1992). Storage conditions of blood samples and primer selection affect the yield of cDNA polymerase chain reaction products of hepatitis C virus. *J. Clin. Microbiol.* **30,** 3220–3224.

Demeke, T., and Adams, R. (1992). The effects of plant polysaccharides and buffer additives on PCR. *BioTechniques* **12,** 333–334.

Diamandis, E. (1993). The role of clinical chemistry in molecular diagnostics. *Clin. Chem. News* **19,** 4.

Duck, G., Alvarado-Urbina, G., Burdick, B., and Collier, B. (1990). Probe amplifier system based on chimeric cycling oligonucleotides. *BioTechniques* **9,** 142–147.

Fahrlander, P. D., and Klausner, A. (1988). Amplifying DNA probe signals: A Christmas tree approach. *Biotechnology* **6,** 1165–1168.

Fahy, E., Kwoh, D. Y., and Gingeras, T. R. (1991). Self-sustained sequence replication (3SR): An isothermal transcription-based amplification system alternative to PCR. *PCR Meth. Appl.* **1,** 25–33.

Farkas, D. H. (1994). Molecular diagnostic testing in 1994 and a glimpse at the future. *Clin. Chem. News* **20,** 4.

Guatelli, J. C., Whitfield, K. M., Kwoh, D. Y., Barringer, K. J., Richman, D. D., and Gingeras, T. R. (1990). Isothermal, *in vitro* amplification of nucleic acids by a multienzyme reaction modeled after retroviral replication. *Proc. Natl. Acad. Sci. USA* **87,** 1874–1878.

Holodniy, M., Kim, S., Katzenstein, D., Konrad, M., Groves, E., and Merigan, T. C. (1991). Inhibition of human immodeficiency virus gene amplification by heparin. *J. Clin. Microb.* **29,** 676–679.

Innis, M. A., Myambo, K. B., Gelfand, D. H., and Brow, M. A. D. (1988). DNA sequencing with *Thermus aquaticus* DNA polymerase and direct sequencing of polymerase chain reaction-amplified DNA. *Proc. Natl. Acad. Sci. USA* **85,** 9436–9440.

Katcher, H. L., and Schwartz, I. (1994). A distinctive property of *Tth* DNA polymerase: Enzymatic amplification in the presence of phenol. *Biotechniques* **16,** 84–92.

Khan, G., Kangro, H. O., Coates, P. J., and Heath, R. B. (1991). Inhibitory effects of urine on the polymerase chain reaction of cytomegalovirus DNA. *J. Clin. Pathol.* **44,** 360–365.

Klinger, J. D., and Pritchard, C. G. (1990). Amplified probe-based assays—Possibilities and challenges in clinical microbiology. *Clin. Microbiol. Newsl.* **12,** 133–135.

Kulski, J. K., and Norval, M. (1985). Nucleic acid probes in diagnosis of viral diseases of man. Brief review. *Arch. Virol.* **83,** 3–15.

Le Guenno, B. (1993). Identifying hantavirus associated with acute respiratory illness: A PCR victory? *Lancet* **342,** 1438–1439.

Leitner, T., Halapi, E., Scarlatti, G., Rossi, D., Albert, J., Fenyo, E. M., and Uhler, M. (1993). Analysis of heterogeneous viral populations by direct DNA sequencing. *BioTechniques* **15,** 120–127.

Liang, T. J., Hasegawa, K., Rimon, N., Wands, J. R., and Ben-Porath, E. (1991). A hepatitis B virus mutant associated with an epidemic of fulminant hepatitis. *N. Engl. J. Med.* **324,** 1705–1709.

Lomeli, H., Tyngis, S., Pritchard, C. G., Lizard, P., and Kramer, F. R. (1989). Use of an immobilized target molecule with Qβ replicase. *Clin. Chem.* **35,** 1826–1831.

Lowe, J. B. (1986). Clinical applications of gene probes in human genetic disease, malignancy, and infectious disease. *Clin. Chim. Acta* **157,** 1–32.

Malek, L. T., Davey, C., Henderson, G., and Sooknanan, R. (1992). Enhanced nucleic acid amplification process. United States Patent #5,130,238.

Miller, D., and Polesky, H. (1988). A salting-out procedure for extracting DNA from human nucleated cells. *Nucleic Acids Res.* **16,** 1215.

Mercier, B., Gaucher, C., Feugeas, O., and Mazurier, C. (1990). Direct PCR from whole blood, without DNA extraction. *Nucleic Acids Res.* **18,** 5908.

Mullis, K. B., and Faloona, F. A. (1987). Specific synthesis of DNA *in vitro* via a polymerase-catalyzed chain reaction. *Meth. Enzymol.* **155,** 335–350.

Omata, M., Ehata, T., Yokosuka, O., Hosoda, K., and Ohto, M. (1991). Mutations in the precore region of hepatitis B virus DNA in patients with fulminant and severe hepatitis. *N. Engl. J. Med.* **324,** 699–704.

Rashtchian, A., Eldrege, J., Ottaviani, M., Abbott, M., Mock, G., Lovern, D., Klinger, J., and Parsons, G. (1987). Immunological capture of nucleic acid hybrids and application to nonradioactive DNA prove assays. *Clin. Chem.* **33,** 1526–1530.

Ray, C. G. (1979). An infectious disease viewpoint of diagnostic virology. *In* "Diagnosis of Viral Infections: The Role of the Clinical Laboratory" (D. Lennette, S. Specter, and K. Thompson, eds.), pp. 241–248. University Park Press, Baltimore.

Ruano, G., Pagliaro, E. M., Schwartz, T. R., Lamy, K., Messina, D., Gaensslen, R. E., and Lee, H. C. (1992). Heat-soaked PCR: An efficient method for DNA amplification with applications to forensic analysis. *BioTechniques* **13,** 266–274.

Saiki, R. K., Scharf, S., Faloona, F. A., Mullis, K. B., Horn, G. T., Erlich, H. A., and Arnheim, N. (1985). Enzymatic amplification of β-globin genomic sequences and restriction site analysis for diagnosis of sickle cell anemia. *Science* **230,** 1350–1354.

Sanger, F., Nicklen, S., and Coulson, A. R. (1977). DNA sequencing with chain-terminating inhibitors. *Proc. Natl. Acad. Sci. USA* **74,** 5463–5467.

Southern, E. M. (1975). Detection of specific sequences among DNA fragments separated by gel electrophoresis. *J. Mol. Biol.* **98,** 503–517.

Tenover, F. C. (1988). Diagnostic deoxyribonucleic acid probes for infectious diseases. *Clin. Microbiol. Rev.* **1,** 82–101.

Urdea, M. S., Running, J. A., Horn, T., Clyne, J., Ku, L., and Warner, B. D. (1987). A novel method for the rapid detection of specific nucleotide sequences in crude biologic samples without blotting or radioactivity: Application to the analysis of hepatitis B in serum. *Gene* **61,** 253–264.

Urdea, M. S., Warner, B., Running, J. A., Kolberg, J. A., Clyne, J. M., Sanchez-Pescador, R., and Horn, T. (1990). Nucleic acid multimer for hybridization assays. WO Patent #8903891.

Walker, G. T., Little, M. C., Nadeau, J. G., and Shank, D. D. (1992a). Isothermal *in vitro* amplification of DNA by a restriction enzyme/DNA polymerase system. *Proc. Natl. Acad. Sci. USA* **89,** 392–396.

Walker, G. T., Fraiser, M. S., Schram, J. L., Little, M. C., Nadeau, J. G., and Malinowski, D. P. (1992b). Strand displacement amplification—In isothermal, *in vitro* DNA amplification technique. *Nucleic Acids Res.* **20,** 1691–1696.

Wiedbrauk, D. L. (1992). Molecular methods for virus detection. *Lab. Med.* **23,** 737–742.

Winberg, G. (1991). A rapid method for preparing DNA from blood, suited for PCR screening on transgenes in mice. *PCR Meth. Appl.* **1,** 72–74.

2

Quality Assurance in the Molecular Virology Laboratory

Danny L. Wiedbrauk
Departments of Clinical Pathology
and Pediatrics
William Beaumont Hospital
Royal Oak, Michigan 48073

Jay Stoerker
Applied Technology
Genetics Corporation
Malvern, Pennsylvania 19355

I. INTRODUCTION

The fields of molecular biology and virology have been interdependent for more than four decades. Molecular methodologies have transformed the study of viruses, viral pathogenesis, and viral diagnostics. Molecular biology itself has been changed by discoveries in the field of virology. Early viral studies helped establish the central dogma of molecular biology: genetic information residing in DNA is used to produce messenger RNA, which in turn is used to make specific proteins. The discovery of reverse transcriptase in two separate viral systems also served to disrupt the central dogma. Today, the molecular biologist uses a toolbox filled with viral artifices, from ligases to reverse transcriptases, DNA modifying enzymes to powerful transcriptional enhancers. It is not surprising therefore, that molecular methodologies are being used to diagnose viral illnesses.

Quality assurance programs constitute an integral part of any diagnostic laboratory and can account for 10–25% of the laboratory's consumable supply costs (Cembrowski and Carey, 1989). Because of the exquisite sensitivity of nucleic acid amplification procedures, quality assurance programs in the molecular virology laboratory are more complicated (and expensive) than those used in other laboratories. The important elements of the molecular virology laboratory's quality assurance program are included in this chapter. However, quality assurance programs are not static, and they will continue to evolve as newer technologies move into the diagnostic laboratory.

II. SPECIMEN QUALITY

Specimen quality is the foundation on which all other procedures are built. Collecting the correct specimen by an appropriate method at the proper time during infection significantly enhances the probability that the laboratory will detect the virus, and improves the clinical relevance of the laboratory result (Wiedbrauk and Johnston, 1993). Poor quality specimens can undermine the value of even the best molecular diagnostic procedures. Because specimen collection is so critical, laboratories must establish specimen handling and adequacy standards that must be met before molecular diagnostic procedures are performed. Laboratories must also assume responsibility for training clinicians, nursing personnel, and laboratory technicians in the proper collection and handling of these specimens.

Specimen storage time and temperatures also have a significant impact on nucleic acid recovery and the efficiency of subsequent diagnostic procedures. Red blood cell lysis can influence polymerase chain reactions (PCR) by inhibition of *Taq* DNA polymerase by heme. Heme can also damage DNA at the elevated temperatures used in PCR (Winberg, 1991). In addition, granulocyte lysis can release proteases and nucleases that degrade viral particles and nucleic acids. Cuypers *et al.* (1992) reported a 1000- to 10,000-fold reduction in hepatitis C virus (HCV) RNA concentrations when whole blood and serum were stored at room temperature for 14 days. However, nucleic acid degradation occurred even more quickly when whole blood was stored at 4°C. This difference was probably due to increased granulocyte lysis at 4°C than at room temperature. Minimizing exposure to blood samples to temperatures ≥23°C is also important because specimens stored at these temperatures have decreased DNA yields (Cushwa and Medrano, 1993). For these reasons, the laboratory must establish storage and transport standards for all specimens. The laboratory should also have a protocol for handling (or rejecting) specimens that do not meet these standards.

Once specimens are received in the laboratory, they should be held in their original containers. Pouring off or splitting specimens provides an opportunity for the introduction of extraneous nucleic acids that could produce a false positive result in a clinically negative patient (McCreedy and Callaway, 1993).

Spectrophotometric evaluation and quantification of nucleic acids are often done to evaluate specimen adequacy in the molecular diagnostic laboratory (Farkas, 1993b). The most common spectrophotometric procedure involves determining the absorbance of the extracted specimen at 260 and 280 nm. The A_{260}/A_{280} ratio of highly purified DNA will approach 2.0; in most laboratories, samples with ratios below 1.7 are considered inadequate for further testing. The absorbance of the extracted specimen should also be determined at 230 nm because contaminants such as salts, sodium dodecyl sulfate (SDS), and some amino acids have peak absorbances at or near this wavelength. A_{230}/A_{260} ratios should be 0.3–0.5. Ratios that are substantially different from these values may indicate contamination with unacceptable levels of proteins and/or salts (Farkas, 1993b). Although useful and easy to perform, optical density measurements have some limitations. For instance, phenol strongly absorbs light at 270 nm, so the presence of phenol in the nucleic acid preparation can distort the A_{260}/A_{280} ratio. In addition, free ribonucleotides and deoxyribonucleotides absorb light at 260 nm; their presence can also falsely elevate the A_{260}/A_{280} ratio.

Careful analysis of nucleic acid yields and absorbance values can help the laboratory establish acceptable ranges for each specimen type. Specimens with values outside these ranges may not provide accurate diagnostic information and should be "cleaned" with 70% ethanol, re-extracted, or rejected. However, there may be compelling reasons for testing specimens that would normally be rejected. In these cases, an "inadequate specimen" comment should be formulated and included in the laboratory report.

III. FACILITIES AND EQUIPMENT

The facility requirements of a molecular virology laboratory depend on the type and number of procedures performed in the facility. Laboratories performing nucleic acid amplification procedures must have at least two rooms or dedicated areas, whereas laboratories that do not perform these procedures can function in a single room. The molecular diagnostic laboratory must accommodate four separate functions: reagent preparation, sample extraction, reaction set-up, and postamplification product detection. The ideal molecular diagnostic laboratory will have a separate room for each function. However, laboratories with limited space can separate these functions by establishing separate work stations in one or two rooms (Dragon *et al.,* 1993; Spadoro and Dragon, 1993). Each work station should have dedicated supplies and equipment. Benchtop hoods with ultraviolet decontamination lights should be used for reagent preparation, sample extraction, and reaction set-up areas (Dragon *et al.,* 1993; Spadoro and Dragon, 1993). More elaborate facility designs have been described for larger laboratories (McCreedy and Callaway, 1993; Sikro and Ehrlich, 1994). Laboratory work surfaces must be smooth and constructed to withstand regular cleansing with dilute acid, bleach, and/or shortwave ultraviolet radiation.

Quality assurance programs must include periodic equipment checks designed to ensure that the equipment is functioning as expected. Laminar airflow hoods should have face airflow velocities >75 ln ft/min at the work opening. Laminar airflow hoods should also be checked by a certified inspector at least once each year. The intensity of the ultraviolet (UV) radiation in laminal flow and static air hoods should be checked at the workbench level at least every 6 months and the ultraviolet bulbs should be replaced when the UV intensity drops below acceptable levels. In rooms with automatic UV ceiling lights, the UV intensity should be checked at several locations in the room to ensure even coverage.

The electrical supply should be checked before installing a thermocycler for two reasons: (1) these devices generally have high amperage consumptions and (2) the cycling of these devices can cause line voltage fluctuations that could damage other equipment. Thermocycling devices should be maintained carefully and checked for performance as part of the overall quality assurance program. Block temperatures should be checked periodically to ensure that the temperatures attained in each block position correspond to the digital readout of the instrument display panel (Linz, 1990; Haff *et al.,* 1991). Thermocyclers should be decontaminated periodically by UV illumination or by swabbing the blocks with a cotton-tipped applicator soaked in 10% household bleach, followed by water-soaked applicators to prevent corrosion.

Centrifuges should be equipped with aerosol contaminant rotors to protect personnel from aerosolized pathogens and to prevent cross-contamination of specimens. Alternatively, centrifuge tubes with compressible O-ring seals may be used (McCreedy and Callaway, 1993). Speed checks should be performed and recorded every 6 months. In addition, the condenser coils on refrigerated centrifuges should be cleaned at least twice each year.

The voltage of electrophoresis power supplies should be checked every 6 months to ensure that the voltage output corresponds with the setting on the power supply. pH meters should be calibrated daily before the first use of the day, using pH standards on either side of the target pH. The pH of the standards should be recorded and these records should be reviewed periodically by the laboratory supervisor. Balances should be kept clean and protected from drafts. Balances should be calibrated every 6 months using National Institute of Standards and Technology (NIST) Class S weight standards. The results of all these calibration checks should be recorded and compared with established tolerance limits. Equipment with performance characteristics outside the tolerance limits should be removed from service until it can be recalibrated.

IV. PIPETTES AND PIPETTING

The most frequent cause of false positive results is the introduction of foreign nucleic acids via contaminated pipettes (Sikro and Ehrlich, 1994). Air displacement pipettes often produce aerosols during pipetting operations that can contaminate the barrel of pipetting devices. The use of positive displacement pipettes or aerosol-resistant pipette tips can significantly reduce nucleic acid contamination. Positive displacement pipettes use a disposable plunger and a molded tip assembly that physically isolate the pipette barrel from the solution. Aerosol-resistant tips contain a hydrophobic microporous filter that is bonded onto the walls of the pipette tip. The microporous filter traps aerosols before they can contaminate the barrel of the pipette. These filters can also prevent contamination of the specimen when a contaminated pipette is used inadvertently. Aerosol-resistant pipette tips can be used with standard pipetting devices and may provide a more economical alternative to positive displacement pipettes. However, aerosol-resistant pipette tips that have microporous filters that are merely inserted (not bonded) into the tip may not provide an air-tight seal and will therefore provide inadequate aerosol protection (McCreedy and Callaway, 1993). The quality and performance of each lot of aerosol-resistant tips should be checked as described by Dragon et al. (1993) before being placed into service. Lots that fail this test should be returned to the vendor.

   A number of precautions must be taken during sample collection, extraction, and nucleic acid amplification steps to minimize cross-contamination between samples. Gloves are a common source of cross-contamination and should be changed frequently. In addition, microcentrifuge tubes should be brought to room temperature and subjected to a quick "pulse" centrifugation before they are opened to minimize aerosolization and sample splatter. Microcentrifuge caps should be opened and closed carefully to minimize aerosolization of the contents. All nonsample components such as dNTPs, primers, buffers, and enzymes should be added to the tube before adding the sample (Kwok, 1990). The "hot-start" PCR procedure described by Chou et al. (1992) has a higher potential for contamination because of the increased number of sample handling steps. However, even this procedure can be organized (e.g., by using wax plugs) to minimize contamination. Finally, all nonessential tubes should be kept closed during sample addition to minimize cross-contamination.

Because accurate measurement of reagent volumes play a critical role in all molecular diagnostic procedures, each pipetting device should be serviced regularly and kept clean. All pipettes, including robotic pipettes, should be checked for accuracy and precision every 3–6 months or after being serviced. Sample handling pipettes that are kept under ultraviolet lights to prevent nucleic acid contamination should be serviced and checked more frequently than other pipetting devices because ultraviolet light and ozone can degrade plastic parts and rubber seals. The accuracy and precision of adjustable pipettes should be determined at the high and low ends of the range and at the most commonly used volumes. Pipettes that fail the precision and accuracy checks must be recalibrated or replaced.

V. BIOCHEMICAL METHODS OF PREVENTING AMPLICON CARRYOVER

Re-amplification of previously amplified sequences (amplicons) is the most frequent cause of false positive results in any nucleic acid amplification procedure. Several physical (Fox *et al.*, 1991; Ou *et al.*, 1991; Sarkar and Sommer, 1991), chemical (Cimino *et al.*, 1991; Isaacs *et al.*, 1991; Aslanzadeh, 1992), and enzymatic (Longo *et al.*, 1990; Thornton *et al.*, 1992) nucleic acid inactivation methods have been developed to control amplicon carryover. Irradiation of PCR components with short-wavelength UV light was the first published method for inactivation of nucleic acids (Fox *et al.*, 1991; Ou *et al.*, 1991; Sarkar and Sommer, 1991). Ultraviolet irradiation creates thymidine dimers and other covalent DNA modifications that destroy the ability of the nucleic acids to act as templates for further amplification. Although these methods are inexpensive and easy to perform, their inactivation efficiencies can vary widely (Dwyer and Saksena, 1991; Sarkar and Sommer, 1991; Persing and Cimino, 1993).

Psoralens and isopsoralens have also been used to inactivate nucleic acids (Cimino *et al.*, 1991; Isaacs *et al.*, 1991; Aslanzadeh, 1992). Psoralens are planar tricyclic compounds that can intercalate between the base pairs of nucleic acids. When activated by longwave (300–400 nm) UV light, psoralens covalently cross-link pyrimidine bases and block template-dependent extension reactions (Cimino *et al.*, 1991). Isopsoralen compounds should be used whenever hybridization-based detection methods are employed because these compounds do not cross-link complementary strands and treated amplicons are capable of efficient and specific hybridization (Cone *et al.*, 1990; Persing and Cimino, 1993).

Hydroxylamine has also been used to inactivate nucleic acids after PCR amplification. Hydroxylamine preferentially reacts with cytosine residues to create covalent adducts that prevent base pairing with guanine residues (Brown and Schell, 1961; Aslanzadeh, 1992). Hydroxylamine inactivation procedures are especially useful for eliminating short G + C products that are difficult to remove by other methods (Persing and Cimino, 1993).

Enzymatic inactivation of amplified nucleic acids can be accomplished by incorporating uracil-*N*-glycosylase (UNG), a DNA repair enzyme, into the PCR reaction mixture (Longo *et al.*, 1990; Thornton *et al.*, 1992). In this procedure, deoxythymidine triphosphate is replaced by deoxyuracil triphosphate (dUTR) in the PCR master mix. The resulting DNA amplicons contain deoxyuracil, the UNG substrate. If dUTP-containing contaminants are introduced into subsequent PCR mixes, UNG will remove the deoxyuracil residues from the DNA backbone. The resulting abasic sites are unstable and undergo alkaline hydrolysis at the pH and temperature conditions found in most PCR protocols. Hydrolyzed amplicons cannot serve as templates for further replication.

Although the previously mentioned approaches are effective for inactivating amplified products, they are not suitable for cleaning up spills or for routine cleansing of surfaces and equipment. Several studies have shown that dilute (10%) household bleach solutions provide the most effective means for inactivating nucleic acids on surfaces (Hayatsu *et al.*, 1971; Prince and Andrus, 1992; Persing and Cimino, 1993). Bleach solutions cause oxidative damage to target, primer, and amplicon nucleic acids and can be used to inactivate RNA and DNA. Prince and Andrus (1992) demonstrated that bleach was significantly more effective in inactivating DNA than hydrochloric acid treatment. Thus, regular use of freshly made 10% bleach solution should be an essential component of the quality control program of every molecular diagnostic laboratory (Persing and Cimino, 1993). Because bleach solutions can damage finishes and corrode metal surfaces, it is important to rinse treated surfaces with autoclaved water or 70% ethanol to remove bleach residues.

Ultraviolet lights can also be used to control surface contamination. However, the UV source should be located within 30–60 cm of the working surface to provide effective cross-linking of DNA targets (Persing and Cimino, 1993). Fairfax *et al.* (1991) reported that the use of UV lights in a biological safety cabinet for 1–2 hr was only marginally effective in reducing contamination by amplified DNA. However, overnight irradiation was effective in eliminating a 200-bp amplicon.

All nucleic acid amplification procedures, except perhaps the cycling probe technology (Chapter 1), are susceptible to false positive results due to amplicon contamination. However, the effectiveness of the nucleic acid inactivation methods described here has yet to be determined with the newer amplification reactions. Although no inactivation method can substitute for good laboratory technique, clearly, amplicon inactivation is an important part of every nucleic acid amplification procedure. Without an inactivation method, the newer amplification protocols will never become widely accepted in the clinical laboratory.

VI. PROTECTIVE CLOTHING

Universal precautions dictate that gloves, laboratory coats, and face shields should be worn whenever the potential exists for spilling or splashing body fluids. In the molecular biology laboratory, laboratory coats should be dedicated for use in each testing area and should remain in those areas, except when being cleaned. Because amplification products can be introduced into an assay by a shirtsleeve dragged across a contaminated benchtop (McCreedy and Callaway, 1993), laboratory coats should be worn at all times while in the laboratory and should be removed when leaving the laboratory. Disposable sleeve protectors can also be used to provide additional protection when working in a laminar airflow hood.

Gloves should be worn to prevent contamination of the laboratory worker and to minimize the chances for contaminating the specimen with foreign nucleic acids and/or nucleases. However, gloves are easily contaminated and are an important source of cross-contamination, especially when handling microcentrifuge tubes. One method of minimizing glove contamination is placing a sterile 2 × 2 gauze over the cap when opening and closing microcentrifuge tubes (a separate gauze should be used for each tube). Aerosol contamination can also be a problem when uncapping vacuum-containing blood tubes. Glove contamination can be minimized if vacuum-containing tubes are uncapped under a gauze pad or with a disposable stopper remover. These procedures also provide some protection if the glass container is cracked or broken. Gauze pads and disposable stopper removers should not be reused.

VII. REAGENTS AND GLASSWARE

Reagent preparation is an important part of any molecular biology laboratory. The reagent requirements of each laboratory will vary depending on whether the laboratory uses diagnostic test kits, commercially prepared reagent sets, or makes reagents from individual components. Whenever possible, all solutions should be autoclaved to minimize bacterial contamination and reduce potential degradation of enzymes and nucleic acids. However, contaminating DNA cannot be removed by autoclaving (Dwyer and Saksena, 1992). Master mixes of nucleic acid amplification reagents and buffers should be aliquoted into single-use volumes to prevent contamination and to improve run-to-run reproducibility. Master mixes can be prepared, aliquoted, and stored under oil at $-15°C$ for up to 2 weeks or at $-70°C$ for months with minimal loss of activity (Civitello *et al.*, 1992). Meticulous record keeping is required; the laboratory must be able to trace all reagents used in a diagnostic test to the original vendor and lot number. Expiration dates must be established for all reagents; expiration dating must not be exceeded except in carefully defined cases (e.g., rare or expensive reagents). If expiration dating is to be extended, the laboratory must establish performance criteria and testing frequencies to ensure that the reagents are functioning as expected.

Reagent water quality should be carefully controlled because water is the largest single component used in the molecular diagnostic laboratory. Deionized or glass-distilled water should be used for all reagents. Water quality should be monitored and recorded daily to ensure that it is suitable for diagnostic procedures (≥ 10 megaohm-cm resistivity). Because reagent water should be sterile, most deionizing systems include a $0.2-\mu m$ filter at or near the dispenser. Flexible-tubing extending from the dispenser can serve as a source of bacterial contamination; care should be taken to position the tubing so that it drains fully after each use. Water sterility should be checked weekly (Farkas, 1993a) using bacteriological media.

All volumetric glassware should be NIST Class A or calibrated against a Class A standard. Broken, chipped, or cracked glassware should be discarded. Washed glassware should be rinsed thoroughly to remove any residual detergents. Glassware should be checked for residual detergent using pH paper to determine the pH of representative pieces of wet glassware. Because most detergents are alkaline, the presence of detergent residue will cause an increase in the normal pH. Some nonionic cleaning agents cannot be detected by this method, so an indicator such as bromosulfone–phthalein must be used (Farkas, 1993a).

General laboratory glassware is often a source of RNase contamination that can plague laboratories testing for, or generating, RNA. In these laboratories, glassware should be baked at 250°C for at least 4 hr (Maniatis *et al.*, 1982). Alternatively, glassware can be treated with a 0.1% diethylpyrocarbonate (DEPC) solution for 12 hr at 37°C. Residual DEPC must be removed from the glassware by heating to 100°C for 15 min or by autoclaving (Maniatis *et al.*, 1982). Most molecular diagnostic laboratories use sterile disposable plasticware whenever possible to prevent contamination.

VIII. PROCEDURAL CONTROLS

Every batch of reactions should contain positive and negative controls. Controls in each PCR procedure should be chosen carefully. Positive controls should amplify weakly to monitor the sensitivity of the reaction and to reduce the potential for amplicon contamination. Two types of negative (no target) controls should be used with each batch of samples: environmental controls and contamination controls. The environmental control is the first tube to receive an aliquot of the master mix and the last tube to be closed before temperature cycling. At least one contamination control (no target control) should be included for every 18–24 samples. A positive result in any of the contamination controls dictates that all reagents must be checked for amplicon or target DNA contamination. A positive result in the environmental control invalidates the test run and the specimens should be tested again.

In addition to positive and negative controls, an amplification control reaction can be set up for each specimen. Amplification controls are used to demonstrate specimen integrity and to protect against false negative results due to inhibitors of the amplification reaction. A resident target gene such as β-globin or the human leukocyte antigen DQα (HLA-DQα) can be used as a control for specimens that contain human genomic DNA. Alternatively, a separate specimen aliquot can be spiked with a low copy number positive control to demonstrate the ability of the reaction to amplify the target nucleic acids. Multiplex PCR procedures that simultaneously detect target nucleic acids and the co-amplification control sequences (e.g., HLA-DQα) are ideal for this purpose; when available, they are extremely efficient and cost effective.

IX. PROFICIENCY TESTING

Proficiency testing is a way of life in the clinical laboratory. Proficiency tests provide an excellent means of determining the testing variability among laboratories (external proficiency tests) and within laboratories (internal proficiency tests). Formal external proficiency testing surveys are not currently available for the molecular virology laboratory. However, some laboratories participate in informal proficiency testing programs by sharing specimens and comparing results.

Internal proficiency testing is an important part of the molecular virology laboratory. Some type of internal testing is required under the Clinical Laboratory Improvement Act of 1988 (CLIA 1988) regulations. Internal proficiency testing can be started by using leftover virology (VR) survey materials from the College of American Pathologists. In the long run, the supervisor should generate blind specimens for testing by the laboratory staff. Additional training may be required for personnel who consistently fail the internal proficiency tests.

X. CONCLUSIONS

Each diagnostic molecular virology program is built around quality control procedures that ensure the accuracy, precision, and reproducibility of the test results. A major difference between an academic research laboratory and the clinical diagnostic laboratory is formality and pervasiveness of the quality control process. Any new diagnostic procedure that moves from a research environment to the diagnostic laboratory must contain quality assurance procedures that ensure that the procedure is performing correctly. Because new diagnostic test methodologies often require new quality assurance procedures, quality assurance programs in the molecular virology laboratory will continue to evolve.

REFERENCES

Aslanzadeh, J. (1992). Application of hydroxylamine hydrochloride for post-PCR sterilization. *Ann. Clin. Lab. Sci.* **22**, 280.

Brown, D. M., and Schell, P. (1961). The reaction of hydroxylamine with cytosine and related compounds. *J. Mol. Biol.* **3**, 709–710.

Cembrowski, G. S., and Carey, R. N. (1989). "Laboratory Quality Management." American Society of Clinical Pathologists, Chicago.

Chou, Q., Russell, M., Birch, D. E., Raymond, J., and Bloch, W. (1992). Prevention of pre-PCR mis-priming and primer dimerization improves low-copy-number amplifications. *Nucleic Acids Res.* **20**, 1717–1723.

Cimino, G. D., Metchette, K. C., Tessman, J. W., Hearst, J. E., and Isaacs, S. T. (1991). Post-PCR sterilization: A method to control carryover contamination for the polymerase chain reaction. *Nucleic Acids Res.* **19**, 99–107.

Civitello, A. B., Richards, S., and Gibbs, R. A. (1992). A simple protocol for the automating of DNA cycle sequencing reactions and polymerase chain reactions. *DNA Sequence* **3**, 17–23.

Cone, R. W., Hobson, A., Huang, M. L., and Fairfax, M. R. (1990). Polymerase chain reaction decontamination: The wipe test. *Lancet* **316**, 686–687.

Cushwa, W. T., and Medrano, J. F. (1993). Effects of blood storage time and temperature on DNA yield and quality. *BioTechniques* **14**, 204–207.

Cuypers, H. T. M., Bresters, D., Winkel, N., Reesink, H. W., Weiner, A. J., Houghton, M., van der Poel, C. L., and Lelie, P. N. (1992). Storage conditions of blood samples and primer selection affect the yield of cDNA polymerase chain reaction products of hepatitis C virus. *J. Clin. Microbiol.* **30**, 3220–3324.

Dragon, E. A., Spadoro, J. P., and Madej, R. (1993). Quality control of the polymerase chain reaction. *In* "Diagnostic Molecular Microbiology: Principles and Application" (D. H. Persing, T. F. Smith, F. C. Tenover, and T. J. White, eds.), pp. 105–121. American Society for Microbiology, Washington, DC.

Dwyer, D. E., and Saksena, N. (1991). Failure of ultra-violet irradiation and autoclaving to eliminate PCR contamination. *Mol. Cell. Probes* **6,** 87–88.

Fairfax, M. R., Metcalf, M. A., and Cone, R. W. (1991). Slow inactivation of dry PCR templates by UV light. *PCR Meth. Appl.* **1,** 142–143.

Farkas, D. H. (1993a). Establishing a molecular biology laboratory. *In* "Molecular Biology and Pathology: A Guidebook for Quality Control" (D. H. Farkas, ed.), pp. 1–38. Academic Press, San Diego.

Farkas, D. H. (1993b). Specimen procurement, processing, tracking, and testing by the Southern blot. *In* "Molecular Biology and Pathology: A Guidebook for Quality Control" (D. H. Farkas, ed.), pp. 51–75. Academic Press, San Diego.

Fox, J. C., Ait-Khalid, M., Webster, A., and Emery, V. C. (1991). Eliminating PCR contamination: Is UV the answer? *J. Virol. Meth.* **33,** 275–383.

Haff, L., Atwood, J. G., DiCesare, J., Katz, E., Picozza, E., Williams, J. F., and Woudenberg, T. (1991). A high-performance system for automation of the polymerase chain reaction. *BioTechniques* **10,** 102–112.

Hayatsu, H., Pan, S. K., and Ukita, T. (1971). Reaction of sodium hypochlorite with nucleic acids and their constituents. *Chem. Pharm. Bull.* (*Tokyo*) **19,** 2189–2192.

Isaacs, S. T., Tessman, J. W., Metchette, K. C., Hearst, J. W., and Cimino, G. D. (1991). Post-PCR sterilization: Development and application to an HIV-1 diagnostic assay. *Nucleic Acids Res.* **19,** 109–116.

Kwok, S. (1990). Procedures to minimize PCR-product carry-over. *In* "PCR Protocols: A Guide to Methods and Applications" (M. A. Innis, D. H. Gelfand, J. J. Sninsky, and T. J. White, eds.), pp. 142–145. Academic Press, San Diego.

Linz, U. (1990). Thermocycler temperature variation invalidates PCR results. *BioTechniques* **9,** 268–293.

Longo, M. C., Berninger, M. S., and Hartley, J. L. (1990). Use of uracil DNA glycosylase to control carry-over contamination in polymerase chain reactions. *Gene* **93,** 125–128.

Maniatis, T., Fritsch, E. F., and Sambrook, J. (1982). "Molecular Cloning: A Laboratory Manual." Cold Spring Harbor Laboratory Press. Cold Spring Harbor, NY.

McCreedy, B. J., and Callaway, T. H. (1993). Laboratory design and work flow. *In* "Diagnostic Molecular Microbiology: Principles and Application" (D. H. Persing, T. F. Smith, F. C. Tenover, and T. J. White, eds.), pp. 149–159. American Society for Microbiology, Washington, DC.

Ou, C. Y., Moore, J. J., and Schochetman, G. (1991). Use of UV irradiation to reduce false positivity in the polymerase chain reaction. *BioTechniques* **10,** 442–446.

Persing, D. H., and Cimino, G. D. (1993). Amplification product inactivation methods. *In* "Diagnostic Molecular Microbiology: Principles and Application" (D. H. Persing, T. F. Smith, F. C. Tenover, and T. J. White, eds.), pp. 105–121. American Society for Microbiology, Washington, DC.

Prince, A. M., and Andrus, L. (1992). PCR: How to kill unwanted DNA. *BioTechniques* **12,** 358–360.

Sarkar, G., and Sommer, S. S. (1991). Parameters affecting susceptibility of PCR contamination to UV inactivation. *BioTechniques* **10,** 590–594.

Sikro, D. A., and Ehrlich, G. D. (1994). Laboratory facilities, protocols, and operations. *In* "PCR-Based Diagnostics in Infectious Disease" (G. D. Ehrlich, and S. J. Greenberg, eds.), pp. 19–43. Blackwell Scientific, London.

Spadoro, J. P., and Dragon, E. A. (1993). Quality control of the polymerase chain reaction. *In* "Molecular Biology and Pathology: A Guidebook for Quality Control" (D. H. Farkas, ed.), pp. 149–158. Academic Press, San Diego.

Thornton, C. G., Hartley, J. L., and Rashtchian, A. (1992). Utilizing uracil DNA glycosylase to control carryover contamination in PCR: Characterization of residual UDG activity following thermal cycling. *BioTechniques* **13,** 180–184.

Wiedbrauk, D. L., and Johnston, S. L. G. (1993). "Manual of Clinical Virology." Raven Press, New York.

Winberg, G. (1991). A rapid method for preparing DNA from blood, suited for PCR screening on transgenes in mice. *PCR Meth. Appl.* **1,** 72–74.

Nucleic Acid Blotting Techniques for Virus Detection

Daniel L. Stoler
*Department of Molecular
and Cellular Biology
Roswell Park Cancer Institute
Buffalo, New York 14263*

Nelson L. Michael
*Department of Retroviral Research
Walter Reed Army Institute
of Research
Rockville, Maryland 20850*

I. INTRODUCTION

The detection of electrophoretically separated nucleic acids immobilized on membranes by molecular hybridization was first described for DNA by Southern (1975) and for RNA by Alwine *et al.* (1977). Southern combined the use of restriction enzyme technology (Kelly and Smith, 1970; Danna and Nathans, 1971), the resolving power of gel electrophoresis, and the ability of nitrocellulose to bind nucleic acids in high salt (Gillespie and Spiegelman, 1965) to create blots of denatured DNA suitable for hybridization to radiolabeled probes. An analogous procedure for the detection of specific RNAs, commonly referred to as Nothern blotting, soon followed from the laboratory of Stark (Alwine *et al.,* 1977). The ability to identify specific nucleic acid sequences within the genome or as part of the pool of transcribed RNAs has proven to be an invaluable tool to all molecular biologists. Southern, Northern, and dot/slot blot procedures are currently widely used for detection of clinically relevant viral nucleic acids. The blotting-based procedures are highly sensitive, reproducible, and relatively inexpensive. In addition, Southern blotting of products of the polymerase chain reaction (PCR) is an important adjunct procedure used to increase further the sensitivity of PCR and to confirm the identity of the amplified DNA.

Since their introductions, the Southern and Northern blotting procedures have evolved with some general alterations and numerous case-specific ones. In this chapter, we discuss techniques to prepare and process nucleic acids from cells and tissues for use in filter hybridization assays. We also discuss a variety of techniques to separate nucleic acid pools to hybridization as well as the preparation of hybridization probes and methods for detection and quantification of hybridization signals. To present detailed descriptions of these procedures for each virus is obviously beyond the scope of a single chapter. We have prepared Table 1 as a source of specific protocols for many clinically relevant viruses. In addition, specific examples of these technologies are shown for the detection of nucleic acids from human immunodeficiency virus type 1 (HIV-1).

II. SAMPLE PREPARATION

Procedures for the isolation of nucleic acids from virus-infected clinical samples are numerous and particular to both virus type and sample source (i.e., tissue, feces, urine, blood, cell, culture, etc.). These methods will not be discussed here, but references listed in Table 1 provide specific methodologies.

TABLE 1
Selected Studies That Have Employed Nucleic Acid Blotting to Detect Clinically Important Viruses

Virus	Reference
Cytomegalovirus (CMV)	Churchill *et al.* (1987), Buffone *et al.* (1988), Landini *et al.* (1990), Shibata *et al.* (1990)
Other Herpes viruses	Desrosiers (1982), Jenkins *et al.* (1982), Aslanzadeh *et al.* (1992), Berneman *et al.* (1992)
Papilloma virus (HPV)	Krzyzek *et al.* (1980), Quick *et al.* (1980), Bloss *et al.* (1990), Burmer *et al.* (1990), Kulski *et al.* (1990)
Hepatitis B virus (HBV)	Brechot *et al.* (1981), Twist *et al.* (1981)
Human immunodeficiency virus (HIV)	Forghani *et al.* (1991), Polonis *et al.* (1991), Golden *et al.* (1992), Herndier *et al.* (1992), Horowitz *et al.* (1992), Taveira *et al.* (1992)
Parvovirus	Clewley (1985), Cunningham *et al.* (1988), Mori *et al.* (1989), Salimans *et al.* (1989), St. Amand *et al.* (1991)
Rotavirus	Street *et al.* (1982), Qian *et al.* (1991)

A. DNA Preparation

Procedures for preparing total DNA from tissues and cells in culture are abundant but generally very similar. Important features in common include rapid isolation and cellular lysis to protect DNA from endogenous nucleases and the enzymatic degradation of cellular proteins. Described here are two procedures that have been used successfully to obtain high molecular weight DNA from tissues and cultured cells; the first, a more traditional procedure, requires 2 days to complete whereas the second can produce DNA suitable for restriction digestion or PCR analysis in 2–3 hr.

1. Method 1

1a. Harvest adherent tissue culture cells by trypsinization or scraping followed by centrifugation. Cells growing in suspension may be centrifuged directly. Wash pelleted cells twice in phosphate buffered saline (PBS) to remove medium and serum proteins.

1b. Mince tissues as finely as possible. These samples may be frozen in liquid nitrogen and crushed with a mortar and pestle or disrupted by dounce homogenization. DNA yields from the former procedure are greater, but the latter produces sufficient material for several restriction digests even from small biopsy samples.

2. Resuspend samples in 10 mM Tris-HCl, pH 8.0, 25 mM EDTA, 100 mM NaCl. Add sodium dodecylsulfate (SDS) (to lyse cells and denature proteins) and proteinase K (to digest the denatured proteins) to final concentrations of 1% and 0.1 mg/ml, respectively. Incubate samples 12–18 hr at 37°C.

3. Extract digested samples twice with phenol:chloroform:isoamyl alcohol (25:24:1) and once with chloroform:isoamyl alcohol (24:1); precipitate with 2 volumes 100% ethanol. Recover white strands of DNA (and contaminating RNA) by centrifugation. Wash pellet with 70% ethanol and dry.

4. Resuspend pellet in 10 mM Tris, pH 8.0, 1 mM EDTA, 1 µg/ml DNase-free ribonuclease (RNase).

5. Repeat organic extractions described in Step 3, followed by ethanol precipitation in the presence of 0.5 volume 7.5 M ammonium acetate.

2. Method 2

This procedure, described by Grimberg et al. (1989), was designed for rapid DNA extraction from whole blood but works well with cultured cells and tissues. High molecular weight DNA suitable for restriction enzyme digestion as well as PCR amplification can be obtained with few manipulations in 2–3 hr, making this a highly attractive technique for the molecular diagnostic laboratory.

1. Release nuclei by adding a Triton X-100-containing solution (0.32 M sucrose, 10 mM Tris-HCl, pH 7.6, 5 mM MgCl$_2$, 1% Triton X-100) to whole blood or pelleted cells. Harvest by centrifugation at 900 g for 5 min.

2. Resuspend nuclei in 10 mM Tris-HCl, pH 8.0, 10 mM NaCl, 10 mM EDTA 1 mg/ml proteinase K. Incubate for 2 hr at 65°C. Jeanpierre (1987) has demonstrated that the activity of proteinase K is far greater when heat-denatured proteins are the substrates for digestion; at 65°C the enzyme self-digests.

3. Use the resultant DNA directly, or subject it to RNase treatment and organic extraction, as described in Method 1.

B. RNA Preparation

As does isolation of DNA, preparation of intact RNA requires rapid separation of RNA from endogenous nuclease activity. RNase is a highly stable protein capable of surviving autoclaving; therefore several precautions should be taken to avoid RNase contamination. Gloves should always be worn when preparing RNA or reagents to be used in the preparation of RNA since skin is a significant source of RNase. Aqueous solutions should be treated with diethylpyrocarbonate (DEPC) which inactivates RNase. Although procedures for removing ribonuclease from glass- and plasticware are relatively simple, disposable pipettes and test tubes are ribonuclease free and require no prior treatment.

Several procedures are commonly used to prepare RNA from tissues and cultured cells. Many rely on guanidinium salts, powerful protein denaturants, to inactivate ribonuclease activity and lyse the cells. The frequently cited method described by Chirgwin et al. (1979), and modifications of it, produce high yields of undegraded RNA from sources rich in ribonuclease. However, this procedure requires many manipulations as well as an overnight ultracentrifugation run, making it less useful to the molecular diagnostic laboratory. Instead, the simple and more rapid procedure of Puissant and Houdebine (1990) is recommended. In this simple one-step procedure, total RNA (high yield and undegraded) is isolated free of other macromolecules using guanidinium thiocyanate, phenol, and chloroform at acid pH.

1. Prepare stock solution of $4\,M$ guanidinium thiocyanate, 25 mM sodium citrate, pH 7, 0.5% sarkosyl in advance. This denaturing solution is stable for up to 3 mo at room temperature. Prior to use, add 2-mercaptoethanol to a final concentration of 0.1 M.
2. Homogenize fresh or frozen tissues at 4°C in 10 ml denaturing solution. RNA can be prepared from monolayer culture by adding denaturing solution directly to the dish.
3. In either case, transfer 5 ml material to a disposable polypropylene tube. Add the following, with vortexing after each addition: 0.5 ml 2 mM sodium acetate, pH 4.0; 5 ml water-saturated phenol; 1 ml chloroform.
4. Centrifuge at 10,000 g for 10 min. Remove the aqueous phase and precipitate RNA with an equal volume of 100% isopropanol at −20°C. At acid pH, DNA and proteins remain in the organic phase and at the interface.
5. Recover the RNA by centrifugation at 3,000 g for 10 min.
6. Vigorously vortex the pellet following the addition of 2 ml 4 M LiCl to dissolve contaminating polysaccharides. RNA, which is insoluble, is again recovered by centrifugation at 3,000 g for 10 min.
7. Dissolve the RNA pellet in 2 ml 10 mM Tris, pH 7.0, 1 mM EDTA, 0.5% SDS and extract with 2 ml chloroform.
8. Centrifuge at 3000 g for 10 min. Collect the upper (aqueous) phase and precipiate with an equal volume of 100% isopropanol in the presence of 0.2 M sodium acetate, pH 5.0.

This entire process can be completed in under 4 hr.

C. Polyadenylated Messenger RNA Preparation

Approximately 1–2% of total RNA is messenger RNA (mRNA); the remainder is composed of ribosomal, transfer, and other assorted small RNAs. For some applications, particularly when the RNA species of interest is present at very low abundance, it is desirable to analyze only mRNA. Removal of nonmessenger RNA species increases the amount of mRNA that can be analyzed by Northern blotting. Messenger RNAs contain polyadenylated [poly (A)] tails that are not present in other RNAs. Under appropriate conditions poly (A)-containing RNAs bind to an oligo-dT cellulose column whereas other RNAs pass through. As in all RNA procedures, RNase-free DEPC-treated solutions are essential, and the use of sterile, disposable plasticware is recommended. Oligo-dT cellulose columns may be purchased from several sources or prepared as described here.

1. Prepare a slurry of oligo-dT cellulose in 1 ml 0.1 M NaOH and pour into a small disposable column. For 2 mg total RNA, use 0.5 g oligo-dT cellulose.
2. Wash the column with 5–15 volumes of water.
3. Equilibrate the column with 10–20 volumes of high salt buffer (10 mM Tris-HCl, pH 7.4, 1 mM EDTA, 0.5 M NaCl).

1. Oligo-dT Cellulose Chromatography

1. Incubate 2 mg total RNA dissolved in 1 ml elution buffer (10 mM Tris-HCl, pH 7.4, 1 mM EDTA) for 5 min at 65°C.
2. Add NaCl to a final concentration of 0.5 M and load the denatured RNA onto the column.
3. Collect the eluant and reload onto the column. Repeat this step two additional times to maximize the binding of mRNA to the column.
4. Wash the column twice with 2 ml high salt buffer and 3 times with 2 ml low salt buffer (10 mM Tris-HCl, pH 7.4, 1 mM EDTA, 0.1 M NaCl).
5. Elute the RNA with 2 ml elution buffer. To obtain highly pure poly(A) + RNA, the selection procedure (Steps 1–4) should be repeated a second time.
6. Recover the RNA by precipitation in 2.5 volumes of ethanol in the presence of 0.3 M sodium acetate.

III. NUCLEIC ACID QUANTIFICATION

Quantification of DNA or RNA in solution is easily accomplished by ultraviolet (UV) spectrophotometry. Concentration and purity are determined by measuring the optical densities of the purified nucleic acid at 260 nm and 280 nm (A_{260} and A_{280}). Nucleic acids exhibit an absorbance maximum at 260 nm; proteins exhibit a maximum at 280 nm. The A_{260} of a 1 mg/ml pure DNA solution is 20; that of RNA is 25. The concentration can be determined by comparing the A_{260} of the nucleic acid sample with that of a 1 mg/ml solution. Highly pure nucleic acid solutions have an A_{260} to A_{280} ratio > 1.8; less pure preparations have a ratio < 1.8. The latter should be repurified before use.

IV. SOUTHERN BLOTTING

Preparing DNA for hybridization to specific probes requires three steps: (1) restriction endonuclease digestion, (2) agarose gel electrophoresis, and (3) Southern transfer.

A. Restriction Digestion

Restriction endonucleases may be purchased from many sources and are often supplied with a concentrated reaction buffer (usually 10×). Activity of the restriction endonuclease is described in units where 1 unit is the amount of enzyme that completely digests 1 µg of a defined DNA substrate in a given amount of time, usually 1 hr. DNA should be digested with several enzymes that produce fragments that are diagnostic for the virus in question. Digest 10 µg DNA in 1× restriction buffer with 2–5 units of restriction endonuclease/µg DNA. Keep the volume of the reactions to a minimum since they will eventually be loaded into the wells of an agarose gel. In addition, the added enzyme should not exceed 10% of the total volume of the reaction. Enzyme storage buffers frequently contain glycerol which, at higher concentrations, may be inhibitory to the reaction. To obtain a detectable signal from a DNA sequence that is present as a single copy in the genome, maximize the amount of target sequence loaded onto a gel. Generally, 10 µg total cellular DNA/restriction digest is sufficient when standard probes (≥500 bp) are labeled to high specific activity (at least 10^9 cpm/mg) and all other conditions are optimized. If the DNA under examination is less complex, for example, purified viral genomic DNA, or the target sequence is present at greater than one copy per genome, proportionately less DNA may be used.

In addition to the samples to be analyzed, other digestions should be performed. If the virus of interest can be cultured *in vitro*, DNA prepared from purified virus stocks of known titer should be digested as well. Alternatively, cloned viral DNA can be used. Inclusion of serial dilutions of such digested DNAs defines the sensitivity of the assay and allows quantification of viral DNA in the sample.

In addition to the digested samples loaded on the gel, one lane should contain molecular weight standards to allow determination of the sizes of restriction fragments in the assayed samples. These standards are readily available from commercial sources.

B. Electrophoresis of DNA Samples

Digested DNA fragments are separated on the basis of size by gel electrophoresis. The most commonly used format, discussed here, is the horizontal agarose gel, but other formats such as vertical acrylamide may also be used. Horizontal electrophoresis chambers, in a variety of sizes and styles, and power supplies are commercially available from a number of sources. Several factors influence the migration of linear DNA fragments through the gel, including:

1. Size of the DNA fragment. Migration through the gel is inversely proportional to the $\log_{10}$ of the size in base pairs of a linear DNA fragment. Larger fragments are retarded while smaller fragments migrate more rapidly.
2. Agarose concentration. This is a critical determinant of the effective range of separation of DNA fragments. For a 0.5% agarose gel, this range is about 1 kilobase (kb)–20 kb; for a 1% gel, the range is 0.5 kb–10 kb. DNA fragments outside the effective range will not be efficiently resolved.
3. Electrophoresis buffer. The two most commonly used buffers are TAE (0.04 M Tris base, 0.02 M glacial acetic acid, 1 mM EDTA) and TBE (0.089 M Tris base, 0.089 M boric acid, 1 mM EDTA). Although DNA fragments migrate about 10% faster in TAE buffered gels than in TBE gels, the difference in resolution is negligible. Note that the buffering capacity of TAE is low, so the buffer requires replacement or recirculation during extended electrophoresis runs. This step is not necessary when using TBE.
4. Applied voltage. At low electric field strengths (<5 V/cm), DNA fragment migration is proportional to the applied voltage. Electrophoresis at elevated field strengths (>5 V/cm) results in proportionally faster migration of higher molecular weight species and loss of resolution. Generally, for maximum resolution, electrophoresis is carried out at 1 V/cm.
5. The intercalating dye ethidium bromide causes a slight reduction in the mobility of linear DNA but no significant loss of resolution. Inclu-

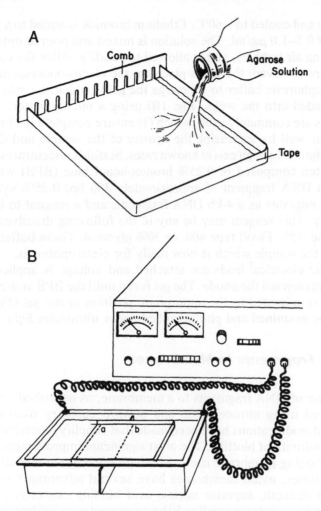

Figure 1 Agarose gel electrophoresis. (A) Melted agarose is poured into the tape-sealed gel plate with comb in place and allowed to solidify. (B) The comb is removed and samples are loaded into the wells (a). Voltage (<5 V/cm) is applied to the gel and electrophoresis proceeds until the bromophenol blue has migrated 1/2 to 3/4 the length of the gel (b).

sion of this dye in the gel enables visualization and photography of the samples and standards after electrophoresis since ethidium bromide fluoresces on exposure to UV light.

Gels are cast on a horizontal plate sealed along the sides and ends with tape. To form the wells, a template or comb is mounted vertically near one end of the plate 1 mm above the glass plate (Fig. 1A). Agarose at the appropriate concentration is melted in electrophoresis buffer (in a flask or bottle) and cooled to ~60°C. Ethidium bromide is added to a final concentration of 0.5–1.0 μg/ml. The solution is mixed and poured onto the gel plates, avoiding air bubbles, and is allowed to solidify. After the comb and sealing tape are removed, the gel is placed in the electrophoresis unit; just enough electrophoresis buffer to submerge the gel is added. Samples and standards are loaded into the wells (Fig. 1B) using a micropipetter. Several loading buffers are commonly used, which (1) ensure complete loading of the sample into the well by increasing the density of the sample and (2) provide dyes that migrate through gels at known rates. Sixfold concentrates of these buffers are often composed of 0.25% bromophenol blue (BPB) which comigrates with a DNA fragment of approximately 300 bp, 0.25% xylene cyanol FF which migrates as a 4-kb DNA fragment, and a reagent to increase sample density. This reagent may be any of the following dissolved in water, 40% sucrose, 15% Ficoll type 400, or 30% glycerol. These buffers are diluted to 1× in the sample which is now ready for electrophoresis.

The electrical leads are attached and voltage is applied so that DNA migrates toward the anode. The gel is run until the BPB and/or xylene cyanol FF have migrated to the appropriate position in the gel (Fig. 1B). The gel may be examined and photographed under ultraviolet light.

C. Transfer of DNA Fragments to Membranes

Transfer of DNA fragments to a membrane, as described by Southern, was achieved using nitrocellulose and upward capillary transfer. Since 1975, several modifications have been made that simplify, accelerate, and increase the sensitivity of blotting. The most significant improvement has been in the immobilizing membrane itself. Although nitrocellulose is still used in many laboratories, nylon membranes have several advantages including greater tensile strength, superior nucleic acid binding capacity, and retention of smaller fragments (as small as 50 bp compared with 500 bp for nitrocellulose). In addition, because of their greater strength, nylon membranes may be reprobed several times whereas nitrocellulose tends to fall apart with repeated probing. Further, during transfer DNA adheres but is not covalently attached. To achieve attachment requires at most a brief exposure to a UV light source for some nylon membranes, compared with 2 hr of baking at 80°C under vacuum for nitrocellulose membranes. Clearly nylon membranes are the preferred material for immobilization. The only reported disadvantage of using nylon is an increased background with some types of nonradioactive hybridization probes. Two types of nylon membranes are available, neutral and positively charged. Either may be used in the following protocols, with only minor alterations for the neutral membranes. Described here are three procedures for transfer of DNA to nylon membranes.

1. Standard Capillary Transfer in Alkaline Buffer

1. Place the gel in a dish with 10 gel volumes of 0.25 M HCl and gently rock. Continue for 10–15 min after the xylene cyanol FF and BPB become green and yellow, respectively. This treatment results in partial depurination and facilitates the transfer of DNA fragments larger than 5 kb. For fragments smaller than 5 kb, this step may be omitted.
2. Drain the dish and add 10 volumes transfer buffer; rock the gel until the dyes have returned to their original color. For a neutral membrane, the transfer buffer is 0.25 M NaOH, 1.5 M NaCl. Increase the NaOH concentration to 0.4 M for positively charged membranes.
3. The transfer stack is assembled as shown in Fig. 2A. Place blotting paper (Whatman 3MM or Schleicher and Schuell GB002) on a platform that is larger and wider than the gel in a large dish so the surface of the platform is covered and the ends of the paper drape over into the dish.
4. Fill the reservoir with transfer buffer to a level below the top of the platform. Allow the wick to absorb the buffer and smooth out any air bubbles with a pipette.
5. Place the prepared gel on the platform and surround it with a waterproof barrier (Parafilm works well for this) so the wick around the gel is completely covered. This precaution prevents buffer flow from the wick directly to the paper towels above the gel. This short circuiting decreases transfer efficiency.
6. Wet a nylon membrane, cut slightly larger than the gel, in distilled H_2O for several minutes and place on top of the gel.
7. Place at least 12 pieces of blotting paper on top of the membrane and smooth out any air bubbles. Stack paper towels (4–5 cm) on top, place a glass plate on the paper towels, and set a weight on the glass. The weight should be sufficient to compress the towels, ensuring good contact between the gel and the membrane. The towels should not make direct contact with the reservoir or the wick.
8. Alkaline transfer is rapid and complete in 2–3 hr. Disassemble the apparatus and mark the membrane with pencil to indicate the position of the wells.
9. Rinse the membrane in 2× SSC (1× SSC: 0.15 M NaCl, 0.015 M Na citrate, adjusted to pH 7.0) to neutralize the membrane and remove adhering bits of agarose.

If positively charged membranes are used, no further manipulations are necessary since DNA is covalently bound to the membrane under alkaline conditions. If neutral nylon membranes are employed, the DNA must be cross-linked to the membrane by exposure to UV light. The optimum time of exposure for cross-linking is dependent on the UV source and must be determined empirically.

2. Downward Capillary Transfer

This procedure reverses the direction of buffer flow in the transfer stack. In the standard procedure, the weight placed above the stack and the weight of wet paper towels can combine to crush the gel and slow the transfer process. The downward capillary transfer stack (Fig. 2B) places little weight on the gel and results in reduction of transfer time to about 1 hr.

1. Prepare the gel and membrane as described for standard transfer.
2. Stack paper towels 3–4 cm high on a level surface.
3. Place 5 pieces of blotting paper cut larger than the membrane over the paper towels. Wet the uppermost piece with transfer buffer.
4. Place the membrane, then the gel on top of the blotting paper. Smooth out any bubbles with a pipette.

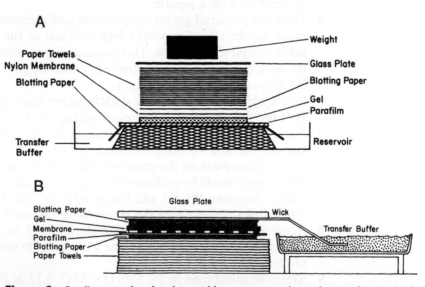

Figure 2 Capillary transfer of nucleic acid from agarose gels to nylon membranes. (A) Standard upward capillary transfer. The direction of fluid flow is from the reservoir up through the gel and membrane and into the paper towels. (B) Downward capillary transfer. Fluid is transferred via the wick down through the gel and membrane. In both cases, nucleic acids are eluted from the gel by capillary action and are immobilized on the membrane. Transfer is more rapid with the downward transfer system; the lack of weight above the transfer stack prevents gel crushing and inhibition of buffer flow.

5. Surround the gel with parafilm, completely covering the paper towels.
6. Saturate 3 pieces of blotting paper, large enough to cover the gel, with transfer buffer and place on top of the stack.
7. Saturate a large piece of blotting paper and place on the stack so the end drapes into a reservoir of transfer buffer.
8. Set a glass plate on top of the stack to prevent evaporation.
9. After 1 hr, disassemble the stack and prepare the membrane as described for standard transfer.

3. Pressure/Vacuum Transfer

An alternative to transfer of nucleic acids from agarose gels to nylon membranes by capillary action involves the use of either positive or negative pressure devices. The gel systems, electrophoresis, and postelectrophoretic gel preparation for pressure transfer are identical to the capillary transfer methods just described up to the point of transfer itself. At this point, the gel is placed on top of a piece of nylon membrane that is placed on a buffer-saturated piece of Whatman 3MM paper on top of a porous support. A sponge saturated with $10 \times$ SSC is then placed over the membrane/gel and the entire sandwich is sealed within the pressure chamber. Positive pressure is applied for 30–60 min to effect nucleic acid transfer. The nylon membrane is then marked for orientation, subjected to UV cross-linking, and processed for hybridization. The advantages of pressure blotting are primarily speed and uniform transfer efficiency across the gel. Positive-pressure blotting requires more effort to set up and involves the expenditure of several thousand dollars for equipment (Stratagene, La Jolla, CA).

V. NORTHERN BLOTTING

As can DNA fragments, RNA molecules can be separated on the basis of size by gel electrophoresis and can be immobilized on membranes by a process referred to as Northern blotting. To achieve tight binding to membranes, the RNA first must be denatured, usually by glyoxal and dimethyl sulfoxide (Thomas, 1980) or formaldehyde (Lehrach et al., 1977; Seed, 1982). Again, nylon membranes are recommended for the reasons previously discussed. The background hybridization may be significantly higher with some nylon membranes and may require the use of nitrocellulose. The two procedures described here work equally well with nitrocellulose or nylon membranes, but covalent attachment of RNA to nitrocellulose requires 2 hr of baking under vacuum after transfer.

A. Electrophoresis of RNA Samples

1. Denaturation with Glyoxal and Dimethyl Sulfoxide (DMSO)

1. Prepare denaturation buffer in advnace [6 M deionized glyoxal, 75% DMSO, 0.1 M NaH$_2$PO$_4$, pH 7.0 (DEPC treated)]. Aliquot and store at $-70°C$. Deionized glyoxal may be prepared by passing a 40% (6 M) solution through a mixed bed resin column until the pH exceeds 5.0.

2. Incubate 10–15 μg total cellular RNA in 15–20 μl denaturation buffer for 1 hr at 50°C. When rare messages are to be detected, up to 3 μg poly(A)+ RNA can be substituted. Samples may be lyophilized prior to addition of denaturation buffer if their volumes exceed 5μl. Chill samples on ice and add loading buffer (see Section IV,B).

As done when analyzing DNA samples, a dilution series of purified viral RNA from stocks of known titer should be included if possible. When using total cellular RNA, a duplicate sample may be included and 28S and 18S ribosomal RNAs may serve as size markers. Alternatively, or when using poly(A)-selected mRNA as samples, commercial preparations of RNA standards can be used. Subsequent to electrophoresis, lanes with standards can be trimmed from the gel, stained with 0.5 μg/ml ethidium bromide, and photographed under UV light. When using nylon membranes, the positions of the ribosomal RNAs can be visualized by exposure of the membrane to ultraviolet light. The ribosomal RNAs appear as dark bands on a bright background, and their positions can be marked on the blot with pencil.

Prior to casting the gel, the gel plate, comb, and electrophoresis chamber should be washed with detergent and rinsed with H$_2$O and ethanol. The equipment may be further treated with 3% hydrogen peroxide to remove any contaminating RNase. The gel is cast and run in much the same way as a DNA gel with the following exceptions: (1) agarose, 1–1.5% depending on message size, is dissolved in 10 mM sodium phosphate, pH 7.0, which is also the running buffer, and (2) ethidium bromide, which reacts with glyoxal is omitted.

3. The gel is run at 3–4 V/cm until the BPB is 1/2 to 3/4 of the way through the gel. Recirculation of the electrophoresis buffer by peristaltic pump is essential to prevent the pH from exceeding to 8.0. Above this pH, glyoxal dissociates from RNA and artifactual migration of the RNA occurs. The gel is ready for transfer on completion of the run. No additional manipulations are necessary.

2. Formaldehyde Denaturation

1. Add formamide and formaldehyde to final concentrations of 50% and 2.2 M, respectively, and running buffer diluted to 1× from a concentrated stock. Two running buffers are commonly used: (a) 10 mM NaH$_2$PO$_4$, pH 7.0, and (b) 0.02 M 3-(N-morpholino)propanesulfonic acid (MOPS), pH 7.0, 8 mM sodium acetate, 5 mM EDTA. pH 8.0. Either may be used in this protocol. Denature RNA samples and standards for 15 min at 60°C.
2. Prepare the gel electrophoresis apparatus as described in Section V,A,1.
3. Melt agarose in water and allow it to cool to 60°C.
4. Add formaldehyde and concentrated running buffer to the agarose solution to final concentrations of 2.2 M and 1×, respectively. Cast the gel in a chemical hood since formaldehyde vapors are toxic.
5. Perform gel loading and electrophoresis as previously described.
6. Remove the gel to a large dish, wash with several changes of distilled H$_2$O to remove formaldehyde, and equilibrate with 10× SSC. The gel is now ready for transfer of the RNA.

B. Transfer of RNA to Membranes

The transfer of RNA to nylon (or nitrocellulose) membranes can be achieved by following the procedures detailed for DNA transfer with the substitution of 10× SSC for the alkaline transfer buffer. High salt transfer is not as rapid as alkaline transfer and should be carried out for 12–24 hr to ensure complete transfer of RNA. RNA binds tightly but noncovalently to nylon and nitrocellulose under these conditions. Ultraviolet cross-linking (see Section IV,C,1) is necessary for both charged and neutral membranes, as is baking at 80°C under vacuum for nitrocellulose blots.

VI. DOT/SLOT BLOTTING DNA AND RNA

Dot and slot blotting are rapid, simple techniques for the quantification of RNA or DNA target sequences without prior electrophoretic separation (Kafatos *et al.*, 1979). These methods differ only in the shape of the immobilized nucleic acid spot deposited on the membrane. As originally performed, nucleic acid was applied to a dry nitrocellulose filter and allowed to dry. The resulting "dots" were variable in size, making accurate estimates of target sequence concentration difficult. Manual spotting of samples has largely been replaced by the use of filtration manifolds.

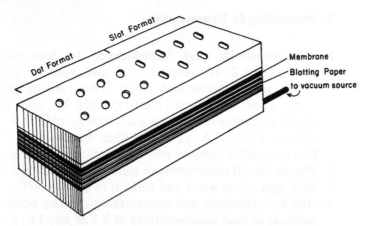

Figure 3 Dot/slot filtration manifold. The apparatus is assembled with a membrane and blotting paper positioned as shown. Buffers and samples are placed in the wells (circular or oval openings at the top) and drawn through the membrane by the applied vacuum. Nucleic acids are retained on the membrane; fluids are collected in the lower chamber of the apparatus. The diagram shows both dot and slot configurations in a single apparatus. However, these manifolds are sold only in one configuration or the other.

The dot/blot apparatus, diagrammed in Fig. 3, supports a membrane and is connected to a vacuum source. Fluid samples loaded into the slots are drawn through the membrane, which retains any nucleic acid in the sample. Filtration manifolds are manufactured by several companies in both dot and slot formats. Procedures for preparation of DNA and RNA dot/slot blots using positively charged nylon membranes are described next.

A. DNA Blots

1. In distilled water, wet one piece each of nylon membrane and blotting paper, precut to fit the manifold.
2. Assemble the manifold according to the manufacturer's specifications with the blotting paper supporting the membrane.
3. Prepare a dilution series of the DNAs to be probed as well as a control DNA in which the abundance of target sequence is known.
4. Denature the DNA by adding NaOH and EDTA, pH 8.0, to a final concentration of 0.4 M and 10 mM, respectively. Boil for 10 min. Immediately chill samples on ice.
5. Apply vacuum and add 100–500 μl water/well (depending on well size).
6. Load the samples when the water has been drawn through the membrane. Rinse each well with 200–500 μl 0.4 N NaOH. Remove the membrane from the apparatus and rinse in 2× SSC.

The alkaline treatment both denature the DNA and covalently attaches it to the membrane, thus eliminating the subsequent denaturation and neutralization required when neutral membranes are used.

B. RNA Blots

1. Assemble the apparatus as described for DNA blots.
2. Denature the RNA as described for sample preparation for Northern blots by using the glyoxal/DMSO or formamide/formaldehyde protocols (Section V,A,1 and 2), or by boiling for 5–10 min and then placing on ice.
3. Prepare a dilution series of RNAs to be probed, as well as controls.
4. Apply the vacuum and rinse wells with 200–500 μl 25 mM NaH_2PO_4, pH 7.0.
5. Add the samples when the wells have drained and wash twice with 25 mM NaH_2PO_4.
6. Remove the filter and cross-link the RNA to the membrane by UV irradiation (Section IV,C,1).

VII. HYBRIDIZATION THEORY

Molecular hybridization is the association between strands of DNA and RNA whose kinetics are governed by the degree of intra- and interstrand relatedness, solvent type, solvent ionic strength, temperature, and time. The basis for this aspect of molecular biology is the structure of DNA first proposed by Watson and Crick (1953). Their work showed that DNA was composed of two antiparallel strands of nucleic acid wound in a double helix with an external polyanionic phosphate backbone and an ordered internal association of the four constituent bases. These bases stack in the vertical plane of the helix by hydrophobic interactions and associate in the horizontal plane according to "base pairing" rules: purine bases, guanine (G) and adenine (A), must base pair with pyrimidine bases, cytosine (C) and thymidine (T), respectively. The rate-limiting step in molecular hybridization is the initial association between two antiparallel strands of nucleic acid that are capable of base pairing (complementary), termed "nucleation" (Flavell *et al.*, 1974). Nucleation is enhanced when the concentration of the two complementary strands is high and when sufficient time is provided for the two strands to find one another. These two parameters, time and concentration of hybridizing strands, are critical to the kinetics of the reaction. In general, hybridization is favored by an increase in one or both of these parameters. In filter hybridization, detection of signal is enhanced by an increasing concentration of immobilized target nucleic acid. However, if quantification of varying amounts of immobilized nucleic acid is desired, the concentration of the probe must be in vast molar excess to the concentration of immobilized nucleic acid. Note that the kinetics of molecular hybridization are 7–10 times slower in filter hybridization than in solution hybridization, presumably as a result of steric effects (Flavell *et al.*, 1974).

The temperature at which two nucleic acid strands are 50% dissociated is termed the "melting temperature" (T_m). The T_m for a given hybrid pair is a function of monovalent cation concentration, percentage of G:C base pairs, and length of the hybrid. Molecular hybrids are stabilized by the presence of monovalent cations since they mitigate the repulsive forces between the polyanionic backbones of two complementary nucleic acid strands. G:C base pairs, possessing three hydrogen bonds per base pair, increase the stability of a hybrid and, thus, the T_m to a greater extent than A:T base pairs, which contain only two hydrogen bonds per base pair. Hybrid length is not critical until the hybrid becomes quite small, as in hybridization using short oligodeoxyribonucleic acid probes. The effects of these factors on the T_m are described by the equation: $T_m = 16.6 \times \log [Na^+] + 0.4$ (% G+C) + 8.15°C − (600/length in base pairs). For a 500-bp probe with a composition of 50% G+C in a typical hybridization buffer with a $[Na^+]$ of 0.825 M, $T_m = 16.6 \times \log [0.825] + 0.4$ (50) + 81.5°C − 600/500 = 98.9°C (Schildkraut, 1965). The optimum hybridization temperature for filter hybridization is typically given by $T_H = T_m - 15$ to 30°C, in the 69–84°C range in the given example. Since these temperatures may be cumbersome to maintain, many investigators prepare hybridization buffers with a proportion of the nonaqueous solvent formamide. The effect of formamide on the melting temperature is given by $dT_m /d\%$formamide = −0.7°C (McConaughy, 1969). Hence, for the hybridization just described, a buffer containing 50% vol/vol formamide would result in a T_m of 63.9°C and a T_H of 34–49°C with an average of 42°C. At these concentrations of formamide, RNA:RNA hybrids are more thermally stable than cognate RNA:DNA hybrids, which are more stable than cognate DNA:DNA hybrids (Casey and Davidson, 1977). This effect will be of practical importance in discussions of RNA probes later in this chapter.

Base mismatch lowers the T_m of a long hybrid by approximately 1°C for every 1.5% of mismatch (Laird *et al.*, 1969). These effects are magnified if the mismatches are clustered and in the middle rather than at the ends of the duplex. Given the relatively broad range for T_H, a small degree of probe–target mismatch can be tolerated in hybridization experiments. Larger degrees of mismatch or the use of mismatched oligodeoxyribonucleic acid probes are more problematic and demand an empirical adjustment of both hybridization and washing conditions. The rate of hybridization can be increased 12- to 13-fold by the addition of 10% dextran sulfate to hybridization reactions (Wetmur, 1971). However, dextran sulfate increases the background in filter hybridization and is both expensive and cumbersome to use. Hybridization is relatively insensitive to pH effects in the range of 5 to 9 (Wetmur and Davidson, 1968). Care must be taken, however, not to perform hybridization experiments with RNA probes or targets with a pH >8.0, especially at higher temperatures, to avoid base-mediated RNA hydrolysis.

VII. HYBRIDIZATION PROBES

A. Probe Types

The choice of probes for molecular hybridization is guided by multiple concerns. Whereas the use of cloned plasmid inserts, restriction fragments, and complementary RNA (cRNA) probes is not dependent on the determination of the nucleotide sequence of the DNA of interest, this information is required for oligodeoxyribonucleic acid- and PCR-derived probes. The most sensitive molecular probes are uniformly labeled cRNAs since they combine a high ratio of labeled atoms/probe length, lack of a competing antiparallel probe strand, and the favorable hybridization kinetics of an RNA probe. The specificity (or stringency) of all hybridization probes is primarily a function of the hybridization and wash conditions. However, the use of whole cloned plasmids has a theoretically lower specificity because of the inclusion of plasmid sequences in the labeled pool that increase the probability of spurious cross-hybridization. This effect can be avoided by purifying the cloned insert away from the plasmid vector before labeling. A list of commonly used molecular probes and their characteristics is given in Table 2. A brief description of the commonly used techniques to prepare molecular hybridization probes, as well as a detailed protocol for each, is presented next.

TABLE 2
Characteristics of Molecular Probes

Probe type	Sequence requirement	Sensitivity	Specificity
Cloned plasmids	No	High	Medium
Restriction/PCR fragments	No	High	High
cRNA	No	Very high	High
Oligodeoxyribonucleotides	Yes	Medium	High

B. Probe Preparation

1. Random Primed Synthesis of Linear DNA Fragments

This technique, first described by Feinberg and Vogelstein (1983), involves the replacement synthesis of labeled deoxyribonucleic acid precursors into linear DNA fragments by a peptide fragment of the *Escherichia coli* DNA polymerase I (Klenow fragment). Circular plasmid DNAs must first be linearized with an appropriate restriction endonuclease prior to labeling. Restriction or PCR fragments, once purified, may be used directly in this method. It is especially critical that Mg^{2+} ions are removed from the DNA sample prior to labeling. A random mix of hexanucleotide primers is utilized that allows the labeling of any double-stranded DNA template. As little as 25 ng template DNA may be labeled to produce probes $>1 \times 10^9$ cpm/μg with 70% incorporation of the labeled deoxyribonucleotide triphosphate precursor. These probes, once denatured, are then suitable for the detection of single copy genes and rare mRNAs in Southern and Northern blotting experiments, respectively, or for the screening of plaque and colony lifts.

1. Mix DNA of interest and 1–5 μg random hexamers into a total of 14 μl water or TE in a microcentrifuge tube.
2. Boil the DNA–primer mix for 2–5 min and plunge into ice.
3. Assemble the following in a separate microcentrifuge tube on ice:

   2.5 μl 0.5 mM d(G,T,C)TP, a mixture of 0.5 mM of each dNTP except dATP

   2.5 μl 10× Klenow/DNA polymerase I buffer (0.5 M Tris-HCl, pH 7.5, 0.1 M MgCl$_2$, 10 mM DTT, 0.5 mg/ml BSA)

   5.0 μl 3000 Ci/mmol [α-^{32}P]dATP (50 μCi)

   1.0 μl Klenow fragment (3–8 units)

4. Add 11 μl reaction mix from Step 3 to the denatured DNA–primer mixture.
5. Incubate at 37°C for 30 min or at room temperature for 2–16 hr.
6. Add 1 μl 0.5 M EDTA, 3 μl 10 mg/ml carrier tRNA, and 100 μl TE.
7. Extract with an equal volume of buffer-saturated phenol.
8. Pass the extracted aqueous layer through a Sephadex G-50 spin dialysis column.
9. Count 1 μl eluate in a scintillation counter.
10. If not used immediately, store the probe at −20°C.

2. Nick Translation of Recombinant Circular Plasmids

This technique, first reported by Rigby *et al.* (1977), is the oldest method of labeling hybridization probes that is still widely used today. Although initially described for the uniform labeling of circular plasmids, the nick translation reaction is suitable for linearized DNA fragments as well. The reaction involves the action of the enzymes DNase I and *E. coli* DNA polymerase I. DNase I is a 3′ endonuclease that introduces single-strand scissions ("nicks") into the DNA duplex, resulting in free 3′ hydroxyl ends. These 3′ hydroxyl ends are then recognized by DNA polymerase I, which extends the nick through the combined action of its 5′ to 3′ polymerase and 5′ to 3′ exonuclease activities. Thus, the polymerase degrades the DNA strand proximally while synthesizing new DNA distally. This neck then progresses, or translates, along the duplex until the polymerase ceases to elongate the chain. Since DNA polymerase I is a poorly progressive enzyme, the length of DNA polymerized on any given run is approximately 400 nucleotides. It is critical to use DNase I sparingly and relatively high concentrations of dNTPs to prevent the resulting labeled DNA from being too small. Since *E. coli* DNA polymerase I is the only DNA polymerase known to possess a 5′ to 3′ exonuclease activity, it is the only enzyme suitable for the nick translation reaction. These concerns, coupled with the relative lability of DNA polymerase I preparations, have led to a gradual decline in the popularity of this labeling technique of DNA duplexes in favor of the random priming technique previously described. However, with careful attention to the reaction conditions, probe specific activities of 1×10^8 cpm/μg can be obtained.

1. Add the following to a 0.6-ml microcentrifuge tube on ice:

   2.5 μl 0.5 mM d(G,T,C)TP, a mixture of 0.5 mM of each dNTP except dATP

   2.5 μl 10× Klenow/DNA polymerase I buffer 0.5 M Tris-HCl, pH 7.5, 0.1 M MgCl$_2$, 10 mM DTT, 0.5 mg/ml BSA)

   10.0 μl 3000 Ci/mmol [α^{32}P]dATP (100 μCi)

   1.0 μl DNase I [freshly diluted 10,000-fold from a 1 mg/ml stock in 20 mM Tris-HCl, pH 7.5, 500 μg/ml BSA (Pentax Fraction V), 10 mM 2-mercaptoethanol]

   1.0 μl *E. coli* DNA polymerase I (5–15 units)

2. Add 0.25 μg DNA in a volume of 8 μl to the reaction mix to bring the final volume to 25 μl.

3. Incubate the reaction at 12–14°C for 15–45 min.

NOTE: *The precise incubation time should be determined empirically for a given batch of DNase I. Once the time to maximal incorporation of free label is determined, this time can be used in successive labeling reactions. Incubation beyond the point of maximal incorporation results in probe degradation with a resultant decrease in specific activity.*

4. Stop the reaction by adding 1 μl 0.5 M EDTA 3 μl 10 mg/ml tRNA, and 100 μl TE.
5. Extract with an equal volume of buffer-saturated phenol.
6. Pass the extracted aqueous layer through a Sephadex G-50 spin dialysis column.
7. Count 1 μl eluate.
8. If not used immediately, store the probe at $-20°C$.

3. *In Vitro* Transcription

Many investigators have come to regard *in vitro* synthesized complementary RNAs (cRNAs) as the most useful molecular probes developed since their description by Melton *et al.* (1984) because of their combination of very high sensitivity and favorable hybridization kinetics. This technology exploits the use of plasmids containing the promoters for highly processive bacteriophage RNA polymerases (T3, T7, and SP6). These promotors are cloned just upstream from a sequence containing multiple restriction sites, facilitating the appropriate ligation of sequences of interest. The resultant recombinant plasmids are then linearized with an appropriate restriction enzyme that cleaves 3' to the cloned insert, before performing "run-off" transcription of the cloned sequences with bacteriophage RNA polymerase and labeled ribonucleotide triphosphate precursors, resulting in the production of cRNA. The plasmid DNA is then selectively degraded with RNase-free DNase; the cRNA probe is purified by phenol extraction and spin dialysis. These probes, when prepared using the protocol given here, yield specific activities on the order of 10^9 cpm/μg. These extremely sensitive hybridization probes should be used within 72 hr of synthesis since they undergo rapid radiolysis. More stable but less sensitive probes may be made by reducing the concentration of labeled radioactive precursor in the reaction.

1. Add the following in order to a 0.6-ml microcentrifuge tube at room temperature (spermidine–DNA complexes precipitate on ice):

   4.0 μl 5$\times$ transcription buffer (200 mM Tris-HCl, pH 7.5, 30 mM MgCl$_2$, 10 mM spermidine, 50 mM NaCl)·

2.0 µl 100 mM DTT
0.8 µl placental RNase inhibitor (20 µl)
4.0 µl 2.5 mM r(A,U,G)TP
2.4 µl 0.1 mM rCTP
1.0 µl form III DNA template in water or TE (0.2–1.0 mg/ml)
5.0 µl 400 Ci/mmol [α³²P]rCTP (50 µCi)
1.0 µl RNA polymerase (10–20 units)

2. Incubate at 40°C for 60 min.
3. Add RNase-free DNase (1 unit/µg DNA template).
4. Incubate at 37°C for 15 min.
5. Extract with an equal volume of buffer-saturated phenol.
6. Process through an RNase-free Sephadex (G-50 spin dialysis column.
7. If not used immediately, store the probe at −20°C.

4. 5′ End-Labeling of Oligodeoxyribonucleotides

The widespread availability of automated phosphoramidite technology to synthesize oligodeoxyribonucleotides has made the use of these reagents as molecular hybridization probes technically and financially feasible. Construction of appropriate oligodeoxyribonucleotide probes requires knowledge of the target nucleotide sequence. However, if this information is available, the creation of oligodeoxyribonucleotide probes avoids the subcloning step necessary for cRNA probes. If the target nucleic acid is not available in a laboratory for the production of nick-translated or random primed probes, an oligodeoxyribonucleotide hybridization probe can be prepared based solely on the knowledge of the nucleotide sequence of the target nucleic acid. These reagents have been especially useful in the recovery of genes encoding proteins for which only amino acid sequence data was available. This information allows the construction of a family of oligodeoxyribonucleotides with a proportion of degenerate bases that correspond to all possible codons for the amino acid sequence. This set of "degenerate oligodeoxyribonucleotides" can then be used to screen bacteriophage cDNA libraries to purify the coding sequences for the protein of interest. These short molecular hybridization probes have also proven useful in the detection of small portions of target DNA or cDNA sequences amplified by PCR. The most common method of labeling oligodeoxyribonucleotides is the addition of a radioactive phosphoryl group from the gamma position of ATP to the free 5′ hydroxyl group of the synthetic oligodeoxyribonucleic acid molecule, a reaction catalyzed by the bacteriophage T4 enzyme polynucleotide kinase. RNA, especially tRNA, is also a suitable substrate for this enzyme.

1. Add to a 0.6-ml microcentrifuge tube on ice:

   0.5 μl 20 μM oligodeoxyribonucleotide (10 pmol)
   2.0 μl 10 × 5′ end-labeling buffer (500 mM Tris-HCl, pH 7.6,
      100 mM MgCl$_2$, 50 mM DTT, 1.0 mM EDTA, pH 8.0)
   5.0 μl 5000 Ci/mmol [γ^{32}P]ATP (50 μCi)
   11.5 μl deionized water
   1.0 μl T4 polynucleotide kinase (8 units)

2. Incubate at 37°C for 30–45 min.
3. Separate phosphorylated oligodeoxyribonucleotide from free label on Sephadex G-25 spin dialysis column. **Caution:** The use of a Sephadex G-50 column will result in a significant retention of labeled oligodeoxyribonucleotide probe on the spin dialysis column.
4. Count 1 μl eluate.
5. If not used immediately, store the probe at −20°C.

NOTE: *Synthetic oligonucleotides are typically constructed with 5′ hydroxyl groups. Therefore, they can be used as substrates for phosphorylation with polynucleotide kinase without prior dephosphorylation.*

IX. FILTER HYBRIDIZATION

Detection of target sequences in Southern, Northern, and dot/slot blots is carried out under essentially identical conditions. Three basic processes are involved: (1) prehybridization, which saturates nonspecific DNA binding sites on the membrane with random DNA and polymers; (2) hybridization, during which specific labeled probes are annealed to target sequences; and (3) washing, to remove unhybridized and imprecisely hybridized probe.

Numerous protocols have evolved since filter hybridizations were first described by Gillespie and Spiegelman (1965). The methods differ with regard to many factors including composition of prehybridization and hybridization buffers, hybridization temperature, and washing procedures.

Two protocols are presented here that are commonly used with nylon or nitrocellulose membranes.

1. Method 1

1. Place the membrane into a heat-sealable bag and wet with a small amount of 2× SSC or prehybridization buffer (6× SSC, 5× Denhardt's reagent, 0.5% SDS, 100 μg/ml denatured sheared salmon sperm DNA;

50% vol/vol formamide may be included). A 50× concentrate of Denhardt's reagent consists of 1% each bovine serum albumin (BSA), polyvinylpyrrolidone, and Ficoll 400.

2. Decant the liquid, and add 15 ml prehybridization buffer for a 10 × 15 cm blot. Proportionally less buffer should be used for smaller blots.
3. Seal the bag and submerge in a 65–68°C (42°C if formamide hybridization buffer is used) shaking water bath for 1 hr.
4. Single-stranded probes may be used without any pretreatment. Denature double-stranded probes before hybridization by boiling for 5 min and immediately chilling on ice. Remove the bag from the water bath.
5. Decant the prehybridization buffer from the bag and add the same volume of hybridization buffer with probe. Hybridization buffer is identical to prehybridization buffer without Denhardt's reagent.
6. Reseal the bag, excluding any air bubbles, and return to the water bath for at least 6 hr or overnight.

Sensitivity and stringency of detection are determined by salt concentration and temperature of the buffer during the washing steps. Low stringency washes, with high salt at lower temperatures, increase sensitivity but may also increase background hybridization. High stringency washes, with low salt at higher temperatures, lower sensitivity but limit hybridization to highly specific probe–target interactions. The following protocol is for high stringency washing. Reduction in stringency may be made by appropriate alterations in temperature and salt concentrations.

1. Remove the membrane from the bag to a dish containing at least 200 ml 2× SSC, 1% SDS; agitate at room temperature for 15 min.
2. Decant the liquid and repeat the wash twice.
3. Wash twice for 15 min at 37°C in 0.1× SSC, 0.1% SDS, and twice for 30 min at 65°C in the same buffer.
4. Air dry the membrane.

2. Method 2

This procedure, first described by Church and Gilbert (1984), is highly sensitive and yields blots with a high signal-to-noise ratio when using nylon membranes.

1. Place the membrane in a heat sealable bag.
2. Wet with water, decant, and add 15 ml hybridization buffer per 10 × 15 cm membrane. Hybridization buffer is composed of 1% BSA, 0.5 M Na$_2$HPO$_4$·7H$_2$O, 1 mM EDTA, 7% SDS, adjusted to pH 7.0 with NaOH.

3. Agitate the membrane in the sealed bag for 5 min in a 65°C water bath.
4. Add probe, denatured as described if necessary.
5. Reseal the bag and incubate overnight at 65°C. In this procedure, stringency is determined solely by the temperature of the washes since salt concentrations remain constant. Stringency may be altered by increasing or decreasing the temperature appropriately.
6. Wash the blot twice for 10 min in at least 200 ml 0.5% BSA, 1 mM EDTA, 40 mM NaH$_2$PO$_4$, pH 7.0, 5% SDS at 65°C.
7. Wash 4 times for 10 min in 1 mM EDTA, 40 mM NaH$_2$PO$_4$, pH 7.0, 1% SDS at 65°C.
8. Air dry the blot.

X. METHODS OF DETECTION AND QUANTIFICATION

A. Signal Detection

The most common method for the detection of molecular hybridization signals is autoradiography. This technique involves the transfer of energy from the radioactive probe to silver grains contained on light-sensitive X-ray film. The bombardment of radioactive emissions onto the film causes these silver grains to darken that portion of the film on subsequent photographic development. Thus, the radioactive signal is transduced into a dark image on a clear background. The methods described in this chapter rely on the use of the strong β-particle emitting isotope ^{32}P. This isotope obviates the need for chemical signal enhancers in the sample to transduce emissions into higher energy photon emissions to expose the film. Thus, this particular type of autoradiography is termed "direct" autoradiography. Direct autoradiography is best performed at room temperature since this is optimal for film response. Exposure times are proportional to the radioactive signal intensity and are best determined empirically. The use of impregnating fluors, termed fluorography, although helpful for weaker β emitters such as ^{3}H, ^{14}C, and ^{35}S, offers no advantage for ^{32}P. However, the use of intensifying screens will boost film response times for ^{32}P probes roughly 10-fold. These image-enhancing screens work optimally at subambient temperatures ($-70°$ to $-80°$C). It is recommended that most ^{32}P-probed Southern or Northern blot experiments be analyzed first by a short exposure to X-ray film in the presence of an intensifying screen at low temperature. Thereafter, optimal film exposures can be obtained with prolonged exposures with an intensifying screen at low temperature or without a screen at ambient temperature. The only real disadvantages of intensifying screens are slight loss of signal resolution due to emission scattering and the high cost of the screens themselves. The two most commonly used Kodak films are XAR-5 and XRP. The former is faster but of lower contrast than the latter.

Technologies have now been developed that allow more rapid detection of radioactive images. These systems either directly sample the radioactive emissions or scan the energy from the sample that has first been transferred to a storage phosphor screen. These signals are then used to create a digital image of the original sample. Both these technologies allow a 10-fold reaction in image production time compared with autoradiography at low temperature in the presence of an intensifying screen. However, the power of these imaging technologies is that they can perform quantitative signal analyses, as will be discussed later.

Autoradiography Protocol

1. Perform all manipulations in a photographic dark room with safelight only. Handle X-ray film by the edges or with gloved hands since fingerprints will appear as images on the developed film.
2. Wrap the radioactive sample tightly with clear plastic film (e.g., Saran wrap).
3. Place the sample up against a piece of X-ray film contained in a light-tight cassette.[1] If an intensifying screen is used, first place the screen in the cassette with the active surface facing up; then place a piece of film against the screen; then place the sample against the film before sealing the cassette. A second intensifying screen can be placed on top of this sandwich to further boost signal speed. However, the second screen's active surface must face the active surface of the first screen.
4. Expose the cassette at ambient temperature if no intensifying screen is used or at −70 to −80°C if an intensifying screen is used.
5. Remove the film from the cassette and develop using commercially available developer and fixative solutions according to the manufacturer's directions before rinsing well with water and air drying. Alternatively, an automated X-ray film developer may be used.

NOTE: *It is advisable to allow cassettes exposed at low temperatures to thaw at room temperature for 30 min before developing the film, to prevent film condensation which can lead to stray signals on the film.*

---

[1] A cassette can be fashioned in the laboratory using stiff cardboard and sealed with light-tight material such as heavy foil. Alternatively, paper, vinyl, or metal cassettes can be purchased commercially.

6. If re-exposure is desired, the cassette, sample, and intensifying screens must all be warmed to ambient temperature and free from condensation to prevent the appearance of water marks on the subsequent autoradiogram.

B. Signal Quantification

Many investigators regard the use of Southern and Northern blot technologies as qualitative tools. They look for the presence or absence of signals or, at best, at faint, strong, and intense signal differences. Older methodologies to quantify autoradiographic images as well as the more recent imaging technologies previously discussed are available. The older methodology, quantitative densitometric scanning, involves the determination of the optical density of autoradiographic signals on X-ray film using a laser scanner. This technique requires rigorous attention to film exposure conditions and the simultaneous use of a series of known standards to determine the linear (dynamic) range of the assay. The X-ray film should be preflashed to an optical density of approximately 0.05 with a shielded light source before autoradiography to improve the dynamic range of the assay at low signal intensities. This technique, under the most ideal conditions and after multiple autoradiographic exposures and densitometer runs, is capable of a dynamic range of only 1.5–2 orders of magnitude. Despite these limitations, laser densitometers are affordable ($10,000 to $15,000) for many laboratories.

Innovations in image processing of radioactive samples involve the use of direct detection of radioactive emissions or the detection of radioactive energy stored in phosphor screens previously exposed to radioactive samples. These technologies are capable of quantifying radioactive samples with dynamic ranges of 3–5 orders of magnitude without the cumbersome constraints of laser densitometry. The price of these instruments ($50,000 to $100,000) puts this technology out of the reach of many investigators.

XI. EXAMPLES OF BLOTTING TECHNOLOGY

A. Southern Blot: Detection of HIV-1 RNAs of Negative Strand Polarity by PCR

The presence of negative strand HIV-1 transcripts produced by H9 cells acutely infected with the HXB3, RF, or MN strain of HIV-1 is shown in Fig. 4. A reverse transcriptase–PCR (RT-PCR) approach was employed with primers specific for a 136-bp fragment of *gag* that can amplify cDNA of

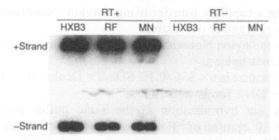

Figure 4 HIV-1 RNA of negative strand polarity as evidenced by RT-PCR. Duplicate 6-ng aliquots of total cellular RNA from H9 cells acutely infected with HXB3, RF, or MN for 3 days were subjected to RT-PCR analysis in the presence (RT+, *left*) or the absence (RT−, *right*) of reverse transcriptase. The upper row of signals result from the use of a negative strand primer (antiparallel to positive strand viral RNA) during the RT step whereas the lower signals result from the use of a positive strand primer (parallel to positive strand viral RNA) during the RT step. The polarity of amplified cDNA is indicated at the left of the figure. The *gag* primers encompass HXB2 sequence positions 1403–1538. The signals were obtained by hybridization with an oligodeoxynucleotide representing an internal sequence within the fragment.

either positive or negative strand orientation, depending on the polarity of the primer chosen for the reverse transcription step. Detection and sequence specificity of the amplified DNA fragment was shown by Southern blotting and hybridization with an internal oligodeoxynucleotide probe. The dependence on RNA as initial template was shown by parallel reactions performed in the absence of reverse transcriptase. RT-PCR reactions performed in the absence of a specific primer during the reverse transcription step to control for the possibility of nonspecific priming by small DNAs contained in the RNA samples were uniformly negative (data not shown). Relative signal intensities in these experiments could not be interpreted since the assay was performed in a strictly qualitative fashion.

Methodology

1. Prepare RNA from acutely infected H9 cells by the RNAzol (Cinna/Biotecz, Friendswood, TX) method.
2. Convert 6 ng RNA to cDNA and amplify using an RT-PCR technique.
3. Electrophorese 20% of the DNA contained in the RT-PCR reaction through a 15% agarose gel in TBE buffer at 6 V/cm for 1 hr. Electrophorese *Hae* III digest of πX174 in parallel to provide size markers.
4. Rock the gel for 20 min in denaturing solution (1.5 M NaCl/0.5 N NaOH); then for 20 min in neutralizing solution (1 M Tris-HCl, pH 7.4/1.5 M NaCl) at room temperature.

5. Set up a capillary transfer using a nylon membrane and 10× SSC; allow to transfer overnight.
6. Mark the nylon blot with indelible ink for orientation and subject to UV cross-linking.
7. Prehybridize in 6× SSC/0.1% SDS/1× Denhardt's/100 μg/ml herring sperm DNA for 30 min at 42°C.
8. Carry out hybridization in the same buffer supplemented with 1.5×10^5 cpm/ml of ^{32}P-labeled oligodeoxyribonucleotide overnight at 42°C.
9. Wash the blot 3 times for 15 min with 1× SSC/0.1% SDS at 37°C, followed by a single wash for 15 min with 0.1× SSC/0.1% SDS at 37°C.
10. Remove excess fluid from the blot and wrap in Saran wrap prior to autoradiography.

B. Northern Blot

Analysis of the induction kinetics of HIV-1 RNA in U1 cells is shown in Fig. 5. Whereas the steady-state levels of spliced RNAs, sized at 4.4 kb and 2.0 kb, were substantial in unstimulated U1 cells, the level of unspliced RNA

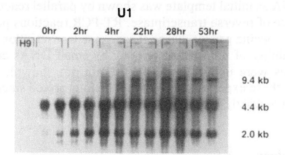

Figure 5 Northern blot analysis of total cellular RNA from U1 cells. These cells are persistently infected with HIV$_{LAI}$. Cells were harvested at the indicated times after exposure to phorbol myristate acetate (PMA); RNA was prepared by a chaotropic salt method. Subsequently, 5 μg RNA was electrophoresed in each of two duplicate lanes through a 1% formaldehyde–agarose gel for each time point shown, transferred to a nylon membrane, and probed with an RNA probe common to the 3′ end of all HIV-1 transcripts. All signals shown were from the same 16-hr exposure of a single nylon membrane. Hybridization signals were quanitfied in a radioanalytic imager (BetaScope; BetaGen, Waltham, MA) and normalized to a beta-actin probe. The apparent molecular weight of each transcript class (shown to the right of the autoradiogram) was determined from the co-migration of a known array of DNA fragments and by the position of rRNAs and tRNAs in the total RNA preparation.

(9.4 kb) was only twice that of the background radioactivity on the blot by radioanalytic imaging. The level of 2.0-kb RNA doubled after 2 hr of phorbol myristate acetate (PMA) stimulation but the amount of the 4.4-kb species did not change significantly from the basal level. As determined by quantitative image analysis, both the spliced transcript class levels increased after 4 hr of stimulation. Unspliced RNA also began to accumulate at this point. Further increases in all species occurred after 22 hr of stimulation. Unspliced RNA accumulated more rapidly than spliced RNA. No further accumulation of 2.0-kb or 4.4-kb RNA is seen after 28 hr of stimulation. Little additional accumulation of HIV-1 RNA occurred at time points between 28 and 53 hr of stimulation. However, cell viability began to fall below 80% after 48 hr. Quantification of the RNA induction kinetics from U1 cells indicated a 3-fold rise in total HIV-1 RNA over the time course as shown in Fig. 5. The 4.4-kb and 2.0-kb RNA size classes rose 3-fold, as well. Unspliced RNA accumulated to a level 25-fold higher than that of unspliced RNA in unstimulated U1 cells.

Methodology

1. Prepare RNA from acutely infected H9 cells by the RNAzol method.
2. Dilute RNA into loading buffer (50% formamide/7% formaldehyde/ 20 mM MOPS, 0.05% BPB), ensuring that the volume of loading buffer is equal to or exceeds the volume of RNA. Heat samples at 50°C for 15 min prior to being placed on ice.
3. Electrophorese samples at ≤6 V/cm until the BPB tracking dye migrates 80% of the distance to the bottom of the gel.
4. Photograph the gel.
5. Rock the gel for 20 min in denaturing solution followed by 20 min in neutralizing solution.
6. Subject the blot to capillary transfer blot for ≥8 hr.
7. Rinse the blot thoroughly in distilled water and UV cross-link.
8. Carry out prehybridization in 50% formamide/5× SSC/5× Denhardt's/20 mM Tris-HCl, pH 7.4/25 μg/ml yeast tRNA for 3 hr at 42°C.
9. Carry out hybridization in the same buffer supplemented with 1.5 × 10^5 cpm/ml of a ^{32}P-labeled cRNA probe overnight at 48°C.
10. Wash the blot twice for 15 min at room temperature with 2× SSC/ 0.1% SDS, followed by 2 washes for 15 min at 50°C with 0.1× SSC/ 0.1% SDS.
11. Remove excess fluid from the blot and wrap in Saran wrap prior to autoradiography.

C. Slot Blot

Analysis of relative tRNA amounts in samples of total cellular RNA from U1 cells is shown in Fig. 6. Total cellular RNA, in descending amounts from various time points following phorbol ester induction, was immobilized onto a nylon membrane using a slot blot manifold. The membrane was then probed with ^{32}P-labeled tRNA, washed, and subjected to autoradiography. All four dilutions revealed slightly different amounts of signal in the original RNA samples when scanned using a direct radioactive imager. The relative amounts of signal, reading left to right on the figure, were 1.1, 1.4, 1.1, 1.2, 1.1, 1.7, 2.3. This calibration allowed for normalization of the HIV-1-specific signals on the Northern blot shown in Fig. 5.

Methodology

1. Dilute total cellular RNA to 0.1 μg/ml using 15$\times$ SSC prior to serial 10-fold dilution.
2. Boil all dilutions for 5 min and then plunge into an ice-water bath for 2 min.
3. Load 50 μl of the cooled dilutions into the appropriate portion of a slot blot manifold containing a nylon membrane and draw through by gentle vacuum suction.
4. Rinse the wells 3 times with 100 μl 15$\times$ SSC.

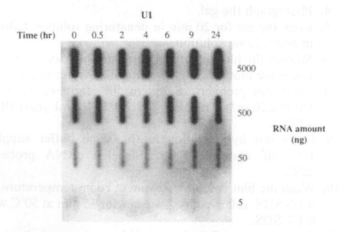

Figure 6 Total cellular RNA from various time points after induction of U1 cells was analyzed by RNA slot blot. The time points are given along the top of the figure and the amount of total cellular RNA loaded is given along the right side of the figure.

5. Remove the nylon membrane from the manifold and cross-link with UV.
6. Carry out prehybridization in 1× SSC, 100 μg/ml salmon sperm DNA, 1× Denhardt's solution at 42°C for 2 hr.
7. Probe the blot with 2 × 10^6 cpm/ml of ^{32}P-labeled tRNA in hybridization buffer overnight at 42°C.
8. Wash the blot 4 times for 30 min in 1× SSC at 37°C, followed by a single wash in 0.1× SSC at 37°C, and subsequent autoradiography and digital scanning.

XII. SOUTHERN BLOTTING AND THE POLYMERASE CHAIN REACTION

Southern and Northern blotting with subsequent nucleic acid hybridization remains the primary method of detection for many viruses in clincial specimens. With the advent of PCR, a second very powerful tool for detection of low abundance nucleic acids became available to the molecular diagnostic laboratory. The high sensitivity of PCR makes it ideally suited to detection of viral nucleic acids in infected materials, where copy number per cell can be extremely low. PCR and Southern blotting are not mutually exclusive procedures. Southern blotting of PCR products is routinely done to increase the sensitivity of detection and to confirm the identity of the amplified DNA. The main barrier to PCR detection of viral genomes is the design of appropriate oligonucleotide primers based on knowledge of viral target nucleic acid sequences. In general, this obstacle is not insurmountable since the nucleotide sequences of many viral pathogens have been determined, and custom synthesized as well as premade primer pairs for virus detection are commercially available. When primers are unavailable, blotting and hybridization techniques can often satisfactorily identify virus-infected material.

REFERENCES

Alwine, J. C., Kemp, D. J., and Stark, G. R. (1977). Method for detection of specific RNAs in agarose gels by transfer to diazobenzyloxmethyl-paper and hybridization probes. *Proc. Natl. Acad. Sci. USA* **74,** 5350–5354.

Aslanzadeh, J., Osmon, D. R., Wilhelm, M. P., Espy, M. J., and Smith, T. F. (1992). A prospective study of the polymerase chain reaction for detection of herpes simplex virus in cerebrospinal fluid submitted to the clinical virology laboratory. *Mol. Cell. Probes* **6,** 367–373.

Berneman, Z. N., Ablashi, D. V., Li, G., Eger-Fletcher, M., Reitz, M. S., Jr., Hung, C. L.,

Brus, I., Komaroff, A. L., and Gallo, R. C. (1992). Human herpesvirus 7 is a T-lymphotropic virus and is related to, but significantly different from, human herpesvirus 6 and human cytomegalovirus. *Proc. Natl. Acad. Sci. USA* **89,** 10552–10556.

Bloss, J. D., Wilczynski, S. P., Liao, S. Y., Walker, J., Manetta, A., and Berman, M. L. (1990). The use of molecular probes to distinguish new primary tumors from recurrent tumors in gynecologic malignancies. *Am. J. Clin. Pathol.* **94,** 432–434.

Boerman, R. H., Arnoldus, E. P. J., Raad, A. K., Peters, A. C. B., Schegget, J., and van der Ploeg, M. (1989). Diagnosis of progressive multifocal leucoencephalopathy by hybridization techniques. *J. Clin. Pathol.* **42,** 153–161.

Brechot, C., Hadchovel, M., Scotto, J., Fonck, M., Potet, F., Vyas, G. N., and Tiollais, P. (1981). State of hepatitis B virus DNA in hepatocytes of patients with hepatitis B surface antigen-positive and -negative liver diseases. *Proc. Natl. Acad. Sci. USA* **78,** 3906–3910.

Buffone, G. J., Demmler, G. J., Schimbor, C. M., and Yow, M. D. (1988). DNA hybridization assay for congenital cytomegalovirus infection. *J. Clin. Microbiol.* **26,** 2184–2186.

Burmer, G. C., Parker, J. D., Bates, J., East, K., and Kulander, B. G. (1990). Comparative analysis of human papillomavirus detection by polymerase chain reaction and virapap/viratype kits. *Am. J. Clin. Pathol.* **94,** 554–560.

Casey, J., and Davidson, N. (1977). Rates of formation and thermal stabilities of RNA:DNA and DNA:DNA duplexes at high concentrations of formamide. *Nucleic Acids Res.* **4,** 1539–1552.

Chirgwin, J. M., Przybyla, A. E., MacDonald, R. J., and Rutter, W. J. (1979). Isolation of biologically active ribonucleic acid from sources enriched in ribonuclease. *Biochemistry* **18,** 5294–5299.

Church, G. M., and Gilbert, W. (1984). Genomic sequencing. *Proc. Natl. Acad. Sci. USA* **81,** 1991–1995.

Churchill, M. A., Zaia, J. A., Forman, S. J., Sheibani, K., Azumi, N., and Blume, K. G. (1987). Quantitation of human cytomegalovirus DNA in lungs from bone marrow transplant recipients with interstitial pneumonia. *J. Infect. Dis.* **155,** 501–509.

Clewley, J. P. (1985). Detection of human parvovirus using a molecularly cloned probe. *J. Med. Virol.* **15,** 173–181.

Cunningham, D. A., Pattison, J. R., and Craig, R. K. (1988). Detection of parvovirus DNA in human serum using biotinylated RNA hybridization probes. *J. Virol. Meth.* **19,** 279–288.

Danna, K., and Nathans, D. (1971). Specific cleavage of simian virus 40 DNA by restriction endonuclease of *Hemophilus influenzae. Proc. Natl. Acad. Sci. USA* **68,** 2913–2917.

Desrosiers, R. C. (1982). Specifically unmethylated cytidylic-guanylate sites in herpesvirus saimiri DNA in tumor cells. *J. Virol.* **43,** 427–435.

Elsner, C., and Dorries, K. (1992). Evidence of human polyomavirus BK and JC infection in normal brain tissue. *Virology* **191,** 72–80.

Feinberg, A. P., and Vogelstein, B. (1983). A technique for radiolabeling DNA restriction endonuclease fragments to high specific activity. *Anal. Biochem.* **132,** 6–13.

Flavell, R. A., Birfelder, E. J., Sanders, J. P., and Borst, P. (1974). DNA–DNA hybridization on nitrocellulose filters. 1. General considerations and nonideal kinetics. *Eur. J. Biochem.* **47,** 535–543.

Forghani, B., Hurst, J. W., and Shell, G. R. (1991). Detection of the human immunodeficiency virus genome with a biotinylated DNA probe generated by polymerase chain reaction. *Mol. Cell. Probes* **5,** 221–228.

Gillespie, D., and Spiegelman, S. (1965). A quantitative assay for DNA–RNA hybrids with DNA immobilized on a membrane. *J. Mol. Biol.* **12,** 829–842.

Golden, M. P., Kim, S., Hammer, S. M., Ladd, E. A., Schaffer, P. A., DeLuca, N., and Albrecht, M. A. (1992). Activation of human immunodeficiency virus by herpes simplex virus. *J. Infect. Dis.* **166,** 494–499.

Grimberg, J., Nawoschik, S., Belluscio, L., McKee, R., Turck, A., and Eisenberg, A. (1989). A simple and efficient non-organic procedure for the isolation of genomic DNA from blood. *Nucleic Acids Res.* **17,** 8390.

Henson, J., Rosenblum, M., Armstrong, D., and Furneauy, H. (1991). Amplification of JC virus DNA from brain and cerebrospinal fluid of patients with progressive multifocal leukoencephalopathy. *Neurology* **41,** 1967–1971.

Herndier, B. G., Shiramizu, B. T., Jewett, N. E., Aldape, K. D., Reyes, G. R., and McGrath, M. S. (1992). Acquired immunodeficiency syndrome-associated T-cell lymphoma: Evidence for human immunodeficiency virus type 1-associated T-cell transformation. *Blood* **79,** 1768–1774.

Horowitz, M. S., Boyce-Jaeino, M. T., and Faras, A. J. (1992). Novel human endogeneous sequences related to human immunodeficiency virus type 1. *J. Virol.* **66,** 2170–2179.

Jeanpierre, M. (1987). A rapid method for the purification of DNA from blood. *Nucleic Acids Res.* **15,** 9611.

Jenkins, F. J., Howett, M. K., Spector, D. J., and Rapp, F. (1982). Detection by RNA blot hybridization of RNA sequences homologous to the *Bg*III-N fragment of herpes simplex virus type 2 DNA. *J. Virol.* **44,** 1092–1096.

Kafatos, F. C., Jones, C. W., and Efstratiadis, A. (1979). Determination of nucleic acid sequence homologies and relative concentrations by a dot hybridization procedure. *Nucleic Acids Res.* **7,** 1541–1552.

Kelly, T. J., and Smith, H. O. (1970). A restriction enzyme from Hemophilus influenzae. II. *J. Mol. Biol.* **51,** 393–409.

Krzyzek, R. A., Watts, S. L., Anderson, D. L., Faras, A. J., and Pass, F. (1980). Anogenital warts contain several distinct species of human papillomavirus. *J. Virol.* **36,** 236–244.

Kulski, J. K., Demeter, T., Mutavdzic, S., Sterrett, G. F., Mitchell, K. M., and Pixley, E. C. (1990). Survey of histologic specimens of human cancer for human papillomavirus type 6/11/16/18 by filter in situ hybridization. *Am. J. Clin. Pathol.* **94,** 561–565.

Laird, C. D., McConaughy, B. L., and McCarthy, B. J. (1969). Rate of fixation of nucleotide substitutions in evolution. *Nature* **224,** 149–154.

Landini, M. P., Trevisani, B., Guan, M. X., Ripalti, A., Lazzarotto, T., and La Placa, M. (1990). A simple and rapid procedure for the direct detection of cytomegalovirus in urine samples. *J. Clin. Lab. Anal.* **4,** 161–164.

Lehrach, H., Diamond, D., Wozney, J. M., and Boedtker, H. (1977). RNA molecular weight determinations by gel electrophoresis under denaturing conditions, a critical reexamination. *Biochemistry* **16,** 4743.

McConaughy, B. L., Laird, C. D., and McCarthy, B. J. (1969). Nucleic acid reassociation in formamide. *Biochemistry* **8,** 3289–3295.

Melton, D. A., Drieg, P. A., Rebagliati, M. R., Maniatis, T., Zinn, K., and Green, M. R. (1984). Efficient *in vitro* synthesis of biologically active RNA and RNA hybridization probes from plasmids containing a bacteriophage SP6 promoter. *Nucleic Acids Res.* **12,** 7035–7056.

Mori, J., Field, A. M., Clewley, J. P., and Cohen, B. J. (1989). Dot blot hybridization assay of B19 virus DNA in clinical specimens. *J. Clin. Microbiol.* **27,** 459–464.

Polonis, V. R., Anderson, G. R., Vahey, M. T., Morrow, P. J., Stoler, D., and Redfield, R. R. (1991). Anoxia induces human immunodeficiency virus expression in infected T cell lines. *J. Biol. Chem.* **266,** 11421–11424.

Puissant, C., and Houdebine, L.-M. (1990). An improvement of the single-step method of RNA isolation by acid guanidinium thiocyanate–phenol–chloroform extraction. *BioTechniques* **8,** 148–149.

Qian, Y., Saif, L. J., Kapikian, A. Z., Kang, S. Y., Jiang, B., Ishimaru, Y., Yamashita, Y.,

Oseto, M., and Green, Y. (1991). Comparison of human and porcine group C rotaviruses by Northern blot hybridization analysis. *Arch. Virol.* **118,** 269–277.

Quick, C. A., Watts, S. L., Krzyzek, R. A., and Faras, A. J. (1980). Relationship between condylomata and laryngeal papillomata. *Ann. Otol. Rhinol. Laryngol.* **89,** 467–471.

Rigby, P. W. J., Dieckmann, M., Rhodes, C., and Berg, P. (1977). Labeling deoxyribonucleic acid to high specific activity in vitro by nick translation with DNA polymerase I. *J. Mol. Biol.* **113,** 237–251.

Salimans, M. M., Holsappel, S., van de Rijke, F. M., Jiwa, N. M., Raap, A. K., and Weiland, H. T. (1989). Rapid detection of human parvovirus B19 DNA by dot hybridization and the polymerase chain reaction. *J. Virol. Meth.* **23,** 19–28.

Schildkraut, C. (1965). Dependence of the melting temperature of DNA on salt concentration. *Bipolymers* **3,** 195–208.

Seed, B. (1982). Attachment of nucleic acids to nitrocellulose and diagonium-substituted supports. *In* "Genetic Engineering: Principles and Methods" (J. K. Setlow and A. Hollaender, eds.), Vol. 4, p. 91. Plenum Publishing, New York.

Shibata, M., Morishima, T., Terashima, M., Kimura, H., Kuzushima, K., Hanada, N., Nishikawa, K., and Watanabe, K. (1990). Human cytomegalovirus infection during childhood: detection of viral DNA in peripheral blood by means of polymerase chain reaction. *Med. Microbiol. Immunol.* **179,** 245–253.

Southern, E. M. (1975). Detection of specific sequences among DNA fragments separated by gel electrophoresis. *J. Mol. Biol.* **98,** 503–517.

St. Amand, J., Beard, C., Humphries, K., and Astell, C. R. (1991). Analysis of splice junctions and in vitro and in vivo translation potential of the small, abundant B19 parvovirus RNAs. *Virology* **183,** 133–142.

Street, J. E., Croxson, M. C., Chadderton, W. F., and Bellamy, A. R. (1982). Sequence diversity of human rotavirus strains investigated by Northern blot hybridization analysis. *J. Virol.* **43,** 369–378.

Taveira, N. C., Ferreira, M. D., and Pereira, J. M. (1992). Detection of HIV 1 proviral DNA by PCR and hybridization with digoxigenin labelled probes. *Mol. Cell. Probes* **6,** 265–270.

Thomas, P. S. (1980). Hybridization of denatured RNA and small DNA fragments transferred to nitrocellulose. *Proc. Natl. Acad. Sci. USA* **77,** 5201–5205.

Twist, E. M., Clark, H. F., Adon, D. P., Knowles, B. B., and Plotkin, S. A. (1981). Integration pattern of hepatitis B virus DNA in human hepatoma cell lines. *J. Virol.* **37,** 239–243.

Vallbracht, A., Lohler, J., Gossman, J., Gluck, T., Petersen, D., Gerth, H.-J., Gencic, M., and Dorries, K. (1993). Disseminated BK type polyomavirus infection in an AIDS patient associated with central nervous system disease. *Am. J. Pathol.* **143,** 29–39.

Watson, J. D., and Crick, F. H. C. (1953). Molecular structure of nucleic acids. A structure of deoxyribose nucleic acid. *Nature* **171,** 737–738.

Wetmur, J. G. (1971). Excluded volume effects on the rate of renaturation of DNA. *Bipolymers* **10,** 601–613.

Wetmur, J. G., and Davidson, N. (1968). Kinetics of renaturation of DNA. *J. Mol. Biol.* **1,** 349–370.

4

In Situ Hybridization

Jeanne Carr
Thomas F. Puckett Laboratory
Hattiesburg, Mississippi 39402

I. INTRODUCTION

In situ hybridization (ISH) was first described in 1969 for localizing specific DNA sequences on chromosomes in cytological preparations of *Xenopus* oocytes (Gall and Pardue, 1969; John *et al.*, 1969). Modifications of the original technique have been described that permit the detection and localization of nucleic acids within tissues from a variety of different organisms.

The principle advantage of ISH over other molecular techniques is that ISH provides information about the location of the target nucleic acids within cells and/or tissues. Pathologists are simultaneously able to look for specific nucleic acids and to study the cellular and tissue morphology of the sample. In addition, ISH can provide results more rapidly than some cell culture methods.

Some viruses are difficult, dangerous, or impossible to grow in cell cultures. For instance, primary human fetal glial cells can be used to isolate JC virus. Unfortunately, after several subcultures these primary cells are no longer permissive for viral infection, making routine culturing of JC virus difficult or impossible. Human immunodeficiency virus (HIV) can be cultivated *in vitro,* but culturing this dangerous organism has no real benefit for most clinical laboratories. Human papillomavirus (HPV) cannot be cultivated *in vitro,* so *in situ* hybridization is especially valuable when HPV infections are suspected. Additionally, ISH can be helpful in cases when the immunological diagnosis of viral infections is difficult (e.g., JC virus) and when there are no reliable, commercially available serologic reagents (Hulette *et al.,* 1991).

The target molecules of ISH are nucleic acids. Nucleic acids are complexed with proteins in the cell; when a tissue is embedded in a complex matrix, the nucleic acids are cross-linked to that matrix. Thus major challenges of ISH are to make target nucleic acid available to the probe; and once proper hybrids are formed, to stabilize them without destroying the cell morphology.

II. GENERAL PROCEDURES

In situ hybridization procedures have several steps:

1. Processing of tissue—fixation, embedding, cutting, and mounting
2. Prehybridization—to increase the accessibility of the target nucleic acids and to decrease hybridization background
3. Hybridization—to allow the probe to bind to the target nucleic acids
4. Washing—to remove unhybridized probe
5. Detection—to observe the signal

ISH methods can use isotopic or nonisotopic detection methods. The isotopic technique employs a probe that has been labeled with a radioactive molecule such as ^{35}S. The nonisotopic technique uses a probe that has been labeled with a nonradioactive molecule such as biotin. Although radioactive probes are still in use and must be used if a quantitative result is desired, nonisotopic probes are more convenient for most clinical laboratories. A diagram of a nonisotopic *in situ* hybridization assay is given in Fig. 1.

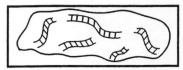

I. Processed tissue sample on microscope slide.

II. Denaturation of nucleic acid into single strands; addition of biotin-labeled probe.

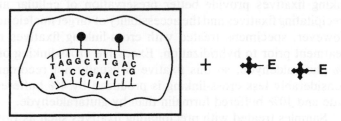

III. Hybridization of probe to homologous target nucleic acid; removal of unbound
probe by washing; addition of enzyme-labeled avidin.

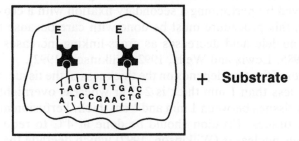

IV. Binding of enzyme-labeled avidin to biotin-labeled probe; removal of unbound
probe by washing; addition of substrate.

V. Examination of tissue for appropriate signal.

Figure 1   Basic outline of a nonisotopic *in situ* hybridization assay.

A. Processing Tissue

1. Fixation

Ideally, the sample should be treated so that the target nucleic acid is made available to the probe without destroying the target nucleic acids or the cellular architecture. Thus, tissues may be snap-frozen or placed in a fixative. Two types of fixative are commonly used with ISH: cross-linking fixatives and precipitating fixatives. Cross-linking fixatives include 4% buffered paraformaldehyde, 10% buffered formalin, and glutaraldehyde. Cross-linking fixatives provide better preservation of cellular architecture than precipitating fixatives and the accessibility of target nucleic acid is also better. However, specimens treated with cross-linking fixatives require protease treatment prior to hybridization. Excessive cross-linking often occurs with 1% glutaraldehyde, so this fixative is usually not recommended for ISH. Considerably less cross-linking is produced by 4% buffered paraformaldehyde and 10% buffered formalin than by glutaraldehyde.

Samples treated with precipitating fixatives such as Bouin's, Carnoy's, acetone, or ethanol–acetic acid do not preserve the target nucleic acids or cellular structures as well as cross-linking agents. Cellular morphology can be preserved by performing a secondary fixation with a cross-linking agent. However, this procedure must be done with care because the accessibility of target nucleic acid decreases as cross-linking increases (Lawrence and Singer, 1985; Lewis and Wells, 1992; Wilkinson, 1992).

Length of fixation depends on the thickness of the tissue. Optimal fixation for tissue less than 1 mm thick is 2–3 hr whereas overnight fixation is suggested for tissues between 1 mm and 1 cm thick. Perfusion may be necessary for larger tissues. Fixation should be done at 4°C to reduce the action of endogenous nucleases (Wilkinson, 1992) which degrade the nucleic acids in the tissues and thereby reduce the signal. Special care is required when the target is RNA because endogenous ribonucleases (RNases) are especially resistant to inactivation. Gloves should always be worn when testing for RNA because hands are replete with RNases. Reagents must also be treated with diethylpyrocarbonate (DEPC) and/or be autoclaved (Lloyd, 1987).

2. Embedding and Cutting

Fresh frozen tissues can be cut with a cryostat and placed on microscope slides. However, frozen tissue sections must be washed to remove water-soluble freezing medium (OCT) and then fixed before hybridization.

Samples originally placed in fixative can be processed routinely following standard laboratory procedures. After embedding in paraffin wax, the sample is cut into 3- to 5-μm slices with a microtome, floated atop an additive-free water bath, and mounted on a microscope slide.

3. Mounting

One of the most common problems with ISH is loss of sample from the slide during the procedure. Coating of microscope slides with materials such as 3-amino-propyltriethoxysilane, polylysine, gelatin, or Elmer's Glue™ (Brahic and Ozden, 1992) can enhance the adherence of the section to the slide.

Samples other than paraffin-embedded tissue may be tested by ISH. Suspension cell cultures can be cytocentrifuged onto treated slides, air dried, and placed into fixative (Brahic and Ozden, 1992). Another method is centrifuging dissociated cells, removing the culture medium, washing twice in saline, and fixing the cells by resuspending them in 10% unbuffered formol–saline overnight. The cells are then centrifuged to form a pellet that is placed in paraffin wax (Coates *et al.*, 1991). If anchorage-dependent cells are grown in flasks, on cover slips, or in multiwell trays (Mougin *et al.*, 1991), the fixative can be added directly to the vessel or well. Alternatively, cells can be trypsinized, centrifuged, washed, and cytocentrifuged onto treated slides (Brahic and Ozden, 1992).

B. Prehybridization

Before hybridization, steps must be taken to increase the accessibility of target nucleic acid and to decrease any background signal that occurs due to nonspecific binding of the probe. First, the paraffin is melted. The temperature and length of incubation must be adequate to melt the paraffin and increase the adherence of the tissue to the slide. The optimum conditions must be determined empirically. However, incubation at 55–65°C for 30 min to overnight is usually sufficient. The slides are allowed to cool before dewaxing and tissue rehydration through graded ethanol rinses. Proteolytic enzymes such as proteinase K, pronase, or pepsin are placed on the mounted tissue. This treatment, with or without the addition of a mild detergent, can reduce protein cross-linking and give the probe easier access to the target nucleic acid.

To reduce background signal, tissue binding sites should be blocked with carrier nucleic acids (e.g., tRNA, salmon sperm DNA), Denhardt's solution, acetic anhydride, or heparin (Sambrook *et al.*, 1989; Wilkinson, 1992). Acetic anhydride acetylates positively charged amino acids, thereby reducing non-specific attachment of the negatively charged probe to these residues (Wilkinson, 1992). When using an enzymatic detection system with tissue samples containing endogenous peroxidases, for example, horseradish peroxidase with liver, the sample can be treated with methanol containing 1% hydrogen peroxide for 30 min to block the endogenous peroxidase activity (Naoumov et al., 1988; Lau et al., 1991). Some tissues such as kidney, liver, and pancreas (Green et al., 1992) contain biotin-binding proteins and can produce nonspecific granular cytoplasmic staining (Naoumov et al., 1988). The following controls should be run to rule out nonspecific staining due to endogenous biotin with these types of tissues: (1) a section of tissue should be treated with RNase before hybridization, (2) a section of tissue should be hybridized with a 100:1 ratio of unlabeled probe to labeled probe, and/or (3) a section of tissue should be hybridized with unlabeled probe only. All these controls should allow a decrease or absence of signal if the signal is target specific (Emson and Gait, 1992). When ^{35}S-labeled probes are used, nonspecific sulfur-binding sites in tissue can be blocked by pretreatment with a mixture of dithiothreitol, iodoacetamide, and N-ethylmaleimide (Zeller and Rogers, 1989).

C. Hybridization

Hybridization conditions vary widely and must be determined experimentally. Commercial ISH kits have helped eliminate much of the trial-and-error process. To achieve an acceptable outcome, however, these protocols may require more adjustments than other commercial kits used in clinical laboratories.

If the target or probe nucleic acid is DNA, it must be denatured before the hybridization of target and probe can begin. If the target is RNA, secondary structure is prominent so this target should be denatured as well. The optimal hybridization temperature is 20–30°C below the melting temperature (T_m).

The T_m is the temperature at which one half of a population of complementary fragments is single stranded or "melted." This temperature is dependent on the guanine–cytosine (GC) content, the length, and the degree of homology of the complementary strands, as well as on the environmental conditions such as the concentration of salts or formamide. The nature of the probe and target also determine the T_m. RNA:RNA hybrids are more stable than RNA:DNA hybrids, which in turn are more stable than DNA:DNA hybrids. The T_m increases with a higher GC content. The GC pairs are held together by three hydrogen bonds and are more stable than adenine–thymine (AT) pairs, which are held together by two hydrogen bonds.

Longer probes and a higher concentration of monovalent cations can increase the T_m. Increasing concentrations of denaturing agents such as formamide destablize the hybrids and decrease the T_m. The use of formamide therefore allows the hybridization reaction to take place at a lower temperature and produces less disturbance of specimen morphology. Note that hybrids are less stable in ISH than in solution hybridizations, presumably because of the steric hindrances imposed by the cross-linking that takes place in the fixed, embedded tissues. The conditions of monovalent cation concentration and temperature for hybridization and washing are referred to as the "stringency" of the protocol (Wilkinson, 1992).

D. Probe Labeling

1. Random Primer Extension

Degenerate oligomeric primers consisting of 6–12 nucleotides containing every combination of the four deoxynucleoside triphosphates in each position can be prepared in the laboratory (Sambrook *et al.*, 1989) or purchased commercially. Such oligonucleotide primers are very heterogeneous and are able to hybridize at many positions along the template to be used for the probe. A DNA polymerase, three unlabeled dNTPs, and one α-labeled dNTP are added. If the template is single-stranded DNA, the Klenow fragment of DNA polymerase I is the polymerase of choice. If the template is single-stranded RNA, reverse transcriptase is added. Any unincorporated precursor may be removed from the completed probe mixture by centrifugation through a small Sephadex G-50 column. This step is usually unnecessary because nearly 90% of the labeled nucleotide is incorporated into the probe (Sambrook *et al.*, 1989).

2. Nick Translation

The double-stranded DNA that is to be used as a probe is "nicked" by the action of pancreatic DNase I. *Escherichia coli* DNA polymerase I, three unlabled dNTPs, and one labeled dNTP are added. The $5' \rightarrow 3'$ exonuclease activity of the DNA polymerase I removes nucleotides beginning at the $5'$ end of each nick. As this takes place, the $3' \rightarrow 5'$ polymerase activity of the enzyme fills in the enlarging gap with free dNTPs, some of which are radiolabeled (Sambrook *et al.*, 1989).

3. Polymerase Chain Reaction

Labeled polymerase chain reaction (PCR) products can be used as ISH probes. In this procedure, oligonucleotide primers that are complementary to the vector sequences flanking the probe sequence are used. *Taq* polymerase, template DNA, and labeled and unlabeled dNTPs are added to the reaction to produce a labeled probe (Sambrook *et al.*, 1989).

4. Synthetic Oligonucleotides

A DNA synthesizer (Applied Biosystems, Foster City, CA) can be used to prepare probes. The instrument employs T4 polynucleotide kinase or terminal deoxynucleotidyl transferase (TdT) to add labeled nucleotides to the ends of oligonucleotides. The sequence of the added nucleotides is determined by the machine operator. The length of the generated probe is usually 20–50 bases.

5. Universal Linkage System

The universal linkage system (ULS) is a nonenzymatic method for nonisotopic labeling of DNA. A platinum derivative acts as a linker between nonradioactive labels (such as biotin, digoxigenin) and the guanine moieties of DNA. The adenine nucleotides are labeled to a lesser extent. This method is reported to produce probes with high labeling densities (Jelsma and Houthoff, 1994).

Several labels have been used with ISH. The most common radioactive labels are ^{32}P, ^{3}H, and ^{35}S. The nonradioactive labels most commonly used are biotin and digoxigenin. Incorporation of biotin or digoxigenin into a probe is done through a modification of the standard nick translation procedure. The random primer method can be employed for labeling with these molecules, but nick translation generally produces smaller probes (Boyle, 1990).

When selecting the type of label to be incorporated into the probe, the advantages and disadvantages of each label should be considered. Early ISH work was done using radioactive probes, which remain the "gold standard" by which other labels are judged. Because of the safety regulations that must be addressed when handling radioisotopes and the shortened detection time needed with nonisotopic probes, the nonisotopic labels have become popular.

Of the radioactive probes, ^{35}S is usually preferred. Although the half-life of ^{35}S is longer than ^{32}P, ^{35}S produces better resolution. Compared with ^{3}H, ^{35}S has lower resolution, but its half-life is shorter and a briefer exposure time is required—approximately 1 week instead of the several weeks required with ³H. As mentioned in the "prehybridization" section, a reducing agent must be included when using ³⁵S-labeled probes (Wilkinson, 1992). Recently ³³P has been reported to give favorable results (McLaughlin and Margolskee, 1993).

Of the nonisotopic labels, biotin is the most commonly used. As stated earlier, some tissues contain endogenous biotin. Although this problem can usually be resolved through proper controls, digoxigenin-labeled probes are being used more frequently.

After hybridization, the unhybridized probe is removed by washing and detection reagents are applied. The stringency of the washing is very important: it must be harsh enough to wash all unhybridized probe away without being so harsh that true target–probe hybrids are destroyed.

Just as steps are taken in the prehybridization portion of the procedure to reduce background signal, a number of steps and reagents can be incorporated after the post-hybridization washes to remove any unhybridized probe that was not removed during washing. If the probe was a riboprobe, RNases can be used to degrade any remaining single-stranded probe. If the probe was DNA, then the slides may be reacted with S1 nuclease. This DNase specifically digests single-stranded DNA.

E. Detection

The method of detection used is determined by the type of label on the probe. If radioactive probes are employed, emulsion autoradiography is used for detection. After hybridization and washing, the slides are dipped into a liquid nuclear track emulsion using a slow and smooth motion so that the emulsion film will be complete and even. The slides are allowed to dry and subsequently are placed in a light-safe slide holder with desiccant. Exposure is carried out at 4°C. Slides are developed in a darkroom. Once developed, the silver grains are seen as white dots on a black background when observed with dark-field microscopy. Strong signals can be observed with transillumination. To enhance the morphology of the specimen, counterstaining with Giemsa, hematoxylin/eosin, or toluidine blue can be performed.

Several detection options are available when nonisotopic probes are used. If the label is biotin, enzyme-conjugated avidin (or streptavidin) molecules are added (Fig. 1). Each avidin molecule has four biotin binding sites. Avidin has a strong affinity for biotin, and the enzyme-conjugated avidin molecules bind readily to the biotin-labeled probe. The sample is washed to remove any unbound avidin–enzyme complex. If the enzyme attached to avidin is alkaline phosphatase, nitroblue tetrazolium (NBT) and bromochloro-indolyl phosphate (BCIP) are added as substrates; a purple-blue precipitate forms in a positive reaction. If the enzyme is horseradish peroxidase, diaminobenzidine (DAB) is the substrate and a positive reaction results in a pink to brick red precipitate. A light microscope is used to read the reaction. Alternatively, biotin-labeled probes can be detected with fluorescently tagged avidin.

If the label is digoxigenin, enzyme-conjugated anti-digoxigenin antibodies are used to detect the probe in the enzyme-mediated reactions already described. Alternatively, the anti-digoxigenin antibodies can be fluorescently tagged (Boyle, 1990).

F. Controls

To validate the conditions for tissue fixation, processing, prehybridization, hybridization, and detection, a control probe that detects human genomic DNA sequences should be included in each batch of reactions. The expected result would be positive staining in all or nearly all of the cells.

To rule out nonspecific staining resulting from hybridization of vector sequences to tissue nucleic acids, a control probe consisting of labeled vector sequences minus the inserts should be run. The expected result would be no positive staining. This control will also be helpful to rule out potential interference (cytoplasmic staining) of endogenous alkaline phosphatase with an alkaline phosphatase detection system. If the target sequences are nuclear, the signal can usually be distinguished from the background (Chang et al., 1992).

Endogenous biotin can cause nonspecific granular cytoplasmic staining (Naoumov et al., 1988). Controls for this problem were outlined in Section II,C. As mentioned previously for endogenous alkaline phosphatase, if the target sequences are nuclear, the target signal may be distinguishable from the background, but elimination of the background sequence is preferable to reading around it. If the target sequences are cytoplasmic RNA, the nonspecific background signal must be eradicated.

All probes should be evaluated for specificity. Cross-reactions have been observed with commercially available biotinylated probes. For instance, some cytomegalovirus (CMV) and herpes simplex virus (HSV) probes have been shown to cross-react with tissues containing *Pneumocystis carinii* (Lloyd, 1987).

An important control to include is a cell culture or tissue sample that has been shown to be positive in the testing system. This control gives continuity to the assay by providing parallel testing materials for validating new lots of reagents.

G. Sensitivity

The sensitivity of an ISH assay depends on many factors. The nature of the probe (i.e., single strandedness versus double strandedness, the label that it carries, etc.) plays an important role in determining the sensitivity of the assay. The organization of the target nucleic acid is also important since a clustered target can be detected more easily than a dispersed one. The stringency of the reactions, the detection system, and individual technique are also important elements.

Some reported sensitivities for ISH with variously labeled probes follow. Radioactive probes can detect 5–10 molecules per 20 μm^3 cell (Wilkinson, 1992). Biotinylated probes can detect approximately 10 copies of HPV per cell (Burns *et al.*, 1987) or less than 5 copies of Epstein–Barr virus (EBV) per cell (Chang *et al.*, 1992) and have been reported to be as sensitive as radioactive probes (Allan *et al.*, 1989). Digoxigenin-labeled probes have reported sensitivities of 2.5–12 copies of HPV per nucleus (Herrington *et al.*, 1991).

III. INSTRUMENTATION

A few instruments are on the market to aid in the performance of ISH assays. Digene Diagnostics, Inc. (Beltsville, MD) has introduced the DTI-1 (Fig. 2), a large heating block with digital temperature control. The heating surface area is 9 ³/₄ × 15 ³/₄ in. The DTI-1 has a hinged cover that seals the unit and minimizes temperature fluctuations. During incubations that require moisture, absorbent pads are provided that can be saturated with deionized water and placed inside the instrument. Triangle Biomedical Sciences (Durham, NC) manufactures the H-IPI-100, which allows staining, rinsing, and incubation within the same unit. Biotek Solutions, Inc. (Santa Barbara, CA) has recently acquired the Code-On Molecular Pathology System. This robotic system handles 60 slides at a time from prehybridization through detection (Hulette *et al.*, 1991).

IV. DETECTION OF HUMAN PAPILLOMAVIRUS BY *IN SITU* HYBRIDIZATION

The protocol presented here is modified from the Digene tissue hybridization kit (Digene Diagnostics, 1991a) and is used in conjunction with the Digene

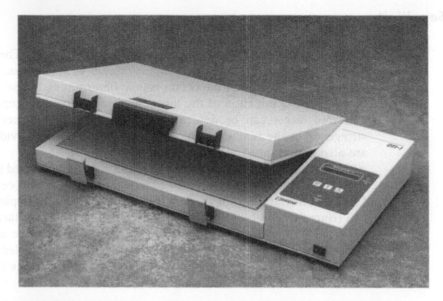

Figure 2   The DTI-1 instrument. Courtesy of Digene Diagnostics, Beltsville, MD.

HPV OmniProbe™ set (Digene Diagnostics, 1991b) and the Digene Vira-Type® *in situ* HPV probe set (Digene Diagnostics, 1991c). The instructions have been reproduced in part with permission from Digene Diagnostics, Inc.

The kit has been optimized for use with routinely processed human tissues that have been fixed with neutral buffered formalin and have been paraffin embedded. Decreased sensitivities have been observed with other fixatives. Optimal results are obtained when tissues have been fixed no longer than 24 hr. Fixation conditions should be standardized to achieve consistent staining results. Proteolytic digestion conditions may need to be modified for different applications such as frozen sections, cytospins, and cytosmears.

A. Protocol

1. Kit Contents

Digestion reagent A: 2 *N* HCl solution
Detection reagent: Alkaline phosphatase conjugate
Substrate 1: 5-bromo-4-chloro-3-indolylphosphate (BCIP) in dimethyl-
    formamide

Substrate 2: Nitroblue tetrazolium (NBT) in dimethylformamide
Batch packs containing:
  Digestion reagent B: A proteolytic enzyme
  Buffers 1, 2, and 3: Wash solutions
Nuclear fast red stain
Treated slides: beveled slides treated with 3-aminopropyltriethoxysaline
  (AES)

2. Materials Required but Not Supplied

Biotinylated HPV DNA probes:
  1. OmniProbe™ for detection of 14 HPV types in fixed, embedded
     tissue (types 6, 11, 16, 18, 31, 33, 35, 42, 43, 44, 45, 51, 52, and
     56)
  2. ViraType® *in situ* HPV probe set for detection of HPV DNA groups
     (types 6/11, 16/18, 31, 33, and 35) in fixed, embedded tissue
  3. Alternative source of biotinylated HPV DNA probes

Neutral-buffered, formalin-fixed, paraffin-embedded tissue
Slide warmer or DTI-1 capable of maintaining 37 ± 2°C
Surface thermometer or Spy-Temp® (Digene Diagnostics, Silver Spring,
  MD)
Water bath capable of maintaining 37 ± 2°C
Oven, incubator, or DTI-1 capable of maintaining 37 ± 2°C
Oven, incubator, or DTI-1 capable of maintaining 58 ± 2°C
Humidity chamber or box (e.g., plastic box lined with wet towels) or
  DTI-1
Staining racks and dishes to hold 25 slides, using 200 ml volumes (Coplin
  jars if smaller batches are being run)
Glass cover slips, 18 × 18 mm (#1)
Slide marker or diamond pen
Xylenes
Absolute ethanol
95% Ethanol
Deionized or distilled water for washes and reconstitution of rea-
  gents
Pipettes able to dispense 500 μl
Graduated cylinders for reagent preparation
Permanent mounting medium miscible with xylenes
Absorbent wipes
Standard light microscope

3. Preparation of Slides

1. Mark AES-treated slides for identification of tissue.
2. Cut serial tissue sections 3–5 μm thick and float them on a deionized water bath *without additives*.
3. Carefully position the sections on the slides, one section per slide. Do not place sections too close to the frosted end of the slide.
4. Air dry the slides.
5. Bake sections on the slides at 58 ± 2°C for at least 30 min, but not longer than 24 hr. It is convenient to place the slides on a metal slide tray and place them into the DTI-1. When incubation is complete, cool the slides to room temperature.

NOTE: *It may be helpful to circle each section with a diamond pen for easier visualization during the assay. If a delay in processing is experienced, store the slides in a dust-free container at room temperature. In our experience, slides stored in this manner are stable for at least 2–3 days.*

4. Equipment Set-Up

1. Heat slide warmer or DTI-1 to 100 ± 2°C. Allow 30 min for the temperature to be reached. A temperature of 100 ± 2°C is important for the best results. A reliable disk-type thermometer is recommended for accurate temperature regulation. If a surface thermometer is not available, Temp-Spy® temperature recorders, specially developed for slide warmers, are available from Digene (Catalog No. 1010–1015).
2. If not using the DTI-1 for hybridization, prepare a humid chamber (petri dish or plastic box lined with clean, wet paper towels) and warm to 37 ± 1°C.

5. Preparation of Reagents

The instructions given here are for 200-ml volumes. Smaller batches using 40-ml volumes can be used if reagents are scaled down accordingly. For buffers 1, 2, and 3, it is recommended to reconstitute as directed, add sodium azide to a final concentration of 0.05%, and store the unused portion at 2–8°C for up to 1 wk. For digestion reagent B, add 80 mg proteolytic enzyme to 40 ml diluted digestion reagent A. Store the unused portion of proteolytic enzyme, desiccated, at 2–8°C.

1. Digestion Reagent A: Dilute digestion reagent A by pipetting 10 ml digestion reagent A into 190 ml deionized water. Mix thoroughly.

2. Buffer 1: Dissolve 1 packet of buffer 1 in 250 ml deionized water. Mix thoroughly.
3. Buffer 2: Dissolve 1 packet of buffer 2 in 1 liter deionized water. Mix thoroughly and allow sufficient time to dissolve completely. Transfer dissolved solution into a staining dish and prewarm solution to $37 \pm 2°C$. Store at 2–8°C if prepared more than 1 hr before use and warm to 37°C before use.
4. Buffer 3: Dissolve 1 packet of buffer 3 in 1 liter deionized water. Mix thoroughly. Occasionally, black particles precipitate that do not interfere with the test.
5. Final digestion solution and substrate solution: These solutions are labile in diluted form and should not be prepared in advance. Directions for preparation are included in the appropriate sections of the procedure.
6. Digene biotinylated DNA probes are provided ready-to-use. If an alternative source of biotinylated DNA is being used, the biotinylated DNA must be diluted in Digene biotin-probe diluent (Catalog No. 4200-1145) to a final concentration of 0.1–1.0 μg/ml. Biotinylated DNA probes added to Digene biotin-probe diluent should not constitute more than 10% of the final volume.

6. Pretreatment of Sections

1. Deparaffinize tissue sections in two changes of xylenes for 5 min each.
2. Immerse sections in two changes of absolute ethanol for 5 min each.
3. Air dry sections (approximately 5–10 min).
4. Prepare working digestion solution: Add 1 packet of digestion reagent B to the diluted digestion reagent A and dissolve thoroughly. Transfer dissolved solution to a staining dish and prewarm the solution to $37 \pm 2°C$.
5. Incubate sections in digestion solution for 15 min at $37 \pm 2°C$.

NOTE: *The digestion time of 15 min is satisfactory in most cases. Variations in tissue processing may require that the incubation time be shortened or lengthened to preserve tissue morphology and achieve acceptable hybridization signal.*

6. Wash sections in 200 ml buffer 1 for 1 min.
7. Dehydrate sections in 95% alcohol for 1 min followed by absolute ethanol for 1 min.
8. Air dry sections (approximately 5–10 min). Use within 24 hr.

7. Hybridization

1. Allow biotinylated probes to come to room temperature. Mix thoroughly.
2. Add 1 drop (approximately 40 μl) probe to the appropriate tissue section and place a glass cover slip over the section and the probe. Be careful to avoid bubble formation in this step. Large sections may require 2 drops of probe and larger cover slips to cover the section adequately.
3. Place slides flat on the prewarmed heating block or DTI-1 and incubate at 100 ± 2°C for 5 min.
4. If using the DTI-1, set the temperature to 37 ± 2°C and open the cover of the instrument. Deionized water may be added carefully to the moisture pad using a pipette to help bring the temperature to 37°C. If using a heating block, take the slides off and place them flat in a humidified incubation chamber. Incubate at 37 ± 2°C for 2–24 hr depending on the concentration of probe; the greater the probe concentration, the shorter the incubation time. Instructions for the HPV OmniProbe™ set state that the incubation time is 2–24 hr (Digene Diagnostics, 1991b). In our laboratory, an overnight incubation (16–18 hr) is convenient. Instructions for the ViraType® *in situ* HPV probe set state that the incubation time is 2 hr (Digene Diagnostics, 1991c). We have successfully increased the incubation time with some tissues to obtain a stronger signal.
5. Prewarm 800 ml buffer 2 to 37 ± 2°C. Distribute equally among four containers.
6. Remove slides from the incubation chamber. To remove cover slips and rinse off excess probe, individually rinse slides with gentle agitation in the first container of buffer 2. Cover slips will float off the slides and sink to the bottom of the staining dish.
7. Place slides in staining rack and wash in three changes of buffer 2 for 3 min each at 37 ± 2°C.

8. Detection

1. Remove slides one at a time from the rack. Carefully wipe away excess liquid from around the outside of each section with an absorbent wipe. Handle only 3–4 slides at a time to ensure that tissue sections do not dry out completely. Complete Step 2 and then proceed to additional slides.
2. Add 1 drop (approximately 40 μl) detection reagent to each tissue section. Larger sections may require 2 drops to cover the section

adequately. Place slides into humidified incubation chamber or DTI-1.

3. Incubate at $37 \pm 2°C$ for 20 min. While the slides are incubating, proceed to the next step.
4. Prepare substrate solution by pipetting 0.5 ml substrate 1 and 0.5 ml substrate 2 into 200 ml buffer 3. Mix thoroughly. Transfer clear, light yellow solution into a staining dish and prewarm the solution to $37 \pm 2°C$.

NOTE: *Use only glass pipettes or polypropylene pipette tips to measure substrates 1 and 2. The solvent in the substrates will dissolve polystyrene.*

5. Remove slides from incubation chamber and re-insert in staining rack. Wash in three 200-ml changes of buffer 3 for 3 min. each.
6. Incubate slides in substrate solution at $37 \pm 2°C$ for 60 min.
7. Wash in three changes of deionized water.
8. Counterstain by immersion in nuclear fast red stain for 30–60 sec.
9. Wash in three changes of deionized water.
10. Dehydrate section in 95% ethanol for 1 min followed by absolute alcohol for 1 min.
11. Clear sections with xylenes for 1 min.
12. Mount sections with permanent mounting medium and view results with a standard light microscope.

B. Quality Control

A positive control probe monitors both the adequacy of tissue preparation and possible procedural errors. The Digene positive DNA control probe (Catalog No. 4200-1096) is available in each Digene probe set or as a separate item. It is specific for human genomic DNA sequences and generates a hybridization signal in human cells. A purplish-blue hybridization signal is produced in the majority of nuclei and verifies that tissue fixation, pretreatment, hybridization, and detection conditions were adequate (Fig. 3a).

A negative control probe aids in the interpretation of weak signals and distinguishes possible nonspecific signal from true signal. The Digene negative DNA control probe (Catalog No. 4200-1097) consists of unrelated DNA sequences and is available in each Digene probe set or as a separate item. Tissues hybridized with the negative DNA control probe should appear pink as a result of the counterstain, with no purple nuclei (Fig. 3b).

Most laboratories find it useful to identify and save HPV-positive cases to serve as controls for future assays. Testing such specimens with the appropriate HPV DNA probe provides an additional control for tissue processing.

C. Interpretation of Sample Results

In general, the presence of target DNA is indicated by a purplish-blue precipitate, primarily in the nuclei of HPV-infected cells (Fig. 3c). Tissue that does not contain target DNA should appear pink because of the nuclear fast red stain.

Not all koilocytes will show the presence of HPV DNA. Generally the strongest signal will be observed in the cells of the outer epithelial layers. In some instances, signal can be apparent down to the basal layer. Most often, the color evenly covers the entire nucleus of the cell. Occasionally, a punctate pattern of nuclear signals may be seen (Fig. 4a).

Specimens tested with the ViraType® *in situ* HPV DNA probes (types 6/11, 16/18, 31/33/35) may demonstrate a signal with more than one group of probes. Since several HPV types are closely related and share regions of sequence homology, tissues that are strongly positive for one HPV group can also exhibit weaker cross-hybridization with other HPV probe groups. This effect is generally seen in tissues containing large amounts of target DNA. Because of this cross-hybridization, it is advisable to use all HPV probe groups during an initial screen of a tissue. In most cases, HPV type can be determined based on the probe group generating the strongest and most abundant signal.

If the signal is equally strong for two of the HPV probe groups and is present in nonoverlapping areas of the epithelial regions, there may be a true multiple infection. Signals of equal intensity with two or three HPV probe groups in the same locus on serial sections suggest that the sample is positive for HPV, but may contain an HPV type not included in the kit (Digene Diagnostics, 1991c). Interpretation of the results of such an assay should be augmented by evaluation of tissue morphology.

D. Troubleshooting

1. Poor or No Signal on a Suspected Positive Sample Indicates Inadequate Tissue Preparation or Procedural Error

Possible causes and remedies include:

1. inadequate proteolytic digestion. If morphology is still acceptable, the digestion step can be increased to 30 min.
2. improper tissue fixation and processing. Extended fixation times (>48 hr) in neutral buffered formalin or any Bouin's fixation generally result in specimens that are inadequate for *in situ* hybridization.
3. improper proteolytic digestion. Ensure that the temperature of the digestion solution is 37°C. Ensure that digestion reagent A was diluted

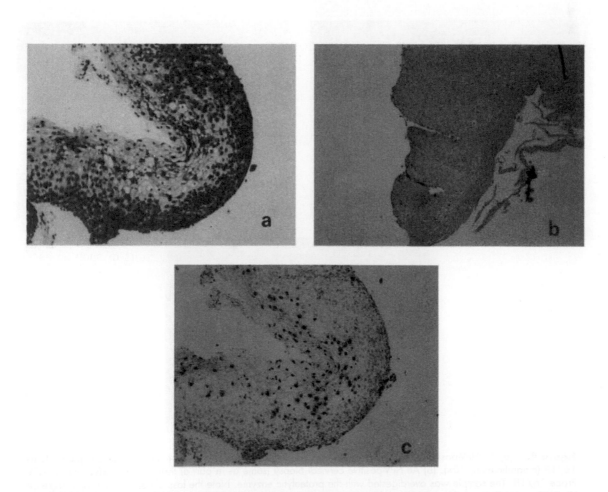

Figure 3. (a) An HPV-positive cervical biopsy tested with the Digene positive control probe (magnification, 69×). (b) An HPV-positive cervical biopsy tested with the Digene negative control probe (magnification, 69×). (c) An HPV-positive cervical biopsy tested with Omniprobe (magnification, 69×).

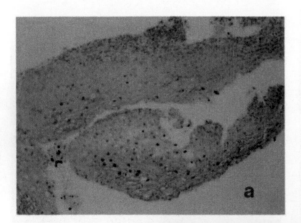
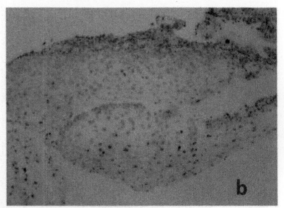

Figure 4. (a) An HPV-positive cervical biopsy tested with ViraType *in situ* HPV Probe 16/18. The sample contains 16/18 (magnification, 70×). (b) An HPV-positive cervical biopsy (same as in part a) tested with ViraType *in situ* HPV Probe 16/18. The sample was overdigested with the proteolytic enzyme. Note the loss of cell structure and decrease in positive signal (magnification, 70×). (c) An HPV-positive cervical biopsy tested with ViraType *in situ* HPV probe 16/18. This tissue section was incubated at 58°C for 1 hr to melt the paraffin and adhere the section to the slide. Incubating an additional section at 65°C for 18 hr eliminated the folding of the sample (magnification, 70×).

correctly and that the entire packet of digestion reagent B was added and fully dissolved.

4. improper denaturation. Ensure that the temperature of the heat block is $100 \pm 2°C$ using an accurate, recently calibrated temperature-measuring device designed for use on a flat surface.

5. improper reagent preparation. Reconstitute and prepare solutions as directed. Do not substitute reagents.

6. improper dehydration and clearing. Use recommended times because specific signal is partially soluble in ethanol and xylenes.

Adequacy of tissue preparation and possible procedural error can be monitored by testing the tissue with the Digene positive DNA control probe.

2. Tissue Destruction and/or Loss of Nuclei or Nuclear Details Can Result from Overdigestion or Overdenaturation

Possible causes and remedies include:

1. slides improperly pretreated. Use only treated slides provided in the kit (available separately from Digene; Catalog No. 4200-1045). Ensure that during prehybridization the temperature and length of incubation of slides is sufficient to adhere the tissue section to the slide (Fig. 4c).

2. overdigestion (Fig. 4b). Reduce the time of the digestion step from 15 min to 10 min or try a digestion time gradient test. Ensure that tissue is fixed in neutral-buffered formalin and processed adequately.

3. excessive denaturation. Check the temperature of the heat block and adjust to $100 \pm 2°C$, if necessary. Denature no longer than 5 min.

3. Diffuse Nonspecific Bluish Background Is Seen Throughout Some Tissues

Possible causes and remedies include:

1. improper substrate conditions. Ensure that substrate solution is prepared as directed. Reduce substrate incubation time to 30 min.

2. improper washing conditions. Ensure that buffer 2 is mixed well and warmed to $37 \pm 2°C$ before use.

3. excessive denaturation. Check the temperature of the heat block and adjust to $100 \pm 2°C$, if necessary. Denature no longer than 5 min.

4. improper hybridization conditions. Ensure that hybridization is done in a humid chamber at $37 \pm 2°C$. Do not allow probe mix to evaporate.

5. probe concentration that is too high. Reduce probe concentration to $0.1–0.2$ $\mu g/ml$ and extend hybridization time to overnight.

Some types of tissue exhibit this type of background more than others, and it tends to appear consistently in specific areas in a specimen. If signal on tissue hybridized with the specific probe looks like that seen on the same tissue hybridized with the Digene negative DNA control probe, the sample should not be considered positive.

Occasional precipitation of substrate crystals causes the appearance of small, blue, rod-shaped crystals. These should not interfere with assay interpretation.

E. Limitations of the Procedure

The Digene tissue hybridization kit is for research use only and is not approved for diagnostic or therapeutic procedures. The kit has been designed to detect DNA in neutral-buffered, formalin-fixed, paraffin-embedded human tissue by the *in situ* hybridization procedure. Other fixatives may give reduced detection sensitivity. Failure to detect target DNA may result from improper sampling, handling, fixation, and/or processing of the specimen, or from the presence of target DNA at levels below the sensitivity of the assay. Therefore, lack of signal does not ensure that no target DNA is present.

V. SELECTED STUDIES

Table 1 lists additional reference papers that discuss *in situ* techniques used for specific viruses and *Chlamydia*. The Bagasra study describing the detection of HIV-1 provirus appeared in the *New England Journal of Medicine* (1992). A subsequent edition of the same journal contained letters to the editor, one of which concerned possible artifacts and suggested modifications (Long *et al.*, 1992). The authors' reply also appears.

VI. *IN SITU* HYBRIDIZATION IN CONJUNCTION WITH OTHER TECHNIQUES

A. Immunohistochemistry and in Situ Hybridization

Several techniques combine nucleic acid detection by ISH and protein detection by immunohistochemistry (IHC) on the same tissue section (Blum *et al.*, 1984; Brahic *et al.*, 1984; Roberts *et al.*, 1988; Porter *et al.*, 1990; Brahic

TABLE 1
Selected Studies That Have Used *in Situ* Hybridization to Detect Viruses or *Chlamydia trachomatis*

Agent	Comment	Reference
Adenovirus	Nonisotopic	Unger *et al.* (1986)
	Nonisotopic, electron microscopy	Puvion-Dutilleul and Pichard (1992)
	Nonisotopic	Nuovo, M. A. *et al.* (1993)
Chlamydia trachomatis	Nonisotopic	Abdelatif *et al.* (1991)
Cytomegalovirus	Nonisotopic	Unger *et al.* (1986)
	Nonisotopic	Coates *et al.* (1987)
	Nonisotopic	Naoumov *et al.* (1988)
	Nonisotopic	Niedobitek *et al.* (1988)
	Nonisotopic	Nuovo *et al.* (1993)
Enterovirus	Nonisotopic	Tracy *et al.* (1990)
	Nonisotopic	Hilton *et al.* (1992)
Epstein–Barr virus	Nonisotopic	Coates *et al.* (1991)
	Isotopic	Hamilton-Dutoit *et al.* (1991)
	Isotopic	Nakhleh *et al.* (1991)
	Nonisotopic	Chang *et al.* (1992)
Hepatitis B virus	Isotopic	Blum *et al.* (1983)
	Nonisotopic	Naoumov *et al.* (1988)
	Nonisotopic	Lau *et al.* (1991)
Hepatitis C virus	Nonisotopic/RT *in situ* PCR	Nuovo *et al.* (1993b)
Herpes simplex virus	Nonisotopic	Dictor (1990)
Human immunodeficiency virus 1	Nonisotopic, peripheral blood mononuclear cells, PCR *in situ* hybridization	Bagasra *et al.* (1992)
	Nonisotopic/PCR *in situ* hybridization	Nuovo *et al.* (1992)
	Isotopic	Smith *et al.* (1992)
	Isotopic	Spiegel *et al.* (1992)
	Nonisotopic/RT *in situ* PCR	Nuovo *et al.* (1993a)
JC virus	Nonisotopic	Hulette *et al.* (1991)
Measles	Nonisotopic	McQuaid *et al.* (1990)
Papillomavirus (human)	Nonisotopic	Cooper *et al.* (1991)
	Nonisotopic	Herrington *et al.* (1991)
Parvovirus B19 (human)	Nonisotopic	Porter *et al.* (1988)
	Nonisotopic	Salimans *et al.* (1989)
	Nonisotopic	Morey *et al.* (1992)

and Ozden, 1992; Chang *et al.*, 1992). The IHC procedure is usually performed first because many proteins will not retain their reactivity after some of the harsh hybridization treatments. Once detection of the protein has been done, the colored precipitate can usually withstand the hybridization conditions. Fixation is once again a prime consideration. Often a fixative that is good for ISH is not good for IHC.

B. Polymerase Chain Reaction and in Situ Hybridization

In efforts to detect viral DNA sequences of low copy number more easily in paraffin-embedded tissue, DNA has been extracted and subsequently amplified by PCR (van den Berg *et al.*, 1989; Telenti *et al.*, 1990; Weiss *et al.*, 1992). Although valuable, this technique loses the advantage of evaluating the types and percentages of cells showing signal, as well as general tissue morphology. However, a technique has been described that places the tissue slide on the hot plate of a thermocycler. The amplification mix is added, topped with cover slips and mineral oil. The cycling process takes place. Afterward, ISH is performed. In this procedure, the advantages of ISH are retained; the technique has improved the sensitivity to 1 copy of HPV per formalin-fixed SiHa cell (Nuovo *et al.*, 1991). An improvement of this technique has been described that uses a single primer pair instead of the multiple primer pairs described previously. This technique, called the "hot-start" method, reduces mispriming, thereby improving specificity (Nuovo *et al.*, 1991; see Chapter 11 for a more detailed discussion).

PCR-amplified RNA can be detected in intact cells using reverse transcription (RT) *in situ* PCR with direct incorporation of digoxigenin-11-UTP. An optimal protease digestion is critical in this method. Pretreatment of the sample with RNase-free DNase will eliminate nonspecific amplification of DNA (Nuovo *et al.*, 1993a,b).

VII. VENDORS

ISH kits are available from Digene Diagnostics, Inc. (Silver Spring and Beltsville, MD), Enzo Diagnostics, Inc. (Syosset, NY), and Oncor, Inc. (Gaithersburg, MD). Individual probes and/or PCR primers are available from Amac, Inc. (Westbrook, ME), Digene Diagnostics, Inc., Enzo Diagnostics, Inc., Genemed Biotechnologies, Inc. (South San Francisco, CA); Oncor, Inc.; Operon Technologies (Alameda, CA); and Synthetic Genetics (San Diego, CA). Radioactivity labeled nucleotides are available from Amersham (Arlington Heights, IN) and Du Pont NEN (Boston, MA).

VIII. CONCLUSION

In situ hybridizaton is labor intensive and has not yet become widely used as a diagnostic technique in clinical laboratories. The ability to localize a viral infection to a specific cell type and to determine the percentage of positive cells within a tissue is an attractive feature of this assay. The possibility of performing IHC and ISH on the same tissue section is an additional advantage in some cases. Probably the most promising developments are those of semi-automated equipment and PCR amplification on the tissue section slide. *In situ* hybridization is sufficiently developed for clinical laboratories to perform the technique and will find its niche in the routine clinical laboratory as more diagnostic applications are developed and kits are approved by the Food and Drug Administration for *in vitro* diagnostic use.

ACKNOWLEDGMENTS

I am grateful to T. F. Puckett, T. G. Puckett, and Linda Eaton for their continued support and to the histology department at T. F. Puckett Laboratory, especially Kim Wright, for processing the tissues. I thank Mary Haddad and the Photo Production Department at William Beaumont Hospital (Royal Oak, MI) for preparation of the graphic. I also thank Meg Nelson and Dina Link at Digene Diagnostics, Inc. I appreciate Daniel Farkas and Dan Wiedbrauk for their assistance and acknowledge Eleta Allen for her technical help. I am grateful to Rhonda Patterson for her work in preparation of the manuscript and photographs.

REFERENCES

Abdelatif, O. M. A., Chandler, F. W., and McQuire, B. S., Jr. (1991). *Chlamydia trachomatis* in chronic abacterial prostatitis: Demonstration by colorimetric *in situ* hybridization. *Hum. Pathol.* **22(1),** 41–44.

Allan, G. M., Todd, D., Smyth, J. A., Mackie, D. P., Burns, J., and McNulty, M. S. (1989). *In situ* hybridization: An optimised detection protocol for a biotinylated DNA probe renders it more sensitive than a comparable ^{35}S-labelled probe. *J. Virol. Meth.* **24,** 181–190.

Bagasra, O., Hauptman, S. P., Lischner, H. W., Sachs, M., and Pomerantz, R. J. (1992). Detection of human immunodeficiency virus type 1 provirus in mononuclear cells by *in situ* polymerase chain reaction. *N. Engl. J. Med.* **326,** 1385–1391.

Blum, H. E., Haase, A. T., and Vyas, G. N. (1984). Molecular pathogenesis of hepatitis B virus infection: Simultaneous detection of viral DNA and antigens in paraffin-embedded liver sections. *Lancet* **ii,** 771–775.

Blum, H. E., Stowring, L., Figus, A., Montgomery, C. K., Haase, A. T., and Vyas, G. N. (1983). Detection of hepatitis B virus DNA in hepatocytes, bile duct epithelium, and vascular elements by *in situ* hybridization. *Proc. Natl. Acad. Sci. USA* **80,** 6685–6688.

Boyle, A. (1990). Enzymatic manipulation of DNA and RNA. *In* "Current Protocols in Molecular Biology" (F. A. Ausubel, R. Brent, R. E. Kingston, D. D. Moore, J. G. Seidman, J. A. Smith, and K. Struhl, eds.), pp. 3.18.1–3.18.7. Greene Publishing and Wiley–Interscience, New York.

Brahic, M., and Ozden, S. (1992). Simultaneous detection of cellular RNA and proteins. *In* "*In Situ* Hybridization: A Practical Approach" (D. G. Wilkinson, ed.), pp. 85–104. IRL Press, New York.

Brahic, M., Haase, A. T., and Cash, E. (1984). Simultaneous *in situ* detection of viral RNA and antigens. *Proc. Natl. Acad. Sci. USA* **81,** 5445–5448.

Burns, J., Graham, A. K., Frank, C., Fleming, K. A., Evans, M. F., and McGee, J. O. (1987). Detection of low copy human papilloma virus DNA and mRNA in routine paraffin sections of cervix by non-isotopic *in situ* hybridisation. *J. Clin. Pathol.* **40,** 858–864.

Chang, K. L., Chen, Y., Shibata, D., and Weiss, L. M. (1992). Description of an *in situ* hybridization methodology for detection of Epstein–Barr virus RNA in paraffin-embedded tissues, with a survey of normal and neoplastic tissues. *Diagnostic Mol. Pathol.* **1(4),** 246–255.

Coates, P. J., Hall, P. A., Butler, M. G., and d'Ardenne, A. J. (1987). Rapid technique of DNA-DNA *in situ* hybridisation on formalin fixed tissue sections using microwave irradiation. *J. Clin. Pathol.* **40,** 865–869.

Coates, P. J., Mak, W. P., Slavin, G., and d'Ardenne, A. J. (1991). Detection of single copies of Epstein–Barr virus in paraffin wax sections by non-radioactive *in situ* hybridisation. *J. Clin. Pathol.* **44(6),** 487–491.

Cooper, K., Herrington, C. S., Strickland, J. E., Evans, M. F., and McGee, J. O. (1991). Episomal and integrated human papillomavirus in cervical neoplasia shown by non-isotopic *in situ* hybridisation. *J. Clin. Pathol.* **44(12),** 990–996.

Dictor, M., Renfjärd, E., and Brun, A. (1990). *In situ* hybridisation in herpetic lesions using a biotinylated DNA probe. *J. Clin. Pathol.* **43(5),** 416–419.

Digene Diagnostics (1991a). "Digene Tissue Hybridization Kit." Digene Diagnostics, Beltsville, MD.

Digene Diagnostics (1991b). "Digene HPV OmniProbe™ Set." Digene Diagnostics, Beltsville, MD.

Digene Diagnostics (1991c). "Digene ViraType® *in Situ* HPV Probe Set." Digene Diagnostics, Beltsville, MD.

Emson, P. C., and Gait, M. J. (1992). *In situ* hybridization with biotinylated probes. *In* "*In Situ* Hybridization: A Practical Approach" (D. G. Wilkinson, ed.), pp. 45–59. IRL Press, New York.

Gall, J. G., and Pardue, M. L. (1969). Formation and detection of RNA–DNA hybrid molecules in cytological preparations. *Proc. Natl. Acad. Sci. USA* **63,** 378–383.

Green, M., Sviland, L., Taylor, C. E., Peiris, M., McCarthy, A. L., Pearson, A. D. J., and Malcolm, A. J. (1992). Human herpes virus 6 and endogenous biotin in salivary glands. *J. Clin. Pathol.* **45(9),** 788–790.

Hamilton-Dutoit, S. J., Delecluse, H. J., Raphael, M., Lenoir, G., and Pallesen, G. (1991a). Detection of Epstein–Barr virus genomes in AIDS related lymphomas: Sensitivity and specificity of *in situ* hybridisation compared with Southern blotting. *J. Clin. Pathol.* **44(8),** 676–680.

Hamilton-Dutoit, S. J., Pallesen, G., Franzmann, M. B., Karkov, J., Black, F., Skinhøj, P., and Pedersen, C. (1991b). AIDS-related lymphoma: Histopathology, immunophenotype, and association with Epstein–Barr virus as demonstrated by *in situ* nucleic acid hybridization. *Am. J. Pathol.* **138(1),** 149–163.

Herrington, C. S., Graham, A. K., and McGee, J. O. (1991). Interphase cytogenetics using biotin and digoxigenin labelled probes: III. Increased sensitivity and flexibility for detecting HPV in cervical biopsy specimens and cell lines. *J. Clin. Pathol.* **44**, 33–38.

Hilton, D. A., Day, C., Pringle, J. H., Fletcher, A., and Chambers, S. (1992). Demonstration of the distribution of Coxsackie virus RNA in neonatal mice by nonisotopic *in situ* hybridization. *J. Virol. Meth.* **40**, 155–162.

Hulette, C. M., Downey, B. T., and Burger, P. C. (1991). Progressive multifocal leukoencephalopathy: Diagnosis by *in situ* hybridization with a biotinylated JC virus DNA probe using an automated histomatic code-on slide and stainer. *Am. J. Surg. Pathol.* **15(8)**, 791–797.

Jelsma, T., and Houthoff, H. J. (1994). Increased labeling of DNA probes for *in situ* hybridization with the universal linkage system (ULS). *J. NIH Res.* **6**, 82.

John, H. A., Birnsteil, M. L., and Jones, K. W. (1969). RNA–DNA hybrids at the cytological level. *Nature* **223**, 582–587.

Lau, J. Y. N., Naoumov, N. V., Alexander, G. J. M., and Williams, R. (1991). Rapid detection of hepatitis B virus DNA in liver tissue by *in situ* hybridisation and its combination with immunohistochemistry for simultaneous detection of HBV antigens. *J. Clin. Pathol.* **44(11)**, 905–908.

Lawrence, J. B., and Singer, R. H. (1985). Quantitative analysis of *in situ* hybridization methods for the detection of actin gene expression. *Nucleic Acids Res.* **13**, 1777–1779.

Lewis, F. A., and Wells, M. (1992). Detection of virus in infected human tissue by *in situ* hybridization. *In* "*In Situ* Hybridization: A Practical Approach" (D. G. Wilkinson, ed.), pp. 121–136. IRL Press, New York.

Lloyd, R. V. (1987). Use of molecular probes in the study of endocrine diseases. *Hum. Pathol.* **18**, 1199–1211.

Long, A. A., Komminoth, P., and Wolfe, H. J. (1992). Correspondence: Detection of HIV provirus by *in situ* polymerase chain reaction. *N. Engl. J. Med.* **327(21)**, 1529–1530.

McLaughlin, Susan K., and Margolskee, R. F. (1993). [33]P is preferable to ^{35}S for labeling probes used in *in situ* hybridization. *Biotechniques* **15(3)**, 506–511.

McQuaid, S., Isserte, S., Allan, G. M., Taylor, M. J., Allen, I. V., and Cosby, S. L. (1990). Use of immunocytochemistry and biotinylated *in situ* hybridisation for detecting measles virus in central nervous system tissue. *J. Clin. Pathol.* **43(4)**, 329–333.

Morey, A. L., Porter, H. J., Keeling, J. W., and Fleming, K. A. (1992). Nonisotopic *in situ* hybridisation and immunophenotyping of infected cells in the investigation of human fetal parvovirus infection. *J. Clin. Pathol.* **45(8)**, 673–678.

Mougin, C., Bassignot, A., Coaquette, A., Bourgeois, A., and Lab, M. (1991). Optimization of *in situ* hybridization for detection of viral genomes in cultured cells on 96-microwell plates: A cytomegalovirus model. *J. Clin. Microbiol.* **29(8)**, 1735–1739.

Nakhleh, R. E., Manivel, J. C., Copenhaver, C. M., Sung, J. H., and Strickler, J. G. (1991). *In situ* hybridization for the detection of Epstein–Barr virus in central nervous system lymphomas. *Cancer* **67**, 444–448.

Naoumov, N. V., Alexander, G. J. M., Feddleston, A. L. W., and Williams, R. (1988). *In situ* hybridisation in formalin fixed, paraffin wax embedded liver specimens: Method for detecting human and viral DNA using biotinylated probes. *J. Clin. Pathol.* **41**, 793–798.

Niedobitek, G., Finn, T., Herbst, H., Gerdes, J., Grillner, L., Landqvist, M., Wirgart, B. Z., and Stein, H. (1988). Detection of cytomegalovirus by *in situ* hybridisation and immunohistochemistry using new monoclonal antibody CCH2: A comparison of methods. *J. Clin. Pathol.* **41**, 1005–1009.

Nuovo, G. J., MacConnell, P., Forde, A., and Delvenne, P. (1991a). Detection of human papillomavirus DNA in formalin-fixed tissues by *in situ* hybridization after amplification by polymerase chain reaction. *Am. J. Pathol.* **139**, 847–854.

Nuovo, G. J., Gallery, F., MacConnell, P., Becker, J., and Bloch, W. (1991b). An improved technique for the *in situ* detection of DNA after polymerase chain reaction amplification. *Am. J. Pathol.* **139,** 1239–1244.

Nuovo, G. J., Margiotta, M., MacConnell, P., and Becker, J. (1992). Rapid *in situ* detection of PCR-amplified HIV-1 DNA. *Diganostic Mol. Pathol.* **1(2),** 98–102.

Nuovo, G. J., Forde, A., MacConnell, P., and Fahrenwald, R. (1993a). *In situ* detection of PCR amplified HIV-1 nucleic acids and tumor necrosis factor cDNA in cervical tissues. *Am. J. Pathol.* **143,** 40–48.

Nuovo, G. J., Lidonnici, K., MacConnell, P., and Lane, B. (1993b). Intracellular localization of polymerase chain reaction (PCR)-amplified hepatitis C cDNA. *Am. J. Surg. Pathol.* **17,** 683–690.

Nuovo, M. A., Nuovo, G. J., Becker, J., Gallery, F., Delvenne, P., and Kane, P. B. (1993). Correlation of viral infection, histology, and mortality in immunocompromised patients with pneumonia: Analysis by *in situ* hybridization and the polymerase chain reaction. *Diagnostic Mol. Pathol.* **2(3),** 200–209.

Porter, H. J., Khong, T. Y., Evans, M. F., Chan, V. T.-W., and Fleming, K. A. (1988). Parvovirus as a cause of hydrops fetalis: Detection by *in situ* DNA hybridisation. *J. Clin. Pathol.* **41,** 381–383.

Porter, H. J., Heryet, A., Quantrill, A. M., and Fleming, K. A. (1990). Combined non-isotopic *in situ* hybridisation and immunohistochemistry on routine paraffin wax embedded tissue: Identification of cell type infected by human parvovirus and demonstration of cytomegalovirus DNA and antigen in renal infection. *J. Clin. Pathol.* **43(2),** 129–132.

Puvion-Dutilleul, F., and Pichard, E. (1992). Segregation of viral double-stranded and single-stranded DNA molecules in nuclei of adenovirus infected cells as revealed by electron microscope *in situ* hybridization. *Biol. Cell* **76,** 139–150.

Roberts, W. H., Sneddon, J. M., Waldman, W. J., Snyder, J. H., and Stephens, R. E. (1988). Cellular localization of CMV infection by simultaneous immunohistochemical staining and DNA hybridization. *Lab. Med.* **19(4),** 240–242.

Salimans, M. M. M., van de Rijke, F. M., Raap, A. K., and van Elsacker-Niele, A. M. W. (1989). Detection of parvovirus B19 DNA in fetal tissues by *in situ* hybridisation and polymerase chain reaction. *J. Clin. Pathol.* **42,** 525–530.

Sambrook, J., Fritsch, E. F., and Maniatis, T. (1989). ''Molecular Cloning: A Laboratory Manual.'' Cold Spring Harbor Laboratory Press, Cold Spring Harbor, New York.

Smith, P. D., Saini, S. S., Raffeid, M., Manischewitz, J. F., and Wahl, S. M. (1992). Cytomegalovirus induction of tumor necrosis factor-α by human monocytes and mucosal macrophages. *J. Clin. Invest.* **90,** 1642–1648.

Spiegel, H., Herbst, H., Niedobitek, G., Foss, H.-D., and Stein, H. (1992). Follicular dendritic cells are a major reservoir for human immunodeficiency virus type 1 in lymphoid tissues facilitating infection of CD4⁺ T-helper cells. *Am. J. Pathol.* **140(1),** 15–22.

Telenti, A., Marshall, W. F., and Smith, T. F. (1990). Detection of Epstein–Barr virus by polymerase chain reaction. *J. Clin. Microbiol.* **28(10),** 2187–2190.

Tracy, S., Chapman, N. M., McManus, B. M., Pallansch, M. A., Beck, M. A., and Carstens, J. (1990). A molecular and serologic evaluation of enteroviral involvement in human myocarditis. *J. Am. Coll. Cardiol.* **22(4),** 403–414.

Unger, E. R., Budgeon, L. R., Myerson, D., and Brigati, D. J. (1986). Viral diagnosis by *in situ* hybridization: Description of a rapid simplified colorimetric method. *Am. J. Surg. Pathol.* **10(1),** 1–8.

van den Berg, F., Schipper, M., Jiwa, M., Rook, R., van de Rijke, F., and Tigges, B. (1989). Implausibility of an aetiological association between cytomegalovirus and Kaposi's sarcoma shown by four techniques. *J. Clin. Pathol.* **42,** 128–131.

Weiss, L. M., Liu, X.-F., Chang, K. L., and Billingham, M. E. (1992). Detection of enteroviral

RNA in idiopathic dilated cardiomyopathy and other human cardiac tissues. *J. Clin. Invest.* **90,** 156–159.

Wilkinson, D. G. (1992). The theory and practice of *in situ* hybridization. *In* "*In Situ* Hybridization: A Practical Approach" (D. G. Wilkinson, ed.), pp. 1–13. IRL Press, New York.

Zeller, R., and Rogers, M. (1989). *In situ* hybridization to cellular RNA. *In* "Current Protocols in Molecular Biology." (F. A. Ausubel, R. Brent, R. E. Kingston, D. D. Moore, J. G. Seidman, J. A. Smith, and K. Struhl, eds.), pp. 14.3.1–14.3.14. Greene Publishing and Wiley–Interscience, New York.

5

Antiviral Susceptibility Testing Using DNA–DNA Hybridization

Richard L. Hodinka

Departments of Pathology and Pediatrics
Children's Hospital of Philadelphia
and School of Medicine
University of Pennsylvania
Philadelphia, Pennsylvania 19104

I. INTRODUCTION

Antiviral therapy for the successful management of viral infections has come of age. Within the last decade, rapid advances have been made in the development of new antiviral agents (for extensive reviews on the subject, see Huraux *et al.*, 1990; Bean, 1992; Keating, 1992; Watts, 1992). Many of these drugs are now available for the treatment and/or prevention of infections caused by herpes simplex virus (HSV), varicella-zoster virus (VZV), cytomegalovirus (CMV), the human immunodeficiency virus (HIV), influenza A virus, respiratory syncytial virus (RSV), and hepatits B and C viruses (HBV, HCV) (Table 1). An important caveat to the increased number and usage of antiviral agents has been the emergence of drug-resistant viruses, particularly in patients with acquired immunodeficiency syndrome (AIDS) or other conditions of immunosuppression such as hemotologic malignacy and organ transplantation. Although still uncommon, drug resistance has been reported after therapy with a number of antiviral agents (Table 2). As more patients fail to respond to appropriate therapy and additional antiviral agents are produced, a definite need arises for the diagnostic laboratory to provide rapid and practical antiviral susceptibiltiy testing to assist the physician in defining drug resistance and choosing appropriate alternative therapies.

Numerous procedures have been described for testing the susceptibility of viruses to antiviral agents (Table 3; comprehensive reviews of various

TABLE 1
Available Antiviral Agents and Their Clinical Indications

Antiviral agent	Clinical indications[a]
Acyclovir	Primary and recurrent herpes, HSV encephalitis, neonatal herpes, treatment and prophylaxis of mucocutaneous HSV infection in immunocompromised host, primary varicella or zoster in immunocompromised host
Vidarabine (Ara-A)	HSV encephalitis, neonatal herpes, herpes keratitis, zoster in immunocompromised host
Trifluridine (TFT)	Herpes keratitis
Idoxuridine (IDU)	Herpes keratitis
Ganciclovir (DHPG)	Treatment of CMV in immunocompromised host, prophylaxis of transplant patients
Foscarnet (PFA)	Acyclovir-resistant HSV, VZV, or CMV infections; CMV retinitis
Zidovudine (AZT)	HIV infection
Didanosine (ddI)	AZT intolerance or AZT treatment failure in HIV infection
Zalcitabine (ddC)	Intolerance or treatment failure with AZT and ddI, used as combination therapy with other antiretroviral agents
Amantadine	Influenza A treatment or prophylaxis
Rimantadine	Influenza A treatment or prophylaxis
Ribavirin	Severe RSV infection
Interferon	Hepatitis B or C chronic liver disease

[a] Abbreviations: HSV, herpes simplex virus; CMV, cytomegalovirus; VZV, varicella-zoster virus; HIV, human immunodeficiency virus; RSV, respiratory syncytial virus.

TABLE 2
Resistance of Viruses to Antiviral Agents

Antiviral agent	Virus[a]	Resistance mechanism	References
Acyclovir	HSV, VZV	Altered or deficient viral thymidine kinase; altered DNA polymerase	Crumpacker et al. (1982), Wade et al. (1983), McLaren et al. (1985), Bean et al. (1987), Norris et al. (1988), Pahwa et al. (1988), Erlich et al. (1989a), Sacks et al. (1989), Birch et al. (1990), Englund et al. (1990), Gateley et al. (1990), Jacobson et al. (1990), Ljungman et al. (1990), Bevilacqua et al. (1991), Kost et al. (1993)
Ganciclovir	CMV	Diminished drug phosphorylation; altered DNA polymerase	Erice et al. (1989), Biron (1991), Drew et al. (1991), Jacobson et al. (1991), Knox et al. (1991)
Foscarnet	HSV, CMV	Altered viral DNA polymerase	Sacks et al. (1989), Birch et al. (1990), Knox et al. (1991), Sullivan and Coen (1991), Safrin et al. (1994)
Amantadine or rimantadine	Influenza A	Altered transmembrane domain of M2 protein	Belsche et al. (1989), Hayden et al. (1989)
Zidovudine or didanosine	HIV	Altered reverse transcriptase	Larder and Kemp (1989), Larder et al. (1989), Rooke et al. (1989), Land et al. (1990), Richman (1990), St. Clair et al. (1991), Gu et al. (1992)

[a] Abbreviations: HSV, herpes simplex virus; VZV, varicella-zoster virus; CMV, cytomegalovirus; HIV, human immunodeficiency virus.

TABLE 3
Selected Publications Describing Methods for Determining the Antiviral Susceptibility of Viral Isolates to Specific Antiviral Agents

Virus[a]	Antiviral agent	Method	Reference
HSV	Acyclovir, Foscarnet	Plaque reduction	Harmenberg et al. (1980)
		DNA hybridization	Swierkosz et al. (1987), Englund et al. (1990)
		Dye uptake	McLaren et al. (1983)
		Enzyme immunoassay	Rabalais et al. (1987)
		Yield inhibition	Prichard et al. (1990)
CMV	Ganciclovir	Plaque reduction	Plotkin et al. (1985), Gerna et al. (1992)
		DNA hybridization	Gadler (1983), Dankner et al. (1990)
		Yield inhibition	Rasmussen et al. (1984), Prichard et al. (1990)
		Antigen synthesis reduction	Telenti and Smith (1989), Pepin et al. (1992)
		Enzyme immunoassay	Tatarowicz et al. (1991)
VZV	Acyclovir, Foscarnet	Plaque reduction	Biron and Elion (1980)
		DNA hybridization	Jacobson et al. (1990)
HIV	AZT, ddI, ddC	p24 antigen inhibition	Averett (1989), Japour et al. (1993)
		RT reduction	Mitsuya et al. (1985), Nakashima et al. (1986a)
		Plaque reduction	Harada et al. (1985), Nakashima et al. (1986b), Chesebro and Wehrly (1988)
		Dye uptake	Montefiori et al. (1988)
		DNA quantification	Averett (1989)
Influenza A	Amantadine, rimantadine	Plaque reduction	Hayden et al. (1980)
		CPE inhibition	Belshe et al. (1989)
		Enzyme immunoassay	Belshe et al. (1989)
RSV	Ribavirin	Plaque reduction	Hruska et al. (1980)
		CPE inhibition	Kawana et al. (1987)
		Enzyme immunoassay	Kang and Pai (1989)

[a] Abbreviations: HSV, herpes simplex virus; CMV, cytomegalovirus; VZV, varicella-zoster virus; HIV, human immunodeficiency virus; RSV, respiratory syncytial virus.

techniques can be found in Newton, 1988; Hu and Hsiung, 1989). In general, all antiviral susceptibility assays require viral replication in a suitable host cell. However, each assay employs a different method for measuring the activity of an antiviral agent against the replication of the virus. The inhibition of virus-induced cytopathic effect (CPE) or plaque formation; a decrease in the production of viral antigens, enzyme activities, or total virus yield; a reduction in viral nucleic acid synthesis; and the inhibition of cell transformation are parameters that can be monitored in an *in vitro* system. This chapter focuses on the use of a newly developed DNA–DNA hybridization assay as it applies to susceptibility testing of HSV isolates.

Acyclovir has been widely used as an effective treatment for and prophylactic against HSV infections. However, resistance of HSV to acyclovir has emerged with the occurrence of chronic, progressive, debilitating disease in immunocompromised patients receiving prolonged or multiple courses of therapy (Table 2). Resistance to acyclovir can result from mutations in the genes encoding the viral thymidine kinase or the viral DNA polymerase (Coen and Schaffer, 1980; Field *et al.*, 1980; Schnipper and Crumpacker, 1980). Most acyclovir-resistant HSV isolates are deficient in, or have altered production of, viral thymidine kinase activity. The frequency with which resistant strains are being found and the increasing morbidity and mortality associated with these virus strains is a problem of growing concern. Foscarnet has been employed successfully as an alternative antiviral agent for treating thymidine kinase-deficient or -altered acyclovir-resistant HSV. However, resistance to this drug has also been documented (Sacks *et al.*, 1989; Birch *et al.*, 1990; Safrin *et al.*, 1994).

Two commonly employed methods for antiviral susceptibility testing of HSV isolates are the plaque reduction and dye uptake assays. The plaque reduction assay has long been considered the standard method used to determine the suceptibiltiy of all cytopathogenic viruses to antiviral agents. In this procedure, cell monolayers are established in 12-well tissue culture plates, infected with 100–200 plaque-forming units (PFU) of virus, and overlaid with medium containing various concentrations of drug and 0.4% agarose, 1–2% methylcellulose, or serum immunoglobulin. Following a 72-hr incubation, the plates are fixed and stained with crystal violet dye. The plaques are then counted; the concentration of drug causing a 50% inhibition in plaque formation represents the 50% inhibitory dose (ID_{50}) for the antiviral agent (Fig. 1). The plaque reduction assay is accurate, reliable, and relatively simple to perform and does not require specialized equipment. It is difficult to automate, however, and is best suited to testing small numbers of specimens. A major disadvantage of this assay is that the enumeration of plaques is tedious, subjective, and time consuming and the number of plaques that can be counted accurately is limited. The infectivity titer of the virus must also be determined before the assay is performed, leading to a delay in reporting results. The dye uptake assay is a colorimetric method that involves

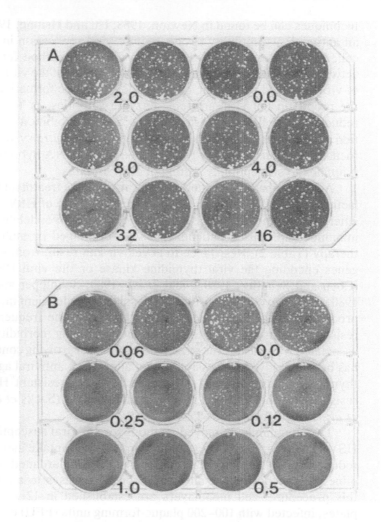

Figure 1 Plaque reduction assay demonstrating acyclovir-resistant (A) and -sensitive (B) HSV isolates on monolayers of Vero cells in 12-well plates. Duplicate wells were overlaid with Eagle's minimum essential medium (EMEM) containing 0.4% agarose and serial 2-fold dilutions of acyclovir (0–64 μg/ml). The cells were incubated for 72 hr, fixed in 10% formalin, and stained with 0.8% crystal violet. Plaques were counted using a dissecting microscope at low power; the ID_{50} values were calculated for each isolate tested. Note the formation of discrete plaques within the wells and the decrease in the number and size of the plaques for the acyclovir-sensitive isolate as the drug concentrations increase. The ID_{50} values were $\geq$32.0 μg/ml for the resistant isolate and 0.19 μg/ml for the sensitive virus.

the quantification of neutral red dye that is preferentially taken up by viable host cells following incubation with virus and various concentrations of antiviral drug. The assay is semi-automated and performed in a 96-well microtiter plate (Fig. 2). The ID_{50} value is the concentration of drug causing

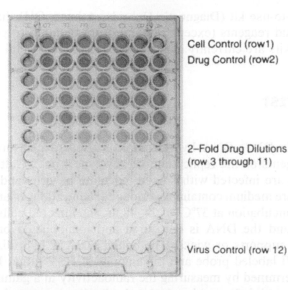

Cell Control (row1)
Drug Control (row2)

2–Fold Drug Dilutions
(row 3 through 11)

Virus Control (row 12)

Figure 2 Dye-uptake assay depicting the preferential uptake of neutral red dye by viable cells relative to cells damaged by viral infection following the incubation of infected cells with various concentrations of acyclovir. Rows 3 through 11 represent serial 2-fold drug dilutions from the highest concentration (row 3) to the lowest (row 11). Row 1 represents a cell control with no virus or drug; row 2 contains cells in the presence of the highest concentration of drug and no virus; and row 12 is a virus control containing cells and virus with no drug. Following a 72-hr incubation of cells and virus with the different drug concentrations, a 0.15% neutral red solution is added to all wells of the plate. The dye taken up by viable cells is then eluted into a phosphate-alcohol buffer and measured in a spectrophotometer. The concentration of drug causing a 50% reduction in viral cytopathic effect (CPE) relative to the cell and virus controls represents the ID_{50} value. (Courtesy of Edgar L. Hill, Burroughs Wellcome Company, Research Triangle Park, NC.)

a 50% reduction in viral CPE compared to cell and virus controls. The assay is reproducible and provides an objective end point. It can be performed on large numbers of clinical isolates and can be adapted for use with different drugs and viruses. However, the infectivity titer of each virus isolate must be determined, and the cost of automating the assay is a factor. The test can be performed manually, but is cumbersome and slow.

A DNA–DNA hybridization assay has been developed that measures the reduction in viral nucleic acid synthesis when virus-infected cells are incubated in the presence of antiviral agent. The assay is simple to perform, is less labor intensive than the conventional methods, and does not require the determination of virus infectivity titers before performing the assay. Like the dye uptake assay, the DNA–DNA hybridization system can be expanded to accommodate multiple clinical isolates and antiviral drugs. The procedure is rapid, providing a reliable determination of the susceptibiltiy of virus isolates to antiviral agents within 48 hrs. Although the hybridization assay has the disadvantage of using a radioiodinated probe, it can be purchased as a ready-to-use kit (Diagnostic Hybrids, Athens, OH) that contain cells, medium, and reagents (except antiviral drug) required to perform the test. Formated kits are currently available for HSV, CMV, and VZV.

II. PRINCIPLE OF THE TEST

Radiometric DNA–DNA hybridization is used to measure the effects of antiviral agents on the replication of HSV type 1 or 2. Cultured cells in 24-well plates are infected with a fixed inoculum of virus and overlayed with tissue culture medium containing various concentrations of an antiviral agent. Following incubation at 37°C for 24–48 hr, the infected cells from each well are lysed and the DNA is denatured and immobilized on separate filter membrane supports or wicks. The bound DNA is hybridized to an HSV-specific, ^{125}I-labeled probe and the amount of HSV target DNA from each well is determined by measuring the radioactivity in a gamma counter. The ID_{50} for the antiviral agent is calculated using linear regression and is defined as the concentration of drug that results in a 50% reduction in DNA hybridization relative to virus-infected control wells containing no drug. The resistance or susceptibility of a particular HSV isolate is established by using an appropriate end point for the antiviral agent studied.

III. COLLECTION AND PREPARATION OF HERPES SIMPLEX VIRUS ISOLATES

To perform the antiviral susceptibility assay, a viable isolate of HSV is required. If necessary, clinical specimens should be collected from dermal, genital, oral–facial, or other appropriate body sites and submitted to a diagnostic laboratory for virus isolation. Otherwise, clinical isolates of HSV can be sent directly to the laboratory for testing. These samples can be submitted as frozen isolates stored at −70°C or as freshly isolated virus demonstrating at least 50% CPE in cultured cells. When submitting fresh isolates, the infected tissue culture tube should be filled completely with an appropriate medium such as modified Eagle's minimum essential medium (EMEM) supplemented with 2 mM glutamine and 2% fetal bovine serum (FBS). The cap should be firmly tightened and covered with parafilm, and the tube should be transported at ambient temperature to the laboratory. Once received in the laboratory, all isolates should be routinely passaged once and regrown to 50–100% CPE in culture tubes of the same cell type. A number of cell types can be used for this purpose including primary rabbit kidney, primary human nenonatal kidney, mink lung, continuous African Green monkey kidney (CV-1 or Vero), continuous human lung carcinoma (A-549), or human diploid fibroblast (MRC-5 or WI-38) cells. CV-1 cells are preferred because this cell line is used to perform the antiviral susceptibility assay. All HSV isolates should be at the first or second passage level; continuous subpassage should be avoided prior to performing susceptibility testing. Aliquots of each isolate should be harvested and frozen at −70°C for future use.

IV. PRELIMINARY CONSIDERATIONS

A. Materials

1. Media and Reagents

Hybriwix™ Herpes (HSV) Antiviral Susceptibility Test Kit (Diagnostic Hybrids, Athens, OH):

1. 3 24-well cell culture dishes containing monolayers of CV-1 cells in EMEM with 10% FBS and gentamicin.
2. 3 27-ml bottles sterile replacement medium containing EMEM with 5% FBS, gentamicin, and amphotericin B.
3. 6 ml DNA wicking agent (lysis solution) containing 0.5 N NaOH and sodium dodecyl sulfate (SDS) in a proprietary formulation; this reagent is used for the single-step lysis of cell monolayers and denaturation of released viral DNA.
4. 36 Hybriwix™ combs, each containing two 9 × 20-mm consecutively numbered wicks.
5. 6 positive and 9 negative control wicks; the negative control wicks are pretreated with a blocking reagent to simulate a negative reaction; purified pSP64 plasmid DNA containing HSV sequences complementary to the HSV DNA probe is absorbed to the positive control wicks.
6. 60 ml wash reagent containing a 20× concentration of 0.3 M sodium chloride, 0.03 M sodium citrate, and 2% SDS in distilled water.
7. 3 3.5-ml vials of HSV probe hybridization agent containing <1.6 μCi ^{125}I-labeled HSV DNA in 50% formamide, 0.5 M NaCl, 0.05 M sodium citrate, and distilled water; the probe is group specific and is produced from a clone of HSV-1 F DNA in bacteriophage M13mp19.
8. Wash container.
9. Radioactive symbol label.

Stock solutions of acyclovir [9-(2-hydroxyethoxymethyl)guanine] (Sigma Chemical Co., St. Louis, MO) and foscarnet (trisodium phosphonoformate) (Sigma), prepared in sterile distilled water at concentrations that are convenient for preparing a range of drug dilutions to be used in the susceptibility assay; final stock concentrations of 640 μg/ml acyclovir and 1800 μg/ml foscarnet are suggested; the concentrations of stock solutions may vary depending on the range of drug concentrations selected for testing and on the antiviral agent employed

EMEM with Earle's balanced salt solution (BioWhittaker, Walkersville, MD), supplemented with 2 mM L-glutamine, 20 mM HEPES, and either 10% (cell growth medium) or 2% (cell maintenance medium) heat-inactivated (56°C, 30 min) FBS (see Hodinka, 1992a, for detailed information on the preparation of cell culture medium and reagents)

CV-1 cells (CCL 70; American Type Culture Collection, Rockville, MD), propagated in the laboratory, as an alternative to purchasing them as part of the commercial susceptibility test kit; stock flasks of these cells can be maintained and used to prepare monolayers in tissue culture tubes and 24-well plates (see Hodinka, 1992b, for details on the serial propagation and maintenance of monolayered cell cultures)

Fresh or frozen virus isolates, including acyclovir and foscarnet-sensitive and -resistant control strains of HSV

Distilled water

2. Supplies

Sterile 1.0-, 5.0-, 10.0-, and 25-ml serological pipettes; disposable plastic pipettes are commercially available and convenient to use

Sterile unplugged pasteur pipettes

Sterile 15-ml and 50-ml polypropylene centrifuge tubes (Corning Glass Works, Corning, NY) for dilution of HSV isolates and antiviral agents

Culture plates, 24-well flat-bottom (Corning Glass Works)

Tissue culture flasks, 80 cm^2 (Nunc, Naperville, IL)

16 × 125-mm Tissue culture tubes (Corning Glass Works)

Forceps

Cryovials (Sarstedt, Newton, NC) for freezing HSV isolates and stock solutions of antiviral agents

0.5% Hypochlorite solution for cleaning work surfaces

Racks designed to hold tissue culture tubes at a 5–7° angle

Safety pipetting devices, protective clothing, latex gloves, infectious-waste disposal containers, and adequate sterilization facilities

Linear regression probability program for calculation of ID$_{50}$ values (Dose-Effect Analysis; Biosoft, Feurgeson, MO; Riacalc; Wallac, Gaithersburg, MD)

Scissors
12 × 55-mm Gamma counting vials (Sarstedt)
^{129}I reference standard (Ludlum Measurements, Sweetwater, TX)

3. Equipment

Inverted and standard light microscopes capable of 100x to 200x magnification
Class II biological safety cabinet
Refrigerator at 2–8°C
Freezer at −70°C
Vacuum source and collection trap containing 0.5% sodium hypochlorite to be used for collecting aspirated infectious waste
Vacuum source and collection trap suitable for collecting low-level radioactive liquid waste
Humidified incubator at 35–37°C with 5% CO_2
Gamma counter
Water baths at 56°C for heat-inactivation of FBS, and 60°C and 73°C for hybridization assay

B. Storage Instructions

   1. On receipt of the Hybriwix™ Herpes (HSV) Antiviral Susceptibility Test Kit, carefully remove the aluminum foil cover from all cell culture dishes, place the plastic lids on the dishes, and incubate at 35–37°C in a humidified atmosphere of 5% CO_2 for 48 hr or until the cell monolayers are confluent. The sterile replacement medium should be stored at 2–8°C and all other reagents and kit accessories should be stored at room temperature. The radioactive probe should be stored in the original container in a specially designated area of the laboratory in accordance with federal and state regulations. The shelf-life of the kit is 2 wks.
   2. The acyclovir and foscarnet stock solutions should be stored in single-use aliquots of 1.5 ml at −70°C no longer than 1 yr. The antiviral agents should be used immediately after thawing. Repeated freezing and thawing or storage under conditions other than those recommended should be avoided.
   3. EMEM with L-glutamine and FBS should be stored at 4°C no longer than 1 mo.
   4. Expiration dates should be indicated on the labels of all media and reagents. The expiration date established by the manufacturer should be observed.

C. Preliminary Preparations

1. Drug Dilutions

1. The dilutions and volumes of drug needed to perform a susceptibility test are dependent on the range of drug concentrations employed and the total number of HSV isolates to be tested. Figures 3 and 4 illustrate the typical ranges of acyclovir and foscarnet, respectively, employed in the HSV hybridization assay and the configurations of the 24-well plates when one clinical HSV isolate and two control virus strains are examined.

2. Sufficient volumes of individual drug dilutions should be prepared

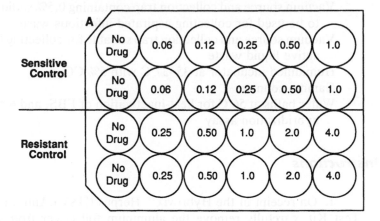

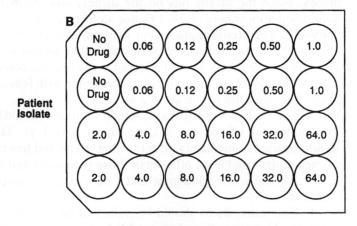

Figure 3 Illustration of the typical ranges of acyclovir (μg/ml) employed in the HSV hybridization assay, and the configuration of the 24-well plates when two control virus strains (A) and one clinical HSV isolate (B) are examined.

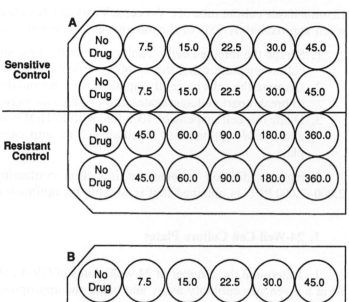

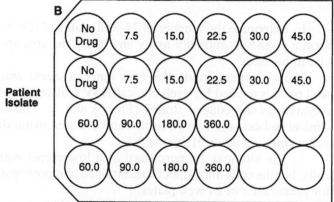

Figure 4 Illustration of the typical ranges of foscarnet (μg/ml)employed in the HSV hybridization assay and the configuration of the 24-well plates when two control virus strains (A) and one clinical HSV isolate (B) are examined.

in sterile EMEM containing 5% FBS and 2 mM L-glutamine (replacement medium) so 1.0 ml of each drug concentration can be added to the appropriate wells of a 24-well plate.

3. All drug dilutions should be prepared immediately before use and each dilution should be mixed thoroughly before making the next dilution.

2. Dilution of Clinical HSV Isolates and Control Strains

The optimum virus inoculum for the susceptibility assay can range from 100 to 10,000 PFU per well. This level can be achieved by growing a clinical HSV isolate or control virus strain to 50–100% CPE in a single 16 ×

125-mm tissue culture tube of CV-1 cells. Virus isolates are usually harvested when approximately 75% of the monolayer is infected.

1. Scrape the infected cell monolayer from the tube surface into 2.0 ml culture medium using the tip of a sterile disposable 1.0-ml pipette.
2. Disperse the cells by gently pipetting them up and down enough times to break apart cell aggregates and produce an even suspension.
3. Dilute this cell suspension 1000-fold with EMEM containing 10% FBS and 2 mM L-glutamine. Inoculate 0.2 ml into each well of 24-well plate.

Using this dilution scheme, an inoculum containing approximately 2,000–3,000 PFU is obtained that is suitable for optimum results.

3. 24-Well Cell Culture Plates

1. Determine the number of 24-well plates of CV-1 cells needed to perform a given susceptibility assay and order the appropriate number of test kits to be used.
2. As a general guide, for each antiviral agent employed, a single 24-well plate is needed for each clinical HSV isolate and an additional plate is required for the control strains (Figs. 3 and 4). Therefore, a total of two clinical isolates and two controls can be tested using the described format and a single 72-assay Hybriwix™ kit.
3. The kits may be purchased at a lower cost without cultured CV-1 cells, but the cells must then be maintained and propagated in the laboratory for preparation of 24-well plates.
4. The cell monolayers of the plates should be examined by microscopy before use to determine the quality of the cells. At the time of virus inoculation, the cell monolayers should be freshly confluent and not overgrown.

D. Preliminary Comments and Precautions

1. Bring all reagents and media to room temperature (20–25°C) before use and return them to appropriate storage temperatures immediately after use.
2. Do not smoke, eat, or drink in areas where specimens, virus isolates, or kit reagents are handled.
3. Take appropriate precautions when handling infectious and radioactive materials. Do not mouth pipette samples or reagents. Avoid contact with broken skin or mucous membranes. Wear disposable gloves, laboratory coats, and other appropriate protective devices. Wash hands thoroughly after handling these materials.

4. Use radioactive material only in designated work areas and cover laboratory bench surfaces with an absorbent material.

5. Use an absorbent material and suitable detergent to clean radioactive spills from involved surfaces and 0.5% sodium hypochlorite to wipe all areas before and after processing of infectious material. Dispose of all infectious and radioactive materials properly.

6. Do not use materials and reagents beyond the assigned expiration dates and do not interchange or mix different lots of reagents.

7. Arrange materials and equipment to provide easy access and to minimize the number of manipulations.

8. Perform accurate pipetting and careful dilution of antiviral drugs and virus isolates for quality results using this procedure. Use separate sterile pipettes when preparing each dilution and during the inoculation of each virus isolate.

9. Use appropriate aseptic technique throughout the procedure.

10. Avoid contact of DNA wicking agent with skin and clothing, since it contains dilute sodium hydroxide (NaOH).

11. Use forceps to manipulate processed Hybriwix™ combs and wicks.

V. ANTIVIRAL SUSCEPTIBILITY ASSAY

A. Virus Inoculation and Drug Addition

1. Obtain a culture tube of HSV-infected CV-1 cells demonstrating 50–100% CPE.
2. Prepare a cell suspension and dilute the infected cells as described in Section IV,C.
3. Remove a 24-well plate of freshly confluent CV-1 cells from the incubator and gently aspirate the medium from each of the 24 wells, taking care not to disturb the cell monolayer.
4. Inoculate the wells with 0.2 ml diluted HSV suspension. Clearly label the plate and wells with an appropriate identification number.
5. Incubate the plate for 60–90 min at 35–37°C in a humidified incubator with 5% CO_2 to allow for virus adsorption. Gently rock the plate back and forth every 15 min to prevent drying of the monolayer.
6. After virus adsorption, gently aspirate the virus inoculum from each well and overlay duplicate wells with 1.0 ml each of the diluted antiviral agent. Two wells should receive 1.0 ml replacement medium to which no drug has been added. These wells serve as untreated controls.
7. Incubate the plate at 35–37°C in the humidified CO_2 incubator for 24–48 hr. Best results are obtained when the CPE in the "no drug"

control wells reaches 50–100% within this time period. When the virus inoculum is prepared as described earlier, the inoculated plates normally require 48 hr to achieve this level of CPE. The monolayers should be examined daily for progression of viral CPE and possible deterioration of the cell monolayers.

B. DNA–DNA Hybridization

See Fig. 5 for an illustration of the hybridization steps.

1. Following the 48-hr incubation, gently but completely aspirate the medium from each well of the 24-well plate. Take care not to disturb the cell monolayers.

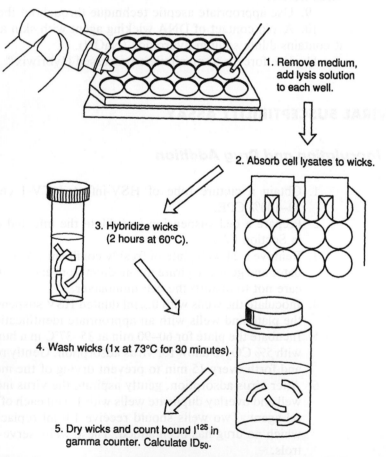

1. Remove medium, add lysis solution to each well.

2. Absorb cell lysates to wicks.

3. Hybridize wicks (2 hours at 60°C).

4. Wash wicks (1x at 73°C for 30 minutes).

5. Dry wicks and count bound I^{125} in gamma counter. Calculate ID$_{50}$.

Figure 5 DNA–DNA hybridization assay for measuring the susceptibility of HSV to antiviral agents.

2. Immediately add two drops of DNA wicking agent (lysis solution) to each well to lyse the cells and virus particles and to denature the DNA. Rock and tap the plate with enough force to ensure that the monolayer in each well is completely covered with the lysis buffer.
3. Using forceps, arrange the Hybriwix™ combs to be used in numerical order on a clean surface or paper towel.
4. Carefully insert the wicks of the combs into their respective duplicate wells and allow the cell lysates to absorb vertically by capillary movement (wicking) onto the wicks. Format the wicks so that increasing wick numbers will correspond to increasing drug concentrations. It is important to place the wicks in the wells so that the DNA from the lysed cells moves upward from the bottom of the wick. The wicks should rest diagonally in the wells and should not stick to the sides of the wells. Wicking is complete when no DNA wicking agent is left in the wells and the Hybriwix™ comb is completely wet. The DNA should be in the lower third of the wick.
5. Once wicking is completed, use forceps to remove the combs from the wells and place them on a clean paper towel. Using scissors, cut the wicks from the combs.
6. Transfer the wicks to a vial of HSV probe hybridization agent (^{125}I-labeled HSV group-specific DNA probe), using one vial for every 24 wicks. Include three negative and two positive control wicks in each hybridization vial.
7. Incubate the hybridization vials for 2 hr at 60 ± 2°C. Also place 114 ml distilled water at 73°C in preparation for Step 12.
8. After hybridization, carefully aspirate the probe hybridization agent from each vial into a trap suitable for collecting low-level radioactive liquid waste.
9. Using a pipette, add 4.0 ml distilled water to cover the wicks in the hybridization vials. Cap the vials and invert several times to rinse the contents.
10. Aspirate this rinse solution into a trap suitable for the collection of radioactive liquid waste.
11. Add 6.0 ml wash reagent to the provided wash container.
12. Carefully transfer the wicks from the vials to the wash container and gently swirl. At this point, add 114 ml preheated (73°C) distilled water to the wash container.
13. Place the wash container in a 73°C water bath for 30 min.
14. Following the 30 min incubation, aspirate the wash into a trap suitable for the collection of radioactive liquid waste.
15. Using forceps, separate the wicks and arrange them in numerical order on a clean paper towel.
16. Transfer the wicks to appropriate vials and count each wick for

2 min in a gamma counter set for a ^{125}I label. Make certain each wick is at the bottom of the vial and that the window of the gamma counter is set to count the entire surface of a wick. The amount of radioactivity present on each wick is dependent on the amount of HSV DNA present after culture in the presence of a specific antiviral agent.

17. Perform appropriate controls for each assay (see Section V,E). Calculate the 50% inhibitory dose (ID_{50}) for each drug and virus isolate tested (see Section V,D).

C. Procedural Notes

1. Care must be taken to use an appropriate virus inoculum in the susceptibility assay. As previously stated, optimum results are achieved when the CPE in the "no drug" control wells is approximately 75% (50–100%) at the end of the incubation period. If a shorter time is required to reach this level of CPE, the HSV inoculum may be too great and the virus may kill the cells before the antiviral agent can be assimilated and have an effect. The opposite effect may occur if the virus inoculum is too weak and too much time is required to achieve the appropriate CPE. The antiviral agent may overwhelm the virus isolate, giving a false impression of the effectiveness of the drug.

2. When aspirating culture medium and reagents from the wells of the cell culture plates, use a sterile disposable pasteur pipette attached to the tubing from a vacuum aspirator and trap. The vacuum should yield gentle aspiration without disturbing the cell monolayer. Hold the 24-well plate at a 20–30° angle and aspirate the medium at the meniscus, directing the tip of the pipette at a 45° angle and moving down the side of each well to the bottom. Completely aspirate all residual material from the wells.

3. Approximately 15 μl DNA wicking agent is used to lyse cells and virus particles within each well. Do not allow this volume to evaporate before placement of the wicks into the wells. By working at a steady rate, multiple plates can be lysed and wicked without a loss in volume of the lysis solution.

4. Strictly adhere to incubation temperatures and times to avoid false results.

5. If probe hybridization cannot be performed immediately after lysis and wicking, allow the wicks to air dry and store them at room temperature. The wicks should be hybridized within 72 hr.

6. The DNA wicking reagent, the 20× wash reagent, and the HSV probe hybridization agent may become turbid or form a precipitate at temperatures below 20°C. If this occurs, warm the reagents in a 37°C water bath until the precipitate dissolves and the solution becomes clear. Precipitation of these reagents will affect the performance of the assay, so close inspection of the solutions before use is important.

7. Instability, deterioration, or toxicity of tissue culture medium, cell monolayers, antiviral agents, and assay reagents should be monitored continuously. Observe medium for microbial contamination; the formation of precipitates, crystals, or sediments; and the ability to support cell growth and viral infection. Determine whether the antiviral agents are cytotoxic to the CV-1 cells. To do this, examine the effect the various concentrations of drug have on normal cell morphology when the drug is incubated with uninfected cells for 48 hr. Define the dose(s) of each antiviral agent required to cause a microscopically visible disruption of normal cell morphology and employ a range of drug concentrations in the *in vitro* susceptibility assay that is not cytotoxic. Cytoxicity in cells treated with drug can also be measured by performing viable cell counts using a 0.4% solution of trypan blue and a hemacytometer. The drug concentration required to reduce the viable cell count by 50% is determined by comparison with cell counts in control cultures without drug.

8. For each assay performed, determine the nonspecific background counts for the gamma counter employed by counting two empty counting tubes for 2 min.

D. Results

1. Calculations

1. Determine the mean reactive value of the nonspecific background for the gamma counter. Subtract this value from the values of each control and specimen to obtain the net values.
2. Determine the mean reactive value of the three negative control wicks.
3. Determine the mean reactive value of each concentration of antiviral agent.
4. Subtract the mean value of the negative controls from the mean value of each concentration of antiviral agent to establish a net reactivity for each drug concentration examined.
5. Then generate a dose–response curve from which the ID_{50} value for the antiviral agent is calculated using a suitable linear regression probability program.

2. Interpretation

The ID_{50} value is defined as the concentration of antiviral drug required to reduce the HSV-specific DNA by 50% compared with the untreated virus controls. In the Clinical Virology Laboratory at Children's Hospital of Philadelphia, HSV isolates with *in vitro* ID_{50} values of ≥ 2 μg/ml for acyclovir and ≥ 100 μg/ml for foscarnet are considered resistant to these antiviral agents. Typically, sensitive HSV-1 isolates have mean ID_{50} values of 0.12 μg/ml and 35.0 μg/ml for acyclovir and foscarnet, respectively, whereas sensitive HSV-2 isolates have mean ID_{50} values of 0.5 μg/ml and 30.0 μg/ml. The range of ID_{50} values for acyclovir-resistant isolates is from 6.0 to 60.0 μg/ml. To date, *in vivo* HSV resistance to foscarnet is rare (Sacks *et al.*, 1989; Birch *et al.*, 1990; Safrin *et al.*, 1994), but this may become a clinically relevant problem as the use of foscarnet against acyclovir-resistant HSV infection increases.

E. Quality Control

The gammma counter used to count the hybridization wicks should be monitored with each assay using a ^{129}I reference source of a known counting rate. A detection efficiency of 75% should be observed for the counter.

The negative and positive control wicks supplied by the manufacturer with each hybridization kit are used to verify the performance of the assay. The negative controls are used to monitor the wash conditions whereas the positive control wicks are used to monitor the hybridization conditions.

1. Typical net negative control values range from 250 to 450 counts per minute (cpm). Two of the three negative controls must have net values that are not in excess of 750 cpm to validate the efficiency of the wash.
2. Typical net positive control values range from 10,000 to 20,000 cpm. Both positive controls must have net values above 7,500 cpm to validate the wash incubation temperature. A substantial decrease in the control values will be observed if the wash temperature exceeds 73°C.

Typical net values for the "no drug" virus controls range from 20,000 to 30,000 cpm.

Acyclovir- and foscarnet-sensitive and -resistant control strains of HSV should be tested with each susceptibility assay. The ID_{50} values for each control should fall within an expected range for the assay to be valid. Alternatively, the ID_{50} values for each control should not exceed two standard deviations from mean values established in previous tests, or the assay should be repeated. Table 4 describes the strains and expected ID_{50} values employed for the hybridization assay. The drug-resistant strains of HSV were originally plaque purified to achieve homogeneous populations of resistant viruses. The resistant strains are generally quite stable and do not routinely revert with subpassage. However, it is recommended that continuous subpassage of the control strains be avoided. These isolates should be frozen at -70°C in single-use aliquots and be regrown only when needed.

TABLE 4
Acyclovir- and Foscarnet-Sensitive and -Resistant HSV Control Strains and Expected ID_{50} Values

Strain designation[a]	HSV type	Expected *in vitro* ID_{50} values (μg/ml)	
		Acyclovir	Foscarnet
F	1	0.25–0.50	30.0–45.0
SLU 360	1	≥2.0	15.0–30.0
PAA^{r}5	1	≥2.0	≥100.0

[a] HSV-1 F and HSV-1 SLU 360 are used when testing clinical isolates against acyclovir. These strains are thymidine kinase-deficient mutants. SLU 360 and HSV-1 PAA^{r}5 are used when testing clinical isolates against foscarnet. PAA^{r}5 is a polymerase mutant. Strains F and SLU 360 were kindly provided by David R. Scholl (Diagnostic Hybrids, Athens, OH), PAA^{r}5 was a gift from Jack Hill (Burroughs Wellcome, Research Triangle Park, NC).

F. Limitations of the Procedure

Numerous methodological factors can influence the results of any antiviral susceptibility assay (Drew and Matthews, 1989) including the DNA–DNA hybridization procedure. These include:

1. the strain of virus examined ("wild" versus "laboratory adapted"). Continuous subpassage of clinical isolates of HSV should be avoided prior to performing susceptibility testing so the isolate tested is representative of the wild-type virus initially identified in a clinical specimen.
2. heterogeneity of the virus population. A given isolate of HSV may be a heterogeneous population containing both susceptible and resistant viruses, which may make *in vitro* results difficult to interpret (Sacks *et al.*, 1989). Also, different body sites from a patient may yield different strains of the same HSV type with dramatically different susceptibility patterns.
3. size of the virus inoculum.
4. type of host-cell system employed, the age and confluency of the cells when used in the assay, and the cell passage level.
5. range of drug concentrations employed.
6. incubation time for virus adsorption and length of incubation of infected cells with drug prior to determination of the ID_{50} value.
7. the use of ID_{50} or ID_{90} drug end points. A consensus has not been reached concerning which value is most appropriate.

8. lack of assay standardization between laboratories, making it difficult to compare results. All assays require the ability of the virus to replicate *in vitro* in a host cell system, but each laboratory may use different parameters of viral replication (e.g., CPE, plaque formation, virus yield, viral antigens, viral nucleic acid, viral enzyme activity, or cell transformation) to measure the activity of an antiviral agent. Comparisons of *in vitro* susceptibility data among different laboratories are valid only if the same assay, virus inoculum, and cell type are used. The definition of a sensitive or resistant isolate may differ for each assay system and for different laboratories performin the same assay. When comparing the activity of different antiviral agents against a virus isolate, the same assay system should also be used.

9. absence of well-characterized, drug-resistant, and -sensitive control strains. These strains are not widely available, but may be obtained as gifts from fellow investigators. Currently, they cannot be purchased through commercial sources.

Certain clinical information should be considered when interpreting the results of *in vitro* antiviral susceptibility assays.

1. Antiviral susceptibility testing appears to be indicated for patients in whom resistance tends to appear the most. These patients normally include immunocompromised individuals, especially transplant recipients and AIDS patients previously treated with anti-HSV agents (Table 3). One report suggests recurrent acyclovir-resistant HSV infection in an immunocompetent individual (Kost *et al.*, 1993) which correlated with clinical failure of antiviral therapy.

2. Clinical situations that favor the development of resistance include long-term suppressive therapy, recurrent intermittent therapy, and the use of less than optimum doses of the antiviral agent.

3. Resistance of HSV isolates to acyclovir is becoming more commonplace, occuring at rates of 2–14% in the immunocompromised host (Wade *et al.* 1983; Englund *et al.*, 1990). In these patients, resistant HSV can be associated with severe progressive disease that does not respond to antiviral treatement (Bean *et al.*, 1987; Norris *et al.*, 1988; Erlich *et al.*, 1989a; Sacks *et al.*, 1989; Gateley *et al.*, 1990; Ljungman *et al.*, 1990).

4. Data suggest that *in vitro* resistance may predict treatment failure and lead to the institution of alternative antiviral therapy (Youle *et al.*, 1988; Chatis *et al.*, 1989; Erlich *et al.*, 1989a,b; Engel *et al.*, 1990; Safrin *et al.*, 1990).

5. Antiviral susceptibility testing may become useful in choosing between continuous high-dose acyclovir and foscarnet to treat acyclovir-resistant HSV. As more antiviral agents are developed and utilized, *in vitro*

results may also guide the physician in the choice of the most appropriate therapy.

6. Results of *in vitro* susceptibility testing are not always predictive of a clinical response. HSV isolates resistant to acyclovir *in vitro* have responded to antiviral treatment (Straus *et al.*, 1984; Englund *et al.*, 1990) and patients with isolates that remain susceptible to the drug *in vitro* have failed therapy (Barry *et al.*, 1985; Erlich *et al.*, 1989a).

7. Acyclovir-resistant clinical isolates of HSV have also been obtained from individuals before therapy (Parris and Harrington, 1982) and from healthy individuals on chronic suppressive therapy (Straus *et al.*, 1984), although no correlation was established between these isolates and the clinical outcome.

8. Acyclovir-resistant isolates may become latent and cause recurrent disease (Erlich *et al.*, 1989b; Safrin *et al.*, 1990). In some patients, recurrences are caused by drug-sensitive isolates after the successful treatment of infection with a resistant virus (Larder and Darby, 1984; Collins, 1988).

9. The absolute ID_{50} value for a single isolate of HSV may be less important than the relative change in ID_{50} values between paired isolates.

10. In general, HSV-2 strains exhibit higher ID_{50} values for acyclovir than HSV-1 strains.

VI. CONCLUSIONS

As the use and number of effective antiviral drugs increases, a definite need will arise for antiviral susceptibility testing, particularly when patients fail to respond to appropriate therapy. The described DNA–DNA hybridization assay is rapid and reproducible, and provides a convenient and objective method for determining the susceptibility of virus isolates to antiviral agents. Although an ^{125}I-labeled probe is currently employed in this procedure, effective nonradiometric probes are being developed and will offer the advantages of being more versatile, being less hazardous to personnel, having a longer shelf-life, and requiring simpler waste disposal. The commercial availability of the hybridization assay and its relative ease of performance should facilitate standardization and the increased use of antiviral susceptibility testing in diagnostic laboratories.

ACKNOWLEDGMENTS

The author is especially grateful to Rachel L. Stetser for excellent technical assistance and helpful discussions.

REFERENCES

Averett, D. R. (1989). Anti-HIV compound assessment by two novel high capacity assays. *J. Virol. Meth.* **23**, 263–276.

Barry, D. W., Nusinoff-Lehrman, S., Ellis, M. N., Biron, K. K., and Furman, P. A. (1985). Viral resistance, clinical experience. *Scand. J. Infect. Dis. (Suppl.)* **47**, 155–164.

Bean, B. (1992). Antiviral therapy: Current concepts and practices. *Clin. Microbiol. Rev.* **5**, 146–182.

Bean, B., Fletcher, C., Englund, J., Lehrman, S. N., and Ellis, M. (1987). Progressive mucocutaneous herpes simplex infection due to acyclovir-resistant virus in an immunocompromised patient: Correlation of viral susceptibilites and plasma levels with response to therapy. *Diagn. Microbiol. Infect. Dis.* **7**, 199–204.

Bleshe, R. B., Burk, B., Newman, F., Cerruti, R. L., and Sim, I. S. (1989). Resistance of influenza A virus to amantadine and rimantadine: Results of one decade of surveillance. *J. Infect. Dis.* **159**, 430–435.

Bevilacqua, F., Marcello, A., Toni, M., Zavattoni, M., Cusini, M., Zerboni, R., Gerna, G., and Palu, G. (1991). Acyclovir resistance/susceptiblity in herpes simplex virus type 2 sequential isolates from an AIDS patient. *AIDS* **4**, 967–969.

Birch, C. J., Tachedjian, G., Doherty, R. R., Hayes, K., and Gust, I. D. (1990). Altered sensitivity to antiviral drugs of herpes simplex virus isolates from a patient with the acquired immunodeficiency syndrome. *J. Infect. Dis.* **162**, 731–734.

Biron, K. K. (1991). Ganciclovir-resistant human cytomegalovirus clinical isolates: Resistance mechanisms and in vitro susceptibility to antiviral agents. *Transplant. Proc.* **23**, 162–167.

Biron, K. K., and Elion, G. B. (1980). *In vitro* susceptibility of varicella-zoster virus to acyclovir. *Antimicrob. Agents Chemother.* **18**, 443–447.

Chatis, P. A., Miller, C. H., Schrager, L. E., and Crumpacker, C. S. (1989). Successful treatment with foscarnet of an acyclovir-resistant mucocutaneous infection with herpes simplex virus in a patient with acquired immunodeficiency syndrome. *N. Engl. J. Med.* **320**, 297–300.

Chesebro, B., and Wehrly, K. (1988). Development of a sensitive quantitative focal assay for human immunodeficiency virus infectivity. *J. Virol.* **62**, 3779–3788.

Coen, D. M., and Schaffer, P. A. (1980). Two distinct loci confer resistance to acycloguanosine in herpes simplex virus type 1. *Proc. Natl. Acad. Sci. USA* **77**, 2265–2269.

Collins, P. (1988). Viral sensitivity following the introduction of acyclovir. *Am. J. Med. (Suppl. 2A)*, **85**, 129–134.

Crumpacker, C. S., Schnipper, L. E., Marlowe, S. I., Kowalsky, P. N., Hershey, B. J., and Levin, M. J. (1982). Resistance to antiviral drugs of herpes simplex virus isolated from a patient treated with acyclovir. *N. Engl. J. Med.* **306**, 343–346.

Dankner, W. M., Scholl, D., Stanat, S. C., Martin, M., Sonke, R. L., and Spector, S. A. (1990). Rapid antiviral DNA–DNA hybridization assay for human cytomegalovirus. *J. Virol. Meth.* **28**, 293–298.

Drew, W. L., and Matthews, T. R. (1989). Susceptibility testing of herpes viruses. *Clin. Lab. Med.* **9**, 279–286.

Drew, W. L., Miner, R. C., Busch, D. F., Follansbee, S. E., Gullett, J., Mehalko, S. G., Gordon, S. M., Owen, W. F., Jr., Matthews, T. R., Buhles, W. C., and DeArmond, B. (1991). Prevalence of resistance in patients receiving ganciclovir for serious cytomegalovirus infection. *J. Infect. Dis.* **163**, 716–719.

Engel, J. P., Englund, J. A., Fletcher, C. V., and Hill, E. L. (1990). Treatment of resistant herpes simplex virus with continuous-infusion acyclovir. *J. Am. Med. Assoc.* **263**, 1662–1664.

Englund, J. A., Zimmerman, M. E., Swierkosz, E. M., Goodman, J. L., Scholl, D. R., and Balfour, H., Jr. (1990). Herpes simplex virus resistant to acyclovir. A study in a tertiary care center. *Ann. Intern. Med.* **112**, 416–422.

Erice, A., Chou, S., Biron, K. K., Stanat, S. C., Balfour, H. H., and Jordan, M. C. (1989). Progressive disease due to ganciclovir-resistant cytomegalovirus in immunocompromised patients. *N. Engl. J. Med.* **320,** 289–293.

Erlich, K. S., Mills, J., Chatis, P., Mertz, G. J., Busch, D. F., Follansbee, S. E., Grant, R. M., and Crumpacker, C. S. (1989a). Acyclovir-resistant herpes simplex virus infections in patients with the acquired immunodeficiency syndrome. *N. Engl. J. Med.* **320,** 293–296.

Erlich, K. S., Jacobson, M. A., Koehler, J. E., Follansbee, S. E., Drennan, D. P., Gooze, L., Safrin, S., and Mills, J. (1989b). Foscarnet therapy for severe acyclovir-resistant herpes simplex virus type-2 infections in patients with the acquired immunodeficiency syndrome (AIDS). An uncontrolled trial. *Ann. Intern. Med.* **110,** 710–713.

Field, H. J., Darby, G., and Wildy, P. (1980). Isolation and characterization of acyclovir-resistant mutants of herpes simplex virus. *J. Gen. Virol.* **49,** 115–124.

Gadler, H. (1983). Nucleic acid hybridization for measurement of effects of antiviral compounds on human cytomegalovirus DNA replication. *Antimicrob. Agents Chemother.* **24,** 370–374.

Gateley, A., Gander, R. M., Johnson, P. C., Kit, S., Otsuka, H., and Kohl, S. (1990). Herpes simplex virus type 2 meningoencephalitis resistant to acyclovir in a patient with AIDS. *J. Infect. Dis.* **161,** 711–715.

Gerna, G., Baldanti, F., Zavattoni, M., Sarasini, A., Percivalle, E., and Revello, M. G. (1992). Monitoring of ganciclovir sensitivity of multiple human cytomegalovirus strains coinfecting blood of an AIDS patient by an immediate-early antigen plaque assay. *Antiviral Res.* **19,** 333–345.

Gu, Z., Gao, Q., Li, X., Parniak, M. A., and Wainberg, M. A. (1992). Novel mutation in the human immunodeficiency virus type 1 reverse transcriptase gene that encodes cross-resistance to 2',3'-dideoxyinosine and 2',3'-dideoxycytidine. *J. Virol.* **66,** 7128–7135.

Harada, S., Koyanagi, Y., and Yamamoto, N. (1985). Infection of HTLV-III/LAV in HTLV-1 carrying cell MT-2 and MT-4 and application in a plaque assay. *Science* **229,** 563–566.

Harmenberg, J., Wahren, B., and Oberg, B. (1980). Influence of cells and virus multiplicity on the inhibition of herpesviruses with acycloguanosine. *Intervirology* **14,** 239–244.

Hayden, F. G., Cote, K. M., and Douglas, R., Jr. (1980). Plaque inhibition assay for drug susceptibility testing of influenza viruses. *Antimicrob. Agents Chemother.* **17,** 865–870.

Hayden, F. G., Belshe, R. B., Clover, R. D., Hay, A. J., Oakes, M. G., and Soo, W. (1989). Emergence and apparent transmission of rimantadine-resistant influenza A virus in families. *N. Engl. J. Med.* **321,** 1696–1702.

Hodinka, R. L. (1992a). Cell culture techniques: Preparation of cell culture medium and reagents. *In* "Clinical Microbiology Procedures Handbook" (H. D. Isenberg, ed.), Vol. 2, pp. 8.19.1–18.9.15. American Society for Microbiology, Washington, D.C.

Hodinka, R. L. (1992b). Cell culture techniques: Serial propagation and maintenance of monolayered cell cultures. *In* "Clinical Microbiology Procedures Handbook" (H. D. Isenberg, ed.), Vol. 2, pp. 8.20.1–8.20.14. American Society for Microbiology, Washington, D.C.

Hruska, J. F., Bernstein, J. M., Douglas, R. G., Jr., and Hall, C. B. (1980). Effects of ribavirin on respiratory syncytial virus *in vitro. Antimicrob. Agents Chemother.* **17,** 770–775.

Hu, J. M., and Hsiung, G. D. (1989). Evaluation of new antiviral agents: I. In vitro perspectives. *Antiviral Res.* **11,** 217–232.

Huraux, J. M., Ingrand, D., and Agut, H. (1990). Perspectives in antiviral chemotherpay. *Fundam. Clin. Pharmacol.* **4,** 357–372.

Jacobson, M. A., Berger, T. G., Fikrig, S., Becherer, P., Moohr, J. W., Stanat, S. C., and Biron, K. K. (1990). Acyclovir-resistant varicella zoster virus infection after chronic oral acyclovir therapy in patients with the acquired immunodeficiency syndrome (AIDS). *Ann. Intern. Med.* **112,** 187–191.

Jacobson, M. A., Drew, W. L., Feinberg, J., O'Donnell, J. J., Whitmore, P. V., Miner, R. D., and Parenti, D. (1991). Foscarnet therapy for ganciclovir-resistant cytomegalovirus retinitis in patients with AIDS. *J. Infect. Dis.* **163,** 1348–1351.

Japour, A. J., Mayers, D. L., Johnson, V. A., Kuritzkes, D. R., Beckett, L. A., Arduino, J. A., Lane, J., Black, R. J., Reichelderfer, P. S., D'Aquila, R. T., Crumpacker, C. S., and the RV-43 Study Group and the AIDS Clinical Trials Group Virology Committee Resistance Working Group. (1993). A standardized peripheral mononuclear cell culture assay for the determination of drug susceptibilities of clinical human immunodeficiency virus type 1 isolates. *Antimicrob. Agents Chemother.* **37**, 1095–1101.

Kang, J., and Pai, C. H. (1989). In situ enzyme immunoassay for antiviral susceptibility testing of respiratory syncytial virus. *Am. J. Clin. Pathol.* **91**, 323–326.

Kawana, F., Shigeta, S., Hosoya, M., Suzuki, H., and DeClercq, E. (1987). Inhibitory effects of antiviral compounds on respiratory syncytial virus replication in vitro. *Antimicrob. Agents Chemother.* **31**, 1225–1230.

Keating, M. R. (1992). Antiviral agents. *Mayo Clin. Proc.* **67**, 160–178.

Knox, K. K., Drobyski, W. R., and Carrigan, D. R. (1991). Cytomegalovirus isolate resistant to ganciclovir and foscarnet from a marrow transplant patient. *Lancet* **337**, 1292–1293.

Kost, R. G., Hill, E. L., Tigges, M., and Straus, S. E. (1993). Brief report: Recurrent acyclovir-resistant genital herpes in an immunocompetent patient. *N. Engl. J. Med.* **329**, 1777–1782.

Land, S., Terloar, G., McPhee, D., Birch, C., Doherty, R., Cooper, D., and Gust, I. (1990). Decreased in vitro susceptibility to zidovudine of HIV isolates obtained from patients with AIDS. *J. Infect. Dis.* **161**, 326–329.

Larder, B. A., and Darby, G. (1984). Virus drug resistance: Mechanisms and consequences. *Antiviral Res.* **4**, 1–42.

Larder, B. A., and Kemp, S. D. (1989). Multiple mutations in HIV-1 reverse transcriptase confer high-level resistance to zidovudine (AZT). *Science* **246**, 1155–1158.

Larder, B. A, Darby, G., and Richman, D. D. (1989). HIV with reduced sensitivity to Zidovudine (AZT) isolated during prolonged therapy. *Science* **243**, 1731–1734.

Ljungman, P., Ellis, M. N., Hackman, R. C., Shepp, D. H., and Meyers, J. D. (1990). Acyclovir-resistant herpes simplex virus causing pneumonia after marrow transplantation. *J. Infect. Dis.* **162**, 244–248.

McLaren, C., Ellis, M. N., and Hunter, G. A. (1983). A colorimetric assay for the measurement of the sensitivity of herpes simplex viruses to antiviral agents. *Antiviral Res.* **3**, 223–234.

McLaren, C., Chen, M. S., Ghazzouli, I., Saral, R., and Burns, W. H. (1985). Drug resistance patterns of herpes simplex virus isolates from patients treated with acyclovir. *Antimicrob. Agents Chemother.* **28**, 740–744.

Mitsuya, H., Weinhold, K. J., Furman, P. A., Clair, H. S., Lehrman, S. N., Gallo, R. C., Bolognesi, D., Barry, D. W., and Broder, S. (1985). 3'-Azido-3'-deoxythymidine (BWA509U): An antiviral agent that inhibits the infectivity and cytopathic effect of human T-lymphotropic virus type III/lymphadenopathy-associated virus in vitro. *Proc. Natl. Acad. Sci. USA* **82**, 7096–7100.

Montefiori, D. C., Robinson, W. E., Jr., Schuffman, S. S., and Mitchell, W. M. (1988). Evaluation of antiviral drugs and neutralizing antibodies to human immunodeficiency virus by a rapid and sensitive microtiter infection assay. *J. Clin. Microbiol.* **26**, 231–235.

Nakashima, H., Matsui, T., Harada, S., Kobayashi, N., Matsuda, A., Ueda, T., and Yamamoto, N. (1986a). Inhibition of replication and cytopathic effect of human T-cell lymphotropic virus type III/lymphadenopathy-associated virus by 3'-deoxythymidine in vitro. *Antimicrob. Agents Chemother.* **30**, 933–937.

Nakashima, H., Koyanagi, Y., Harada, S., and Yamamoto, N. (1986b). Quantitative evaluation of the effect of UV irradiation on the infectivity of HTLV-III (AIDS virus) with HTLV-1-carrying cell lines, MT-4. *J. Invest. Dermatol.* **87**, 239–243.

Newton, A. A. (1988). Tissue culture methods for assessing antivirals and their harmful effects. *In* "Antiviral Agents: The Development and Assessment of Antiviral Chemotherapy" (H. J. Field, ed.), Vol. I, pp. 33–66. CRC Press, Boca Raton, FL.

Norris, S. A., Kessler, H. A., and Fife, K. H. (1988). Severe, progressive herpetic whitlow caused by an acyclovir-resistant virus in a patient with AIDS. *J. Infect. Dis.* **157,** 209–210.

Pahwa, S., Biron, K., Lim, W., Swenson, P., Kaplan, M. H., Sadick, N., and Pahwa, R. (1988). Continuous varicella-zoster infection associated with acyclovir resistance in a child with AIDS. *J. Am. Med. Assoc.* **260,** 2879–2882.

Parris, D. S., and Harrington, J. E. (1982). Herpes simplex virus variants resistant to high concentrations of acyclovir exist in clinical isolates. *Antimicrob. Agents Chemother.* **22,** 71–77.

Pepin, J., Simon, F., Dussault, A., Collin, G., Dazza, M., and Brun-Vezinet, F. (1992). Rapid determination of human cytomegalovirus susceptibility to ganciclovir directly from clinical specimen primocultures. *J. Clin. Microbiol.* **30,** 2917–2920.

Plotkin, S. A., Drew, W. L., Felsenstein, D., and Hirsch, M. S. (1985). Sensitivity of clinical isolates of human cytomegalovirus to 9-(1,3-dihydroxy-2-propoxymethyl)guanine. *J. Infect. Dis.* **152,** 833–834.

Prichard, M. N., Turk, S. R., Coleman, L. A., Engelhardt, S. L., Shipman, C., and Drach, J. C. (1990). A microtiter virus yield reduction assay for the evaluation of antiviral compounds against human cytomegalovirus and herpes simplex virus. *J. Virol. Meth.* **28,** 101–106.

Rabalais, G., Levin, M. J., and Berkowitz, F. E. (1987). Rapid herpes simplex virus susceptibility testing using an enzyme-linked immunosorbent assay performed in situ on fixed virus-infected monolayers. *Antimicrob. Agents Chemother.* **31,** 946–948.

Rasmussen, L., Chen, P. T., Mullenax, J. G., and Merigan, T. C. (1984). Inhibition of human cytomegalovirus replication by 9-(1,3-dihydroxy-2-propoxymethyl)guanine alone and in combination with human interferons. *Antimicrob. Agents Chemother.* **26,** 441–445.

Richman, D. D. (1990). Zidovudine resistance of human immunodeficiency virus. *Rev. Infect. Dis. (Suppl. 5),* **12,** S507–S512.

Rooke, R., Tremblay, M., Soudeyns, H., DeStephano, L., Yao, X. J., Fanning, M., Montaner, J. S., O'Shaughnessy, M., Gelmon, K., and Tsoukas, C. (1989). Isolation of drug-resistant variants of HIV-1 from patients on long-term zidovudine therapy: Canadian Zidovudine Multi-Centre Study Group. *AIDS.* **3,** 411–415.

Sacks, S. L., Wanklin, R. J., Reece, D. E., Hicks, K. A., Tyler, K. L., and Coen, D. M. (1989). Progressive esophagitis from acyclovir-resistant herpes simplex. Clinical roles for DNA polymerase mutants and viral heterogeneity? *Ann. Intern. Med.* **111,** 893–899.

Safrin, S., Assaykeen, T., Follansbee, S., and Mills, J. (1990). Foscarnet therapy for acyclovir-resistant mucocutaneous herpes simplex virus infection in 26 AIDS patients: Preliminary data. *J. Infect. Dis.* **161,** 1078-1084.

Safrin, S., Kemmerly, S., Plotkin, B., Smith, T., Weissbach, N., De Veranez, D., Phan, L. D., and Cohn, D. (1994). Foscarnet-resistant herpes simplex virus infection in patients with AIDS. *J. Infect. Dis.* **169,** 193–196.

Schnipper, L. E., and Crumpacker, C. S. (1980). Resistance of herpes simplex virus to acycloguanosine: Role of viral thymidine kinase and DNA polymerase loci. *Proc. Natl. Acad. Sci. USA* **77,** 2270–2273.

St. Clair, M. H., Martin, J. L., Tudor-Williams, G., Back, M. C., Vavro, C. L., King, D. M., Kellam, P., Kemp, S. D., and Larder, B. A. (1991). Resistance to ddI and sensitivity to AZT induced by a mutation in HIV-1 reverse transcriptase. *Science* **253,** 1557–1559.

Straus, S. E., Takiff, H. E., Seidlin, M., Bachrach, S., Lininger, L., DiGiovanna, J. J., Western, K. A., Smith, H. A., Nusinoff-Lehrman, S., Creagh-Kirk, T., and Alling, D. W. (1984). Suppression of frequently recurring genital herpes. *N. Engl. J. Med.* **310,** 1545–1550.

Sullivan, V., and Coen, D. M. (1991). Isolation of foscarnet-resistant human cytomegalovirus patterns of resistance and sensitivity to other antiviral drugs. *J. Infect. Dis.* **164,** 781–784.

Swierkosz, E. M., Scholl, D. R., Brown, J. L., Jollick, J. D., and Gleaves, C. A. (1987).

Improved DNA hybridization method for detection of acyclovir-resistant herpes simplex virus. *Antimicrob. Agents Chemother.* **31,** 1465–1469.

Tatarowicz, W. A., Lurain, N. S., and Thompson, K. D. (1991). In situ ELISA for the evaluation of antiviral compounds effective against human cytomegalovirus. *J. Virol. Meth.* **35,** 207–215.

Telenti, A., and Smith, T. F. (1989). Screening with a shell vial assay for antiviral activity against cytomegalovirus. *Diagn. Microbiol. Infect. Dis.* **12,** 5–8.

Wade, J. C., McLaren, C., and Meyers, J. D. (1983). Frequency and significance of acyclovir-resistant herpes simplex virus isolated from marrow transplant patients receiving multiple courses of treatment with acyclovir. *J. Infect. Dis.* **148,** 1077–1082.

Watts, D. H. (1992). Antiviral agents. *Obstet. Gynecol. Clin. North Am.* **19,** 563–583.

Youle, M. M., Hawkins, D. A., Shanson, D. C., Evans, R., Oliver, N., and Lawrence, A. (1988). Acyclovir-resistant herpes in AIDS treated with foscarnet. *Lancet* **ii,** 341–342.

6

Quantification of Viral Nucleic Acids Using Branched DNA Signal Amplification

Judith C. Wilber
Nucleic Acid Systems
Chiron Corporation
Emeryville, CA 94608
and Department of Laboratory Medicine
University of California, San Francisco
San Francisco, California 94143

Mickey S. Urdea
Nucleic Acid Systems
Chiron Corporation
Emeryville, California 94608

I. INTRODUCTION

As antiviral therapy becomes more commonplace, the clinical laboratory
will require routine methods with which to monitor the efficacy of therapy

and the development of resistance to antiviral agents. Because chronic viral infections are often treated with prolonged courses of potentially toxic drugs, following the effect of the treatment at the level of virus infection in individual patients is often important. One of the challenges of the clinical virology laboratory is the quantification of virus in a sample. Most quantitative strategies involve molecular amplification of the target nucleic acid (target amplification) followed by detection of the amplified product with standard, and relatively insensitive, methods. Although target amplification techniques such as the polymerase chain reaction (PCR) are extremely sensitive, these methods require extraction and purification of the target materials and extreme care must be used to limit cross-contamination. Once the target has been amplified, it is difficult to relate the resultant number of copies to the physiological level of virus because the steps involved often have unknown or variable efficiencies.

Signal amplification methods do not alter the number of target molecules and are therefore more amenable to quantitative analysis. In addition, viral nucleic acids are measured directly in clinical samples; the resulting signal is thus directly proportional to the concentration of specific nucleic acid in the patient sample. Many techniques for labeling probes for Southern blots or *in situ* hybridization (e. g., fluorescence, radioactivity, chemiluminescence, electrochemiluminescence) can be considered signal amplification techniques (Rashtchian *et al.*, 1987; Blackburn *et al.*, 1991; Hoyland and Fremont, 1991; Wiedbrauk, 1991; Wolcott, 1992), but these methods usually require high concentrations of target nucleic acids.

To increase the sensitivity of direct detection, branched oligodeoxyribonucleotides (bDNA) molecules were developed by Urdea and others (Urdea *et al.*, 1987,1991; Urdea, 1993) at Chiron Corporation (Emeryville, CA). These molecules are used as amplifiers to incorporate as many as 1755 alkaline phosphatase molecules onto each nucleic acid target molecule. The bDNA signal amplification system is highly reproducible, can utilize relatively crude specimens (no purification of target), and requires no special handling procedures beyond those needed for standard laboratory practices.

II. DESCRIPTION OF THE PROCEDURE

Human immunodeficiency virus type 1 (HIV-1) RNA levels in plasma, hepatitis C virus (HCV) RNA in serum, and hepatitis B virus (HBV) DNA in serum can be quantified using a solid phase nucleic acid hybridization assay (Quantiplex™) based on bDNA signal amplification technology, shown in Fig. 1. RNA or DNA is detected in a sample based on a unique solution-phase sandwich hybridization assay coupled with signal amplification employing

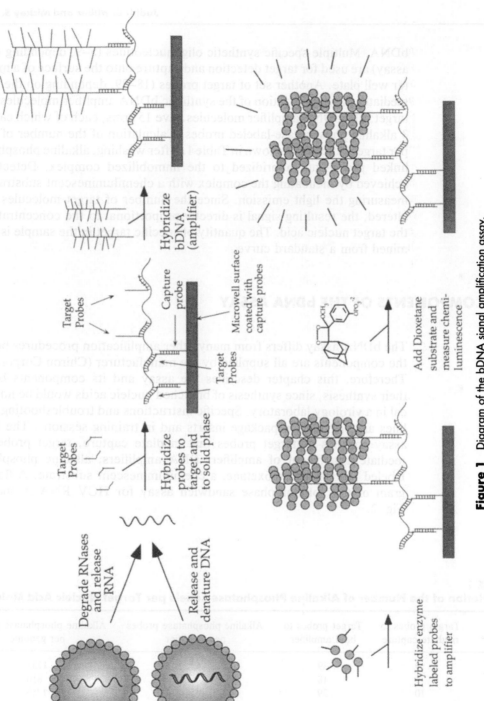

Figure 1 Diagram of the bDNA signal amplification assay.

bDNA. Multiple specific synthetic oligonucleotides (5–9, depending on the assay) are used for target detection and capture onto the surface of a microtiter well plate. Another set of target probes (18–39, depending on the assay) mediates the hybridization of the synthetic bDNA amplifier molecules to the target. The bDNA amplifier molecules have 15 arms, each of which can bind 3 alkaline phosphatase-labeled probes. Calculation of the number of labels per target molecule is shown in Table 1. After washing, alkaline phosphatase-linked probes are hybridized to the immobilized complex. Detection is achieved by incubating the complex with a chemiluminescent substrate and measuring the light emission. Since the number of target molecules is not altered, the resulting signal is directly proportional to the concentration of the target nucleic acid. The quantity of specific target in the sample is determined from a standard curve.

III. COMPONENTS OF THE bDNA ASSAY

The bDNA assay differs from many other amplification procedures because the components are all supplied by the manufacturer (Chiron Corporation). Therefore, this chapter describes the assay and its components but not their synthesis, since synthesis of branched nucleic acids would be impractical in a virology laboratory. Specific instructions and troubleshooting guidelines are provided in package inserts and in training sessions. The bDNA assay consists of target probes that mediate capture, target probes that mediate attachment of amplifier, DNA amplifiers, alkaline phosphatase-labeled probes, and dioxetane, a chemiluminescent substrate. A flow diagram of the solution-phase sandwich assay for HCV RNA is shown in Fig. 2.

TABLE 1
Calculation of the Number of Alkaline Phosphatase Labels per Target Nucleic Acid Molecule

Assay[a]	Target probes to mediate capture	Target probes to bind amplifier	Alkaline phosphatase probes per amplifier	Alkaline phosphatase probes per genome
HBV	9	39	45	1755
HCV	9	18	45	810
HIV	10	39	45	1755

[a] Abbreviations: HBV, hepatitis B virus; HCV, hepatitis C virus; HIV, human immunodeficiency virus.

Add 50 μl specimen plus 150 μl diluent with HCV probes

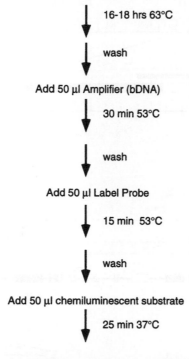

16-18 hrs 63°C

wash

Add 50 μl Amplifier (bDNA)

30 min 53°C

wash

Add 50 μl Label Probe

15 min 53°C

wash

Add 50 μl chemiluminescent substrate

25 min 37°C

Measure light emission and quantitate
HCV RNA equivalents/ml in specimens
using standard curve.

Figure 2 Flow diagram of the laboratory procedure for the bDNA signal amplification assay for HCV RNA.

A. Target Probes

The target probes are 50-base oligonucleotides, one portion (20–40 bases) of which is complementary to the target nucleic acid sequence while the second portion (20 bases) is used to capture the probe–target complex to an oligonucleotide-modified microtiter well plate or to attach the probe–target fragment to the bDNA amplifier. Several (5–50) target and capture probes can be used. The number of probes used in the current assays is shown in Table 1.

All the target probes are designed to hybridize to the most conserved regions of the target nucleic acids. Figure 3 shows the placement of the two

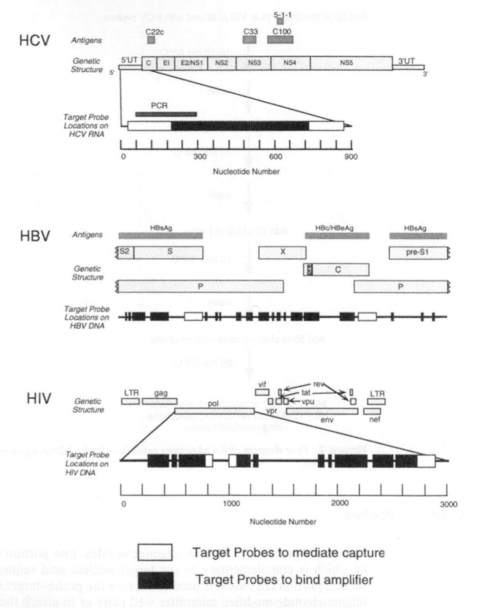

Figure 3 Location of target probes for hepatitis C (HCV), hepatitis B (HBV), and human immuno-deficiency virus (HIV) bDNA signal amplification assays. The HCV diagram shows the most commonly used region for RT-PCR primers for comparison.

types of probe for each of three viral targets. For HCV, the probes are located in the 5' untranslated region and a portion of the core region; for HBV in conserved regions throughout the genome; and for HIV, in the reverse transcriptase (*pol*) region.

B. Branched DNA Amplifiers

Branched DNA amplifiers were initially synthesized by chemical cross-linking of oligonucleotides containing three alkylamine functions. However, the method was of limited utility because structural constraints precluded maximal probe binding. The use of "branching monomers" during the chemical synthesis of oligodeoxyribonucleotides proved to be a more flexible method (Urdea, 1991). Branching monomers are phosphoramidite reagents containing at least two protected hydroxyl groups. In general, a primary linear fragment is synthesized and then tailed with several appropriately spaced branching monomers. Several simultaneous secondary syntheses are then initiated from the branch points. Branched DNAs containing several hundred nucleotides can be constructed in this manner.

Large-branched oligonuceotides for signal amplification have been synthesized using a combination of solid-phase chemistry and enzymatic ligation methods. The amplifier—containing a maximum of 45 alkaline phosphatase probe-binding sites—is produced by synthesizing a bDNA with 15 branches that is then combined with a complementary linker that is in turn complementary to a branch extension or "arm," each of which has 3 binding sites for an alkaline phosphatase-labeled probe (3 sites × 15 branches = 45 labels). The amplifiers are assembled by ligation, then analyzed by capillary electrophoresis. On average, 13 of the 15 arms are incorporated into each branched oligonucleotide. Although the bDNA amplifiers are difficult to fabricate, they are easy to use.

C. Alkaline Phosphatase-Labeled Probes

Alkaline phosphatase probes are synthesized so part of the probe is complementary to a portion of the arms of the branched DNA. Three hybridization sites are located on each branch, yielding 45 alkaline phosphatase probes per bDNA molecule. Finally, alkaline phosphatase catalyzes the dephosphorylation of dioxetane (Lumigen, Detroit, MI), a reaction that produces visible light, the intensity of which is measured with a luminometer. An example of a computer readout is given in Fig. 4.

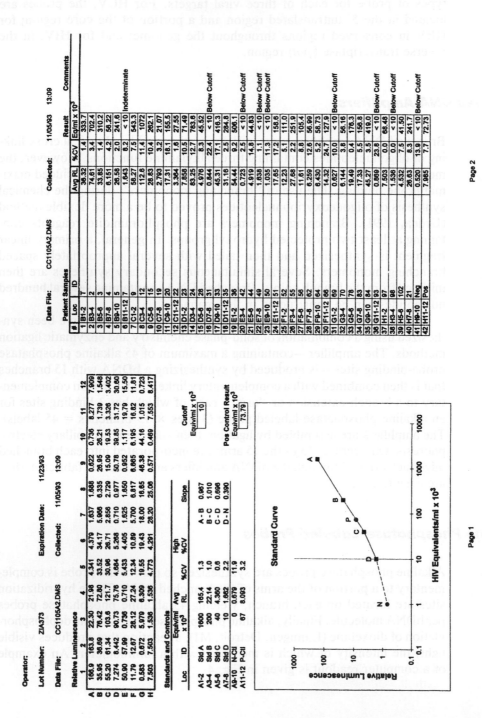

Figure 4 Example of a luminometer printout with calculated results. RL, Relative luminescence.

D. Molecular Standards

Extensive work has been done to determine accurately the number of target molecules in each of the standards used to generate a standard curve. Cloned and extensively purified target nucleic acids are quantified by several methods including phosphate determination, incorporation of $[\alpha\text{-}^{32}P]$guanosine, and determining the absorbance of both intact and digested transcripts. Once the standards have been quantified, they are tested in the bDNA assay and the number of relative light units per attomole of target nucleic acid is determined. An "equivalent" is the amount of virus that generates the same amount of light as a single molecule of the standard. Any comparisons of absolute molecular sensitivity with other assays such as PCR would have to use exactly the same standard to compare numberical results.

E. Protocol for Hepatitis C RNA

Specific details are in the package insert.

1. Pipette 50 μl patient serum into each of two wells of a 96-well plate coated with capture probes. Pipette standards and controls into the appropriate wells. Add lysis diluent containing proteinase K, detergent and two sets of oligonucleotide probes. (Target probes with extensions mediate the capture of the target RNA to the microtiter well and target probes with extenders mediate binding of the target to the bDNA amplifiers). Incubate at 63°C overnight.
2. Wash plate with buffer A. Add bDNA amplifiers and incubate at 53°C for 30 min.
3. Wash plate with buffer A. Add alkaline phosphatase probes and incubate at 53°C for 15 min.
4. Wash plate with buffer A followed by buffer B. (The second buffer does not contain SDS, which will interfere with the enzyme–substrate reaction.)
5. Add dioxetane substrate. Incubate at 37°C for 25 min in the luminometer. Read relative luminescence.

F. Interpretation of Results

Clinical cutoff values for each of the assays were set by assessing the signal generated by a large number of negative samples and determining the level above which the assay could reliably distinguish a negative from a positive sample. Currently, the standard curve for HCV is 3.5×10^5 to 5.7×10^7

RNA equivalents/ml (the HCV assay uses 50 μl per well); for HBV, the curve is 7×10^5 to 4.5×10^9 DNA equivalents/ml (the HBV assay uses 10 μl per well); for HIV, the curve is 1×10^4 to 1.6×10^6 RNA equivalents/ ml (the HIV assay uses 1 ml per well in a centrifuged pellet). The sensitivities are determined by studying populations of individuals known to be infected with each particular virus, as discussed subsequently. Sensitivity comparisons of populations are much more useful than molecular comparisons, since every laboratory uses different standards. The specificity of all the assays is greater than 98%.

IV. USES OF bDNA ASSAYS

A. Hepatitis C Virus RNA

HCV is the major cause of parenterally transmitted non-A, non-B hepatitis (Choo et al., 1989). Until the virus was characterized, diagnosis was made by exclusion of all other known causes of hepatitis. Antibody against HCV is found in over 80% of patients with well-documented non-A, non-B hepatitis. Chronic HCV is characterized by fluctuating alanine aminotransferase (ALT) levels and recognizable changes in liver histology, which may lead to cirrhosis and hepatocellular carcinoma (Alter et al., 1990). Therapy with interferon-α has been approved by the Food and Drug Administration (FDA) and can be effective in eliminating the virus. However, less than half of HCV-infected individuals respond to treatment, and relapse is common (David et al., 1989).

A bDNA-amplified test was developed for monitoring the level of the genomic RNA at the conserved 5' untranslated region and a portion of the core gene in individuals infected with HCV. In a study of chronic HCV patients, the bDNA assay detected and quantified the virus in 650/706 (92%) of those detected by RT-PCR (Li et al., 1993). The quantity of HCV RNA in serum may be useful in monitoring both the disease state and the effectiveness of interferon therapy. Preliminary studies using bDNA to quantify HCV RNA during therapy indicate that there is a correlation between the level of RNA before the initiation of therapy and the prognosis. In other words, the greater the amount of RNA before therapy, the poorer the response (Lau et al., 1993).

B. Hepatitis B Virus DNA

Two studies were performed to determine the correlation of the HBV DNA assay with HBV e antigen (HBeAg) in infected patients. In the first study, with samples from patients in the United States, 53 of 56 specimens (95%) were positive. The second study, conducted with samples from Japanese patients, showed a 100% correlation (42 of 42) with HBeAg. In a comparison with the Abbott liquid hybridization assay, the bDNA assay detected 130/142 (92%) whereas the Abbott test detected 117/142 (82%) (Hendricks *et al.*, 1992). In HBeAg-negative, anti-HBe positive, HBsAg-positive patients, the bDNA assay detected 26/32, whereas a dot blot assay detected only 12 (Habersetzer *et al.*, 1993). The quantity of HBV DNA has been shown to be useful in monitoring interferon treatment and in predicting response to therapy (Perrillo *et al.*, 1990).

C. Human Immunodeficiency Virus RNA

In the HIV RNA assay (Pachl *et al.*, 1994), plasma samples (1 ml) are concentrated by centrifugation in a table-top refrigerated microfuge at 23,500 *g*, and then processed as shown in Fig. 2. In one study, 110 HIV antibody-positive samples were tested, and 79% were reactive; 93% of the 56 samples from patients with CD4 counts <200 were positive for HIV RNA, 67% of the 48 patients with CD4 counts of 200–500 were reactive, and 50% of the patients with CD4 counts >500 were reactive (Urdea *et al.*, 1993). The specificity of the test is 98.9%. Figure 5 shows that the assay is very reproducible, which is very important in quantitative studies.

Reproducibility of Inter-Assay Quantitation bDNA Assay for HIV RNA

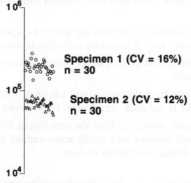

Specimen 1 (CV = 16%) n = 30

Specimen 2 (CV = 12%) n = 30

HIV RNA Equivalents/mL

Figure 5 Reproducibility of 2 samples run in 30 separate runs by branched DNA signal amplification for HIV RNA.

A significant number of HIV-infected individuals have high CD4 levels and high viral load. In a study using the bDNA assay to measure HIV RNA in the plasma of individuals enrolled in a large cohort study, researchers showed that patients with high levels of HIV RNA progressed to AIDS more rapidly than those who maintained low levels of RNA throughout the 30- to 60-mo follow-up period (Gupta *et al.*, 1993). Therefore, it may be useful to stratify patients by viral load rather than CD4 level for entry into clinical trials. The bDNA assay is currently being used in many laboratories to monitor changes in levels of HIV RNA for several antiviral treatment trials.

V. SAMPLE COLLECTION AND STABILITY

Sample collection procedures are shown in Table 2. Most of the assays have been developed for serum or plasma, which are not extracted before

TABLE 2
Sample Collection and Handling for Quantitative bDNA Assays for RNA[a]

HIV

1. 5 ml *plasma* is required—this assures the ability to repeat any problem specimens and to do other tests if necessary (absolute minimum is 2 ml).

2. Blood should be collected in sterile tubes with anticoagulant—EDTA is preferred, acid citrate dextrose (ACD) or heparin is acceptable.

3. Remove plasma aseptically from the centrifuged collection tube within 4 hours of collection.

4. Dispense 1 ml aliquots (measured precisely) into sterile screw-capped tubes (1.5 ml Sarstaedt tubes with outer threads and rubber O-ring).

HCV

1. 0.5–1.0 ml *serum*—this ensures the ability to repeat any problem specimens and to do other tests if necessary (absolute minimum is 0.2 ml).

2. Blood should be collected in sterile serum separator tube (SST) with no anticoagulants.

3. Remove serum from the cells within 1 hour of collection by centrifuging the SST. (If SST is not used, serum must be removed from the clot within 1 hour.)

4. Remove serum aseptically from the centrifuged SST within 4 hours of collection and dispense into sterile screw-capped tubes (1.5 ml Sarstaedt tubes with outer threads and rubber O-ring).

Storage

1. Freeze immediately at −70 C after dispensing into sterile storage tubes.

2. After thawing, keep on ice during use.

3. Avoid repeated thawing and refreezing.

placement in the well. Extraction protocols using guanidine hydrochloride, sodium sarcosyl, and ethanol have been developed to quantify viral RNA (HCV or HIV) in lymph node tissue, liver tissue, and peripheral blood mononucleocytes (Cox, 1968).

Since these assays are designed to be quantitative, it is especially important to preserve the physiological level of virus in each sample. RNA is very susceptible to degradation by the high levels of RNase in blood. Blood should be collected in sterile tubes; the serum or plasma should be separated from the cells within 4 hr of collection and then stored at −70°C (Cuypers *et al.*, 1992; Davis *et al.*, 1994).

VI. CONCLUSION

The bDNA signal amplification system has been used to detect and quantify a variety of viruses in clinical samples. Because of the simple steps and format, many samples (42–126 patient results) can be processed by one individual in 24 hr. The assay is very reproducible (interlaboratory CVs <30%) relative to other nucleic acid quantification methods. The dynamic range of each assay includes at least 90% of the detectable population without dilution. Because the viral nucleic acids remain at their physiological levels, the direct probe method is inherently quantitative and is not subject to the amplicon carryover problems associated with target amplification methods.

REFERENCES

Alter, M. J., Hadler, S. C., Judson, F. N., Mares, A., Alexander, W. J., Hu, P. Y., and Miller, J. K. (1990). Risk factors for acute non-A non-B hepatitis in the United States and association with hepatitis C infection. *J. Am. Med. Assoc.* **264,** 2231–2235.

Blackburn, G. F., Shah, H. P., Kenten, J. H., Leland, J., Kamin, R. A., Link, J., Peterman, J., Powell, M. J., Shah, A., Talley, D. B., Tyagi, S. K., Wilkins, E., Wu, T.-G., and Massey, R. J. (1991). Electrochemiluminescence detection for development of immunoassays and DNA probe assays for clinical diagnostics. *Clin. Chem.* **37(9),** 1534–1539.

Choo, Q.-L., Kuo, G., Weiner, A. J., Overby, L. R., Bradley, D. W., and Houghton, M. (1989). Isolation of a cDNA clone derived from a blood-borne non-A, non-B viral hepatitis genome. *Science* **244,** 359–364.

Cox, R. A. (1968). The use of guanidinium chloride in the isolation of nucleic acids. *Meth. Enzymol.* **12B,** 120–129.

Cuypers, H. T. M., Bresters, D., Winkel, N., Reesink, H. W., Weiner, A. J., Houghton, M., van der Poel, C. L., and Lelie, P. N. (1992). Storage conditions of blood samples and primer selection affect the yield of cDNA polymerase chain reaction products of hepatitis C virus. *J. Clin. Microbiol.* **30(12),** 3220–3224.

Davis, G. L., Balart, L. A., Schiff, E. R., Lindsay, K., Bodenheimer, H. C., Perrillo, R. P., Carey, W., Jacobson, I. M., Payne, J., Dienstag, J. L., VanThiel, D. H., Tamburro, C., Lefkowitch, J., Albrecht, J., Meschievitz, C., Ortego, T. J., Gibas, A., and the Hepatitis Interferon Treatment Group (1989). Treatment of chronic hepatitis C with recombinant interferon alfa. *N. Engl. J. Med.* **321(22)**, 1501–1505.

Davis, G., Lau, J., Urdea, M., Neuwald, P., Wilber, J., Lindsay, K., Perrillo, R., and Albrecht, J. (1994). Quantitative detection of hepatitis C virus RNA with a solid-phase signal amplification method: Definition of optimal conditions for specimen collection and clinical application in interferon-treated patients. *Hepatol.* **19(6)**, 1337–1341.

Gupta, P., Kokka, R., Neuwald, P., Kern, D., Rinaldo, C., and Mellors, J. (1993). Expression of HIV-1 RNA in plasma correlates with the development of AIDS: A multicenter AIDS cohort study (MACS). 9th International Conference on AIDS, Berlin, Germany.

Habersetzer, F., Zhang, X., Cohard, M., Bizollin, T., Chossegros, P., Berthillon, P., Rougier, P., Chevallier, M., Urdea, M., Zarski, J., and Trépo, C. (1993). Efficacy of interferon alpha in HBV chronic active hepatitis appreciated by branched DNA amplification. *Hepatology* **18(4)**, 234A.

Hendricks, D., Hoo, B., Neuwald, P., Urdea, M., Campbell, C., and Perrillo, R. (1992). Use of a new signal amplification assay for the quantification of hepatitis B virus DNA in sera of patients undergoing therapy with alpha interferon. *Hepatology* **16**, 542.

Hoyland, J. A., and Freemont, A. J. (1991). Investigation of a quantitative post-hybridization signal amplification system for mRNA-oligodeoxyribonucleotide in situ hybridization. *J. Pathol.* **164**, 51–58.

Lau, J. Y. N., Davis, G. L., Kniffen, J., Qian, K.-P., Urdea, M. S., Chan, C. S., Mizokami, M., Neuwald, P. D., and Wilber, J. C. (1993). Significance of serum hepatitis C virus RNA levels in chronic hepatitis C. *Lancet* **341**, 1501–1504.

Li, X., de Medina, M., LaRue, S., Shao, L., and Schiff, E. R. (1993). Comparison of assays for HCV RNA. *Lancet* **342**, 1174–1175.

Pachl, C., Todd, J., Kern, D., Sheridan, P., Fong, S.-J., Stempien, M., Hoo, B., Besemer, D., Yeghiazarian, T., Irvine, B., Kolberg, J., Kokka, R., neuwald, P., and Urdea, M. (1994). Rapid and precise quantification of HIV-1 RNA in plasma using a branched DNA (bDNA) signal amplification assay. *J. AIDS. (in press)*.

Perrillo, R. P., Schiff, E. R., Davis, G. L., Bodenheimer, H. C., Lindsay, K., Payne, J., Dienstag, J. L., O'Brien, C., Tamburro, C., Jacobson, I. M., Sampliner, R., Feit, D., Lefkowitch, J., Kuhns, M., Meschievitz, C., Sanghvi, B., Albrecht, J., Gibas, A., and the Hepatitis Interferon Treatment Group (1990). A randomized, controlled trial of interferon alfa-2b alone and after prednisone withdrawal for the treatment of chronic hepatitis B. *N. Engl. J. Med.* **323**, 295–301.

Rashtchian, A., Eldredge, J., Ottaviani, M., Abbott, M., Mock, G., Lovern, D., Klinger, J., and Parsons, G. (1987). Immunological capture of nucleic acid hybrids and application to nonradioactive DNA probe assay. *Clin. Chem.* **33(9)**, 1526–1530.

Urdea, M. S. (1993). Synthesis and characterization of branched DNA (bDNA) for the direct and quantitation detection of CMV, HBV, HCV, HIV. *Clin. Chem.* **39**, 725–726.

Urdea, M. S., Running, J. A., Horn, T., Clyne, J., Ku, L., and Warner, B. D. (1987). A novel method for the rapid detection of specific nucleotide sequences in crude biological samples without blotting or radioactivity: Application to the analysis of hepatitis B in human serum. *Gene* **61**, 253–264.

Urdea, M. S., Horn, T., Fultz, T. J., Anderson, M., Running, J. A., Hamren, S., Ahle, D., and Chang, C-A. (1991). Branched DNA amplification multimers for the sensitive, direct detection of human hepatitis viruses. *Nucleic Acids Res.* **24**, 197–200.

Urdea, M. S., Wilber, J. C., Yeghiazarian, T., Todd, J. A., Kern, D. G., Fong, S.-J., Besemer, D., Hoo, B., Sheridan, P. J., Kokka, R., Neuwald, P., and Pachl, C. A. (1993). Direct and quantitative detection of HIV-1 RNA in human plasma with a branched DNA signal amplification assay. *AIDS (Suppl 2)* **7,** S11–S14.

Wiedbrauk, D. L. (1992). Molecular methods for virus detection. *Lab. Med.* **23(11),** 737.

Wolcott, M. J. (1992). Advances in nucleic acid-based detection methods. *Clin. Microbiol. Rev.* **5(4),** 370–386.

7

Detection Methods Using Chemiluminescence

Irena Bronstein and Corinne E. M. Olesen

Tropix, Inc.
Bedford, Massachusetts 01730

I. GENERAL INTRODUCTION

Chemiluminescence, the emission of light from a chemical reaction, has been studied extensively for many decades. Chemiluminescent processes constitute a very special class of chemical reactions in which products (or intermediates) are produced in electronically excited states that are very short-lived and rapidly decay with concomitant emission of light. Similar chemiluminescent reactions, called bioluminescence, occur in nature in species as diverse as the firefly (*Photinus pyralis*), marine bacteria (*Vibrio harveyi*), and others. Most chemiluminescence reactions involve oxidations of a variety of organic compounds as well as naturally occurring materials, resulting in the generation of light-emitting excited states. This phenomenon was first described with synthetic organic compounds in 1877 (Radziszewski, 1877).

Chemiluminescent reactions do not produce very high intensity light signals because of many efficient quenching processes that compete with the radiative decay of the excited states. Nevertheless, chemiluminescence has been used effectively as a very sensitive detection system in many applications (Carter and Kricka, 1982; Harber, 1982; Kricka and Carter, 1982), largely because no background light signals are generated since the emitting excited state is created in a dark chemical reaction (compared to scattered excitation light in fluorescence.) Therefore, in theory, every photon detected is a true signal of the assay. This feature of chemiluminescent molecules—coupled with long shelf-life, elimination of hazards associated with the use of radioisotopes, and their detectability at 10^{-21} moles (detection of alkaline phosphatase with chemiluminescent dioxetane substrate; Kricka, 1992)—makes them ideal as a reporter system for immunoassays and DNA probe hybridization assays.

In this chapter we describe the use of various chemiluminescence methodologies for the detection of viruses in DNA hybridization assays. A short discussion of instrumentation used in chemiluminescence measurement is also included.

II. CHEMILUMINESCENCE METHODS

A. Dioxetanes

Dioxetanes are four-membered cyclic peroxides that have been implicated as short-lived unstable intermediates in oxidation reactions that result in chemiluminescence (McCapra, 1966). Thus, 1,2-dioxetanes differ from most other chemiluminescence systems because these compounds do not require oxidation to emit light. Recently developed 1,2-dioxetanes that can be activated to luminesce by enzymes have been used successfully for bio-analyte detection. Dephosphorylation of adamantyl- and derivatized adamantyl-1,2-dioxetane phosphate substrates, such as AMPPD® [disodium 3-(4-methoxyspiro{1,2-dioxetane-3,2′-tricyclo[3.3.1.1^{3,7}]decan}-4-yl)phenyl phosphate] and CSPD® [disodium 3-(4-methoxyspiro{1,2-dioxetane-3,2′-(5′-chloro)tricyclo[3.3.1.1^{3,7}]decan}-4-yl)phenyl phosphate] by alkaline phosphatase results in the formation of a destablized anion that fragments further to form an excited state of methyl *meta*-oxybenzoate anion that emits light at

477 nm (Fig. 1; Bronstein *et al.*, 1989a,1991; Bronstein and Dimond, 1990; Bronstein and Sparks, 1992).

1,2-Dioxetane substrates for alkaline phosphatase are widely used in DNA hybridization assays (Bronstein, *et al.*, 1990; Pollard-Knight *et al.*, 1990b; Tumolo *et al.*, 1992). DNA probes are labeled with alkaline phosphatase either indirectly, with a biotin or hapten label followed by binding streptavidin– or antibody–alkaline phosphatase conjugates, or directly by covalent bonding to enzyme (oligonucleotide probes). Biotin has been the most popular ligand for indirect labeling, but hapten labels other than biotin have also been employed including digoxigenin, fluorescein, and 2,4-dinitrophenyl. Dioxetane-based chemiluminescent indirect labeling and detection systems for DNA hybridization assays, as well as for immunoassays and DNA sequencing, are widely available from many commercial suppliers. With a nick-translated biotinylated DNA probe, as little as 380 fg (7.9×10^4 copies) of target pBR322 DNA can be detected on a Southern blot (Bronstein *et al.*, 1990). Using digoxigenin-labeled random-primed DNA probes with a membrane-based assay and photographic film detection, a sensitivity level of 10–50 fg of target DNA was obtained for the detection of purified cytomegalovirus (CMV) or parvovirus B19 DNA (Musiani *et al.*, 1991a).

Direct labeling of oligonucleotide probes with alkaline phosphatase (Jablonski *et al.*, 1986) is possible with systems from Promega (Madison,

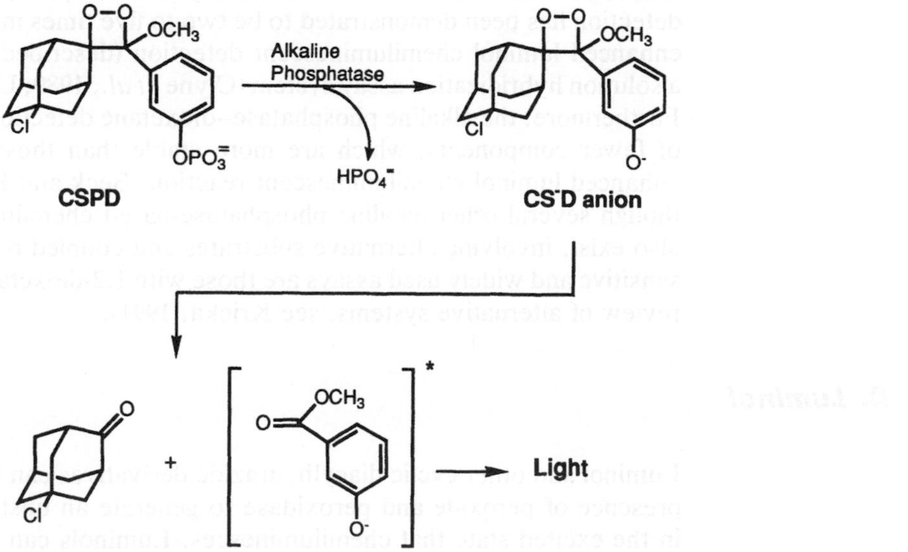

Figure 1 Chemiluminescent decomposition of CSPD® 1,2-dioxetane triggered by enzymatic dephosphorylation.

WI) and Cambridge Research Biochemicals (Wilmington, DE). The detection of a single copy gene in 0.25 μg human genomic DNA with Southern blot analysis has been achieved with an alkaline phosphatase-labeled oligonucleotide probe (Cate *et al.*, 1991).

Comparisons have shown that the sensitivity of alkaline phosphatase–dioxetane chemiluminescence detection is comparable to or better than that of ^{32}P-based detection. In a human genomic Southern blot analysis of the tissue plasminogen activator gene, the sensitivity achieved with an alkaline phosphatase-labeled oligonucleotide probe was 12-fold higher than that achieved with the same [32]P-5'-end-labeled probe, and the speed of detection was enhanced 40-fold with the alkaline phosphatase-labeled probe (Cate *et al.*, 1991). Slot blot hybridization of human serum samples with the alkaline phosphatase-labeled AmpliProbe® system (ImClone Systems, New York) showed a higher sensitivity in the detection of hepatitis B virus (HBV) relative to ^{32}P-labeled nick-translated probes (Yang *et al.*, 1991). Similar sensitivities were obtained with an indirect digoxigenin-labeled probe and a random-primed ^{32}P-labeled probe in a dot blot hybridization assay for amplified human immunodeficiency virus type 1 (HIV-1) DNA (Zachar *et al.*, 1991).

The sensitivity of alkaline phosphatase–dioxetane chemiluminescence detection has been shown to be superior to other nonisotopic systems based on colorimetric detection in membrane-based hybridization assays (Bronstein and Kricka, 1989; Bronstein and Voyta, 1989; Bronstein *et al.*, 1989c; Musiani *et al.*, 1991a,1992). Finally, alkaline phosphatase–dioxetane detection has been demonstrated to be two to five times more sensitive than enhanced luminol chemiluminescent detection (described subsequently) in a solution hybridization assay system (Clyne *et al.*, 1989; Urdea *et al.*, 1990). Furthermore, the alkaline phosphatase–dioxetane detection system consists of fewer components, which are more stable than those required for an enhanced luminol chemiluminescent reaction (Beck and Köster, 1990). Although several other alkaline phosphatase-based chemiluminescent assays also exist, involving alternative substrates and coupled reactions, the most sensitive and widely used assays are those with 1,2-dioxetane substrates (for review of alternative systems, see Kricka, 1991).

B. Luminol

Luminol and other cyclic diacylhydrazide derivatives can be oxidized in the presence of peroxide and peroxidase to generate an unstable intermediate in the excited state that chemiluminesces. Luminols can be used as direct chemiluminescent labels or as the chemiluminescent detectors of a peroxidase enzyme label (Kricka, 1991). Activation of luminol chemiluminescence with horseradish peroxidase (HRP) using an enhanced luminol system (enhanced chemiluminescence, ECL) has been done in DNA hybridization assays (Matthews *et al.*, 1985; Durrant *et al.*, 1990; Durrant, 1992) and immunoassays (reviewed by Bronstein and Sparks, 1992; Whitehead *et al.*, 1983; Thorpe *et al.*, 1985; Kricka *et al.*, 1987).

DNA probes can be labeled indirectly with HRP by binding streptavidin–HRP or anti-hapten–HRP conjugates or covalently by direct enzyme conjugation with oligonucleotides and longer double-stranded DNAs. The detection of single copy genes in 0.5 μg human genomic DNA has been reported with indirectly labeled probes (Simmonds *et al.*, 1991). Direct HRP-labeled DNA probes have been used for both membrane-based DNA hybridization assays (Pollard-Knight *et al.*, 1990a; Simmonds *et al.*, 1991) and solution-phase hybridization assays (Urdea *et al.*, 1990). Detection of a single-copy gene on a Southern blot of <2 μg human genomic DNA, with a sensitivity of <1 amol target DNA, has been demonstrated (Pollard-Knight *et al.*, 1990a) using direct HRP-labeled probes 0.3–5.1 kb in length. Similar sensitivity (1 amol target DNA) was also reported by Durrant *et al.* (1990). Both indirect and direct HRP labeling systems for nucleic acids and detection systems for HRP-catalyzed chemiluminescent reactions (ECL gene detection system) are available from Amersham (Arlington Heights, IL).

C. Acridinium Esters

Acridinium esters (AE) are direct chemiluminescent labels for antibodies (Weeks *et al.*, 1983) and DNA probes (Septak, 1989; Nelson and Kacian, 1990; Nelson *et al.*, 1992), in contrast to dioxetane and luminol systems, in which an enzyme label catalyzes the chemiluminescent reaction. *N*-Methyl acridinium esters react with hydrogen peroxide under basic conditions to yield an excited state *N*-methylacridone which emits light at 430 nm (reviewed by Nelson and Kacian, 1990). Oligonucleotide DNA probes can be labeled convalently with AEs by reaction of modified *N*-hydroxysuccinimide-AE with a primary alkyl amine on a linker arm that was previously incorporated during oligonucleotide synthesis (Nelson and Kacian, 1990). Preparation of AE-labeled oligonucleotide probes has also been described by Septak (1989). The AE label does not affect probe hybridization characteristics; relatively large amounts of clinical specimen material may be used without interfering with hybridization and detection of AE-labeled probes.

Probe hybridization and detection reactions are performed in solution, using either separation or nonseparation formats. In a separation or heterogeneous assay, hybridized probe may be separated and detected by selective binding to microspheres, which can be separated from solution magnetically. In a nonseparation or homogeneous format, also termed a hybridization protection assay (HPA), the ester bond of the unhybridized probe can be hydrolyzed by differential chemical hydrolysis, thus rendering its AE label nonchemiluminescent, whereas the AE label of the hybridized probe is minimally affected (Nelson and Kacian, 1990). This type of assay is possible because hybridization provides an intercalation site for the AE label, thereby protecting the AE molecule residing in the hybridized region from hydrolysis (Arnold *et al.*, 1989).

The sensitivity of this detection system is approximately 5×10^{-19} mol AE-labeled oligonucleotide, and the linear dynamic range is greater than four orders of magnitude (Nelson and Kacian, 1990). Similar sensitivities for the detection of an amplified *gag* sequence (4 HIV proviral copies per 150,000 cells) were achieved with colorimetric, chemiluminescence and ^{32}P-labeling methods (Ou *et al.*, 1990; Rapier *et al.*, 1993). Schmidt (1991) was able to detect 0.05 fmol target HIV-1 DNA with AE-labeled *gag* probes and obtained greater sensitivity with chemiluminescence than with the same ^{32}P-end-labeled probe in a dot blot hybridization assay.

D. Electrochemiluminescence

Electrochemiluminescence is a process in which the excited state products are generated via an electrochemical reaction (Faulkner and Glass, 1982). Electrochemiluminescence occurs when specific metal chelates such as ruthenium (II) tris(bipyridyl) [$Ru(bpy)_3^{2+}$], utilized as labels, undergo a series of chemical reactions at an electrode surface. Electrochemiluminescent labels for DNA hybridization assays have been utilized in a highly sensitive, simple, and versatile assay system. Oligonucleotide probes, synthesized with a free 5'-amino group, are readily labeled with $Ru(bpy)_3^{2+}$–NHS ester (Blackburn *et al.*, 1991; Kenten *et al.*, 1991). Alternatively, oligonucleotide probes may be labeled during synthesis by incorporating labeled phosphoramidites (Kenten *et al.*, 1992; DiCesare *et al.*, 1993).

Electrochemiluminescent labels are relatively small molecules (~1000 dalton) that are extremely stable and may be coupled to nucleic acids, haptens, or proteins without affecting immunoreactivity or hybridization characteristics. The dynamic range for detection of these labels has been reported to be over six orders of magnitude (Blackburn *et al.*, 1991). These advantages, compared with other nonisotopic detection methods, provide potential wide utility in automated nonradioactive clinical diagnostic assays, including both DNA hybridization and immunoassay formats. A disadvantage of electrochemiluminescence, however, is a need for specialized instrumentation that can induce generation of electrochemically-excited states coupled with sensitive light detection.

Blackburn *et al.* (1991) used electrochemiluminescence detection with a DNA probe assay to quantify polymerase chain reaction (PCR)-amplified HIV-1 *gag* sequences. Double-stranded biotinylated PCR product was captured on streptavidin-coated microparticles and treated with alkali.

$Ru(bpy)_3^{2+}$-labeled oligonucleotide probe was then hybridized to the particle-bound DNA, washed, and quantified. A linear response was generated over the range of 50 to 2000 gene copies, and the detection of less than 10 copies of the HIV-1 *gag* was attained. An automated system for electrochemiluminescence quantification of PCR products (QPCR System 5000; Perkin-Elmer Corporation, Norwalk, CT) has been developed (DiCesare *et al.*, 1993) and is used for detection of viral disease. This system provides detection limits of 10–200 amol and a linear dynamic range greater than three orders of magnitude. The system has been used for the detection of HIV-1 over a range of 3 to 10^6 copies of target DNA (Wages *et al.*, 1993).

Because of the electrogeneration of the emitting species, which requires contact of the metal chelate label with an electrode, it is difficult to envision that simple membrane-based blotting assays that can be imaged on film could be designed using electrochemiluminescence.

E. Bioluminescence

Bioluminescent reactions, a special class of chemiluminescent reactions that occur in nature and are catalyzed by a luciferase or photoproteins, offer an alternative method for luminescence detection of protein and DNA (Kricka, 1991). Two bioluminescent reaction systems have been used for DNA hybridization assays, both of which are coupled enzymatic reactions. One system, used for membrane-based DNA hybridization, couples the production of D-luciferin from D-luciferin-*O*-phosphate, catalyzed by alkaline phosphatase (as a direct or indirect label) and the oxidation of D-luciferin, catalyzed by firefly luciferase, with concomitant light emission (Hauber and Geiger, 1987,1988; Hauber *et al.*, 1988,1989; Geiger, 1992). The other system, used with both membrane-based and solution hybridization assays, couples reactions catalyzed by glucose-6-phosphate dehydrogenase (G6PDH), NAD(P)H : FMN oxidoreductase, and marine bacterial luciferase to produce the light (Balaguer *et al.*, 1989a,b,1991a,b; Nicolas *et al.*, 1990,1992). Although bioluminescence-based DNA detection systems have not become as widely used as chemiluminescence systems for DNA hybridization assays, they do offer another alternative for sensitive nonradioactive biomolecule detection.

III. INSTRUMENTATION FOR CHEMILUMINESCENCE ASSAYS

A wide spectrum of instruments is currently available for recording and quantifying chemiluminescent signal intensities. These instruments, known as luminometers, use a light detector that consists of a photomultiplier tube in photon counting mode, positioned close to the light source (microtiter plate or tube) to maximize photon collection efficiency. Among commercially available luminometers, semi-automated tube instruments such as the AutoClinilumat LB952T (Berthold/EG&G, Wallac, Inc., Gaithersburg, MD) and microtiter plate readers such as the ML 1000 (Dynatech Laboratories, Chantilly, VA) are most popular (reviewed by Bronstein and Kricka, 1990; Stanley, 1992a,b,1993b).

Chemiluminescence signals originating from blotting experiments performed on membranes can be detected by imaging on X-ray or instant photographic films. These films offer simple, convenient, and inexpensive detectors of chemiluminescence that can be used successfully for qualitative determinations and some signal quantification. Camera luminometers that house instant photographic film are suitable for the detection of light emission from blots and microtiter plate wells, and are available from Amersham, Analytical Luminescence Laboratory (San Diego, CA), Dynatech Laboratories, and Tropix, Inc. (Bedford, MA).

Finally, photon-counting cameras are available and are most suitable for the detection and accurate quantification of low-light signals. This instrumentation usually consists of a light detector such as a silicon target, silicon diode array, or charge-coupled device (CCD) coupled to a lens system, a controller, and a digital image processor. Since most of these camera systems are capable of imaging in two dimensions, micro- and macroscopic luminescent specimens can be analyzed spatially and temporally. The Argus-100/CL (Hamamatsu Corporation, Photonic Microscopy, Inc., Oak Brook, IL) is a photon-counting imaging device that has been used in the detection of blotted proteins (Hauber et al., 1988). The Star I CCD cooled camera system (Photometrics Ltd., Tucson, AZ) exhibits very low dark current background and a wide dynamic range and has been used successfully to detect protein and nucleic acid analytes in solution and on membranes (Martin and Bronstein, 1993, 1994.).

IV. CHEMILUMINESCENCE ASSAYS FOR VIRUS DETECTION

The combination of DNA hybridization assays with chemiluminescence detection methods has enabled the development of rapid, sensitive, quantitative, nonradioactive assays that are amenable to automation. DNA hybridization technology is becoming accepted as a reliable clinical laboratory technique for the identification of infectious organisms and has fueled the need for more rapid, sensitive, and automated assay formats. Culture assay methods are laborious, time-consuming, and costly, and sometimes impossible to use. Antigen-based detection assays including fluorescent antibody and immunoassay techniques, although faster and automatable, are often less sensitive than culture techniques. With the advent of technologies such as PCR, DNA probe methods offer rapid, easy, and highly sensitive assay formats. DNA hybridization assays using radioactive labels are sensitive and are easily quantified, but health, environmental, disposal, and cost concerns render these systems less than ideal as widely used clinical assays.

Chemiluminescence methods for the detection of viral agents as well as other microorganisms have become widely used (Table 1), and continued development will certainly expand their applications in research and clinical diagnostic tests. More traditional immunoassays have also been developed and used with chemiluminescence for the detection of various viral antigens and the assessment of immune status with respect to viruses (selected references in Table 1). A survey of commercially available products that incorporate chemiluminescence or bioluminescence techniques and reagents for specific assays and nonspecific detection systems is available (Stanley, 1993a,b).

A. DNA Hybridization Assay Formats

Several DNA hybridization assay formats including membrane-based, solution, and *in situ* hybridization have been coupled with chemiluminescence for the detection of viruses and other infectious agents. Membrane-based chemiluminescent hybridization assays have employed either 1,2-dioxetane substrates for alkaline phosphatase or the enhanced chemiluminescence reaction of luminol and HRP, and are imaged on X-ray or photographic films or imaged directly and quantified using a CCD camera system. Solution hybridization assays are performed with 1,2-dioxetanes, luminol, and AE labels, and the emitted light signal is measured in a luminometer. Electrochemiluminescent labels are also used for solution hybridization assays and are detected with an instrument combining an electrochemical flow cell, a potentiostat, and a photomultiplier tube. *In situ* hybridization has been performed using both 1,2-dioxetanes and enhanced luminol with either photographic film detection or a CCD camera system.

B. Chemiluminescence Detection Systems

1. Dioxetanes

Alkaline phosphatase–dioxetane chemiluminescence systems have been used in a wide variety of DNA hybridization assays for detection of infectious

TABLE 1
Selected Studies That Have Used Chemiluminescence to Detect Viruses, *Chlamydia trachomatis*, and Other Microorganisms

Agent	Assay[a]	Reference
Barley yellow dwarf virus	MH	Fouly et al. (1992) (DX)
Bluetongue virus	PCR/H	Akita et al. (1993)
Bovine enteric coronavirus	MH	Collomb et al. (1992) (LU)
Bovine immunodeficiency-like virus	IA	Jacobs et al. (1992)
Bovine leukosis virus	IA	Miliukiene et al. (1991)
Bursal disease virus	H	Akin et al. (1993) (AP)
Chicken anemia virus	PCR/MH	Tham and Stanislawek (1992a,b) (DX)
Cytomegalovirus	MH	Musiani et al. (1991a,1992) (DX); Yang et al. (1991) (DX)
Dengue virus	PCR/H	Henchal et al. (1991) (DX)
Enterovirus (poliovirus)	MH	Fuchs et al. (1993) (DX)
Epstein–Barr virus	MH	Yang et al. (1991) (DX)
	PCR/MH	Vlieger et al. (1992) (LU)
Feline infectious peritonitis virus	MH	Martinez and Weiss (1993) (AP)
Grapevine closterovirus	IA	Pollini et al. (1993) (LU)
Hepatitis B virus	MH	Bronstein et al. (1989c) (DX); Farmar and Castaneda (1991) (DX); Yang et al. (1991) (DX)
	PCR/MH	Escarceller et al. (1992) (DX)
	SH	Urdea et al. (1987,1990) (LU; LU, DX)
	IA	Khalil et al. (1991a,b) (AE); Bouveresse and Bourgeois (1992) (AP); Boxall (1992) (LU); McCartney et al. (1993) (LU)
	IM	Ireland and Samuel (1989) (LU); Robertson et al. (1991) (AE); Weare et al. (1991) (AE)
Hepatitis C virus	PCR/H	Geiger and Caselmann (1992)
	IA	Khalil et al. (1991b) (AE)
Herpes simplex virus	MH	Bronstein and Voyta (1989) (DX)
	ISH	Bronstein and Voyta (1989) (DX)
	PCR/H	Puchhammerstoeckl et al. (1993) (AP)
	IA	Pronovost et al. (1981) (ILU); Dalessio and Ashley (1992) (LU)
Human immunodeficiency virus	PCR/MH	Conway et al. (1990) (AP); Zachar et al. (1991) (DX)
	PCR/SH	Ou et al. (1990) (AE); Blackburn et al. (1991) (EL); Schmidt (1991) (AE); Schmidt and Gschnait (1991); Gudibande et al. (1992) (EL); Kenten et al. (1992) (EL); Suzuki et al. (1992) (DX); Rapier et al. (1993) (AE); Wages et al. (1993) (EL)
	PCR	Bettens et al. (1991) (DX)
	BH	Ishii and Ghosh (1993) (AP)

Organism	Format[a]	Reference (CL method[b])
	ISH	Bronstein et al. (1989b) (DX)
	RT	Suzuki et al. (1993) (DX)
	IA	Khalil et al. (1991b) (AE); Jacobs et al. (1992)
Human papilloma virus	PCR/SH	Balaguer et al. (1991b) (BL); Kenten et al. (1991) (EL)
	MH	Sarkar et al. (1993)
	ISH	Hawkins and Cumming (1990) (LU)
Human T-cell leukemia virus	IA	Khalil et al. (1991b) (AE)
	IM	Kuroda et al. (1992) (LU)
Human T-cell lymphotropic virus	IA	Papsidero et al. (1992) (LU)
Influenza virus	MH	McKimm-Breschkin (1992) (DX)
	IM	Arenkov et al. (1991)
Lentivirus	RT	Cook et al. (1992) (DX)
Parvovirus	MH	Musiani et al. (1991a,b) (DX)
	IA	O'Neill and Coyle (1992) (LU)
Potato virus Y, potato spindle tuber viroid	MH	Welnicki and Hiruki (1993) (AP)
Potato, pome fruit viroid	MH	Podleckis et al. (1993) (AP)
Respiratory syncytial virus, rotavirus	IM	Hornsleth et al. (1988) (LU)
Rubella virus	IM	Chanteloup et al. (1992) (AP)
Varicella zoster virus	PCR/H	Eis-Hubinger et al. (1992) (AP)
Bacteria	MH	Gustaferro and Persing (1992) (LU)
	?	Daly et al. (1991)
Chlamydia trachomatis	SH	Clyne et al. (1989) (DX); Urdea et al. (1989) (DX); Gratton et al. (1990) (AE); Mercer et al. (1990) (AE); Iwen et al. (1991) (AE); Jang (1992) (AE); Scieux et al. (1992) (AE)
	IA	Neman-Simha et al. (1991); Dumornay et al. (1992) (AE); (1992a,b) (AE)
Mycobacteria	SH	Bull and Shanson (1992) (AE)
	SD/SH	Donahue et al. (1993) (DX)
	PCR/MH	Sritharan and Barker (1991) (DX)
Neisseria gonorrhoeae	SH	Urdea et al. (1989) (DX); Vlaspolder et al. (1993) (AE)
Plasmodium falciparum	PCR/MH	Barker et al. (1992) (DX)
Toxoplasma gondii	PCR/MH	Stauber et al. (1991) (LU)

[a] Assay formats include: BH, bead-based hybridization; H, hybridization; IA, immunoassay; IM, immunity; ISH, in situ hybridization; MH, membrane-based hybridization; PCR, polymerase chain reaction; RT, reverse transcriptase; SD, strand displacement; SH, solution hybridization.

[b] Chemiluminescent (CL) methods employed (if known): AE, acridinium ester; AP, alkaline phosphatase (most likely with dioxetane substrate); BL, bioluminescence; DX, dioxetane; EL, electrochemiluminescence; ILU, isoluminol; LU, enhanced luminol.

agents. Membrane-based hybridization assays have been used for the detection of HBV (Bronstein *et al.*, 1989c; Yang *et al.*, 1991; Escarceller *et al.*, 1992), herpes simplex virus (HSV-1) (Bronstein and Voyta, 1989), CMV (Musiani *et al.*, 1991a,1992; Yang *et al.*, 1991), HIV-1 (Zachar *et al.*, 1991), and other viral agents (Fouly *et al.*, 1992; Tham and Stanislawek, 1992a,b; Fuchs *et al.*, 1993). Solution hybridization assays include those for HBV (Urdea *et al.*, 1990), HIV-1 (Suzuki *et al.*, 1992), and *Chlamydia* (Clyne *et al.*, 1989; Urdea *et al.*, 1989). *In situ* hybridization assays have been performed with both HSV-1 infected cells (Bronstein and Voyta, 1989) and HIV-infected cells (Bronstein *et al.*, 1989b). Finally, assays for retroviruses based on the detection of reverse transcriptase activity can be coupled with chemiluminescence detection by measuring the enzymatic incorporation of digoxigenin-labeled nucleotides with anti-digoxigenin alkaline phosphatase and a dioxetane substrate (Suzuki *et al.*, 1993).

Commercially available detection systems incorporating dioxetanes include the AmpliProbe® system (ImClone Systems) for membrane-based hybridization assays for HBV, CMV, and EBV (Yang *et al.*, 1991), the Hybrid Capture™ System HBV DNA Assay (Murex Diagnostics Ltd., Kent, UK), and solution hybridization assay systems for *Chlamydia trachomatis* and HBV detection (Chiron Corporation; Clyne *et al.*, 1989; Urdea *et al.*, 1989,1990).

2. Luminol

DNA hybridization assays using the ECL system with direct HRP-labeled probes include detection of bovine enteric coronavirus in a slot blot hybridization assay (Collomb *et al.*, 1992) and a solution-phase hybridization assay for HBV DNA (Urdea *et al.*, 1987,1990). *In situ* hybridization for detection of human papillomavirus (HPV) type 16 has been performed with an indirect labeled probe (Hawkins and Cumming, 1990). ECL systems have also been used for the immunoassay detection of several viruses, including grapevine closterovirus (Pollini *et al.*, 1993) and parvovirus B 19 (O'Neill and Coyle, 1992).

3. Acridinium Esters

DNA hybridization assays incorporating AE-labeled probes have been developed for detection of several infectious agents from clinical samples, including *C. trachomatis, Neisseria gonorrhoeae*, fungal pathogens, mycobacteria, and several common bacterial pathogens (Nelson and Kacian, 1990). These assay systems, called PACE 2™ and ACCUPROBE™, are available commercially through Gen-Probe, Inc. (San Diego, CA). The Gen-Probe system for screening for *Chlamydia* has been compared with both culture and nonculture antigen detection methods including enzyme immuno-assays and immunofluorescent antibody tests (Gratton *et al.*, 1990; Mercer *et al.*, 1990; Iwen *et al.*, 1991). The PACE 2™ system can provide a rapid, reliable alternative to culture and immunoassay methods for the detection of *Chlamydia* from cervical samples (Iwen *et al.*, 1991). Solution hybridization (hybridization protection) assays with AE-labeled probes have been used for the detection of PCR-amplified HIV-1 DNA (Ou *et al.*, 1990; Schmidt, 1991; Rapier *et al.*, 1993).

In addition, AEs have also been used to label antibodies that have been incorporated into automated immunoassay formats for the detection of infectious agents and antibody screening from clinical samples (Khalil *et al.*, 1991a,b).

4. Electrochemiluminescence

Electrochemiluminescence detection has been used in both manual (Blackburn *et al.*, 1991; Gudibande *et al.*, 1992; Kenten *et al.*, 1992) and automated (QPCR System 5000; Wages *et al.*, 1993) post-PCR amplification DNA hybridization assays for the detection of HIV-1 and HPV (Kenten *et al.*, 1991).

5. Bioluminescence

Detection of asymmetric amplified papillomavirus sequences using solution-phase hybridization with a G6PDH-labeled oligonucleotide and solid-phase capture has been performed using a bioluminescence assay (Balaguer *et al.*, 1991b).

V. CHEMILUMINESCENCE DETECTION PROTOCOLS

A. Hepatitis B Virus

Hepatitis B "core antigen" DNA, immobilized on nylon membrane, is hybridized with an alkaline phosphatase-labeled oligonucleotide probe. Hybridized probe is then detected with the 1,2-dioxetane substrate AMPPD (Bronstein *et al.*, 1989c).

1. Materials

Hepatitis B core antigen plasmid DNA and alkaline phosphatase-labeled probe, included in a SNAP® Hybridization System, and GeneScreen Plus™ nylon membrane were obtained from NEN/DuPont (Boston, MA). AMPPD and CSPD are from Tropix.

2. Target DNA Preparation and Probe Hybridization

HBV "core antigen" (HBVc) plasmid DNA (100 ng; 1.2×10^{10} copies) was dissolved in 25 μl sterile deionized H_2O and serially diluted with 0.3 M NaOH to produce target DNA samples ranging in concentration from 4.88×10^3 to 0.98×10^8 copies/μl. Blots were prepared as described here:

1. Incubate diluted DNA samples at room temperature for 15 min to denature, and spot 1 μl of each dilution onto dry membrane strips (1 $\times$ 8 cm).
2. Rinse blots with 2 M NH_4OAc and then with 0.6 M NaCl, 0.08 M sodium citrate, pH 7.0.
3. Prehybridize with 3 ml hybridization buffer [0.75 M NaCl, 0.075 M sodium citrate (5X SSC), 0.5% bovine serum, 0.5% polyvinylpyrrolidone, 1% sodium dodecyl sulfate (SDS), pH 7.0] for 15 min at 55°C.
4. Hybridize with hybridization buffer containing 1.0 nM alkaline phosphatase-labeled oligonucleotide probe for 30 min at 55°C.
5. Wash sequentially for 5 min each in:

1X SSC, pH 7.0, 1% SDS at room temperature
1X SSC, pH 7.0, 1% Triton X-100 at 55°C
1X SSC, pH 7.0, at room temperature

3. Chemiluminescence Detection

1. Wash hybridized blots with 0.1% bovine serum albumin (BSA), 0.05 M sodium carbonate, pH 9.5.
2. Saturate blot with 100 μl 1.6 mM AMPPD in 0.1% BSA, 0.05 M sodium carbonate, 1.0 mM $MgCl_2$, pH 9.5.

NOTE: *Alternatively, an improved buffer (0.1 M diethanolamine, 1.0 mM $MgCl_2$, pH 10.0) can be substituted for this wash, using 0.25 mM AMPPD or CSPD in this buffer for substrate incubation.*

3. Place blots in a plastic pouch and image light emission in a camera

luminometer with Polaroid Instant Black and White Type 612 (ASA 20,000) photographic film.

NOTE: *Alternatively, blots can be imaged on standard X-ray film.*

4. Digitize photographic film image using a black and white RBP Reflectance Densitometer (Tobias Associates, Inc., Ivyland, PA).

4. Results

Figure 2 shows a time course of the chemiluminescent DNA hybridization assay for HBVc antigen DNA. Serial dilutions of plasmid DNA were hybridized with alkaline phosphastase-labeled oligonucleotide probe, incubated with chemiluminescent substrate, and imaged on photographic film. Each photograph corresponds to a 30-min exposure. With this chemiluminescence assay, 1.18×10^6 copies of HBVc DNA can be detected within 30 min of substrate incubation. After a 2-hr incubation, 4.39×10^4 copies can be detected. In contrast, with the colorimetric bromochloroindolyl phosphate/nitro blue tetrazolium (BCIP/NBT) substrate system, 9.8×10^7 and 1.07×10^7 copies can be detected after 30 min or 2 hr of substrate incubation,

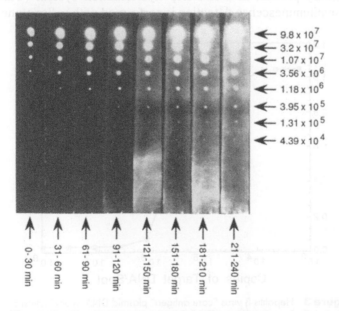

Figure 2 Chemiluminescent detection of hepatitis B "core antigen" plasmid DNA with AMPPD substrate in alkaline phosphatase-based DNA hybridization assay. Reprinted with permission from Bronstein *et al.* (1989c).

respectively (results not shown). Quantitative results were obtained by measuring reflection densities from the imaged photographic film using a black and white reflection densitometer (Fig. 3). These values could be used to establish a dose–response curve for the reflection densities as a function of HBVc plasmid concentration, from which HBVc DNA levels in clinical specimens could be determined. Use of the improved chemiluminescence detection protocol, incorporating the diethanolamine substrate buffer and CSPD chemiluminescent substrate, results in even greater sensitivity for DNA hybridization assays and would increase the sensitivity of this HBV DNA assay. Imaging and quantification of this membrane-based assay with rapidly evolving CCD camera systems will likely provide even greater sensitivity and a greater linear dynamic range than that achieved with densitometry.

5. Summary

Chemiluminescent detection of HBV DNA has also been performed with the AmpliProbe® system (ImClone Systems). This signal amplification probe system incorporates multiple target-specific primary and multiple secondary probes, alkaline phosphatase-labeled oligonucleotides that hybridize to the primary probes, in a two-step hybridization system (Yang *et al.*, 1991). Chemiluminescence detection is performed with a dioxetane substrate. With

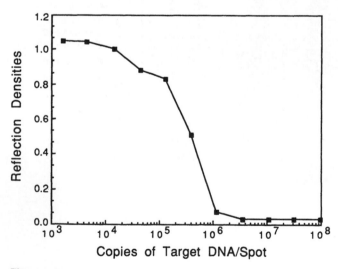

Figure 3 Hepatitis B virus "core antigen" plasmid DNA hybridization assay. Reflection density vs. number of copies of target DNA. Densitometric analysis of the Polaroid instant black and white photographic film image [0.00 (white)–2.00 (black)]. Reprinted with permission from Bronstein *et al.* (1989c).

this system, 0.4 pg (1×10^5 copies) purified target HBV genomic DNA can be detected in a chemiluminescent slot blot assay (Farmar and Castaneda, 1991; Yang et al., 1991). Identical assays performed with serum samples (25 μl) demonstrated that this chemiluminescent DNA hybridization system has the same specificity and sensitivity as immunoassays and is more sensitive than a ^{32}P-labeled nick-translated probe (Yang et al., 1991).

Escarceller et al. (1992) report the use of digoxigenin-labeled probes, anti-digoxigenin alkaline phosphatase, and AMPPD for the direct detection of HBV sequences in human serum samples. These investigators achieved a limit of sensitivity of 2–5 pg, which was equivalent to that obtained with both colorimetric detection and a ^{32}P-labeled probe. These researchers also used digoxigenin-labeled oligonucleotide primers for PCR amplification of HBV DNA purified from human serum, followed by immunological detection of the digoxigenin label (as described), a method that can be used in conjunction with alternatively labeled primers for multiple amplifications.

A chemiluminescent assay incorporating a solution-phase hybridization of synthetic oligonucleotides to target DNA, followed by solid-phase capture, labeling, and detection with either HRP or alkaline phosphatase-labeled oligonucleotides and chemiluminescent substrates has been used to achieve the detection of 0.2 pg (6×10^4 copies) HBV DNA in human serum samples in 4 hr. This solution DNA hybridization method includes novel labeling and amplification schemes and has been performed with both polystyrene bead and microtiter well capture systems (Urdea et al., 1987,1990).

Chemiluminescence techniques have also been used in the development of automated enzyme immunoassay systems for the detection of HBV in human sera (Khalil et al., 1991a,b; Bouveresse and Bourgeois, 1992).

B. Herpes Simplex Virus

Two chemiluminescent DNA hybridization assays for HSV, dot blot hybridization and in situ hybridization, are described here as originally reported by Bronstein and Voyta (1989). In these assays, HSV-1 plasmid DNA, immobilized on nylon membrane, or HSV-1-infected Vero cells, fixed and mounted on microscope slides, were hybridized with an alkaline phosphatase-labeled HSV-1 oligonucleotide probe and detected with AMPPD.

1. Materials

HSV-1 plasmid DNA and alkaline phosphatase-labeled oligonucleotide probe, included in a SNAP® Hybridization System, and GeneScreen Plus nylon membrane were obtained from NEN/DuPont. HSV-1-infected Vero cells were provided by Drs. J. Kershner and E. Jablonski (Molecular Biosystems, San Diego, CA). AMPPD and Emerald™ luminescence-amplifying material are from Tropix.

2. Dot Blot Hybridization and Chemiluminescence Detection

This membrane hybridization protocol is similar to that described for HBV detection.

1. Serially dilute HSV-1 plasmid DNA in 0.3 M NaOH, denature, and spot 1-μl aliquots onto dry membrane strips.
2. Prehybridize blots with hybridization buffer (0.5% BSA, 0.5% polyvinylpyrrolidone, 1% SDS) for 15 min at 55°C.
3. Hybridize with hybridization solution (containing alkaline phosphatase-labeled HSV-1 oligonucleotide probe) for 30 min at 55°C.
4. Wash sequentially for 5 min each in:

2X SSC, 1% SDS at room temperature
1X SSC, 1% Triton X-100 at 55°C
1X SSC, 1% Triton X-100 at room temperature
1X SSC at room temperature

5. Wash hybridized blots with 0.05 M sodium carbonate/bicarbonate, 1 mM MgCl$_2$, pH 9.5 (substrate buffer).
6. Saturate blot with 1.6 mM AMPPD (in substrate buffer) for 5 min.

NOTE: *As described for HBV detection, the diethanolamine buffer and 0.25 mM AMPPD or CSPD can be substituted in Steps 5 and 6 for increased sensitivity.*

7. Image blots with Polaroid Type 612 Instant Black and White film.

NOTE: *Alternatively, blots can be imaged on X-ray film.*

3. *In Situ* Hybridization and Chemiluminescence Detection

1. Infect Vero cells with HSV-1 (MacIntyre strain) for 1 hr at room temperature.
2. Harvest cells with trypsin/versene after the addition of 2% fetal calf serum at 0, 2, 4, 6, 8, 10, 12, 24, and 48 hr.
3. Pellet cells, fix in 95% ethanol, and mount on glass microscope slides.

4. Treat mounted slides with 0.2 M HCl for 2 min, rinse with deionized water, and immerse in 70% ethanol. Prior to hybridization, remove slides from ethanol and dry.

5. Immerse slides in 0.1% BSA, 5X SSC for 15 min at 70°C. Treat with 0.3 M NaOH for 1 min at room temperature. Rinse with phosphate-buffered saline (PBS).

6. Hybridize cells with the alkaline phosphatase-labeled HSV-1 oligo-nucleotide probe at a concentration of 5 nM in 0.1% BSA, 5X SSC for 20 min at 60°C.

7. Wash slides briefly in hybridization buffer at 60°C, and then extensively with 1X SSC at 50°C.

8. Wash with 0.05 M sodium carbonate/bicarbonate, 1 mM MgCl$_2$, pH 9.5 (substrate buffer).

9. Incubate with 0.8 mM AMPPD, 10% Emerald in substrate buffer for 5 min.

10. Place slides in a camera luuminometer and expose to Polaroid Type 612 Instant Black and White film.

4. Results

With the dot blot hybridization assay for HSV-1 plasmid DNA, detection limits achieved with the chemiluminescent substrate AMPPD are 1.3×10^5 and 1.4×10^4 copies of target HSV-1 DNA, with a 30-min exposure performed 1 hr after substrate addition and a 45-min exposure performed 4 hr after substrate addition, respectively (results not shown). The sensitivity achieved with AMPPD is 25- to 100-fold higher than that obtained with the colorimetric BCIP/NBT substrate system (results not shown). Fig. 4 shows the time course of viral infection assayed by *in situ* DNA hybridization with chemiluminescence detection. Use of the AMPPD chemiluminescent substrate enables the detection of HSV-1-infected cells within 6 hr postinfection. Again, with this assay format, CCD detection and imaging may provide even greater sensitivity than that achieved with photographic film.

5. Summary

In situ hybridization with chemiluminescence detection has also been used to detect HIV-infected cells (Bronstein *et al.*, 1989b) and HPV type 16 in a cervical carcinoma cell line (Hawkins and Cumming, 1990). The latter protocol involved the use of biotinylated HPV 16 DNA probes (Enzo Diagnostics, New York), a streptavidin–HRP conjugate, and ECL detection re-

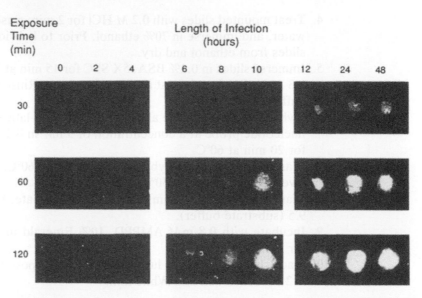

Figure 4 Chemiluminescent detection of *in situ* DNA hybridization of herpes simplex virus I-infected Vero cells: time course of infection. Reprinted from Bronstein and Voyta, *Clinical Chemistry* (1989), **35**, 1856–1857, Courtesy of the American Association for Clinical Chemistry, Inc.

agents coupled with a CCD imaging system. Detection of fewer than 10 HPV-positive cells (containing 600 copies of HPV 16 DNA per cell) among 10,000 HPV-negative cells on a single slide was achieved. However, this detection level is not necessarily the limit of sensitivity; with improved optical instrumentation, *in situ* hybridization coupled with CCD detection may provide a valuable diagnostic tool for the rapid and automated identification of viral sequences within cells.

VI. CONCLUSION

Chemiluminescence detection technologies combined with DNA hybridization methods provide rapid, sensitive, nonradioactive, automatable assay formats for the clinical diagnosis of infectious agents, as well as for research use. Rapidly evolving chemiluminescent enzyme substrates and labels, techniques, and assay and detection instrumentation, coupled with continued advances in DNA hybridization technologies, will further refine and improve the specificity and sensitivity of chemiluminescent DNA detection methods, bringing them into more widespread use.

ACKNOWLEDGMENTS

We are very grateful to Larry Kricka, Chris Martin, John Voyta, and Alison Sparks for editorial assistance.

REFERENCES

Akin, A., Wu, C. C., Lin, T. L., and Keirs, R. W. (1993). Chemiluminescent detection of infectious bursal disease virus with a PCR-generated nonradiolabeled probe. *J. Vet. Diagn. Invest.* **5**, 166–173.

Akita, G. Y., Glenn, J., Castro, A. E., and Osburn, B. I. (1993). Detection of bluetongue virus in clinical samples by polymerase chain reaction. *J. Vet. Diagn. Invest.* **5**, 154–158.

Arenkov, P. I., Berezin, V. A., and Starodub, N. F. (1991). Chemiluminescence fiber optic immunosensor for detecting antibodies to the influenza virus. *Ukr. Biokhim.* **63**, 99–103.

Arnold, L. J., Hammond, P. W., Wiese, W. A., and Nelson, N. C. (1989). Assay formats involving acridinium-ester-labeled DNA probes. *Clin. Chem.* **35**, 1588–1594.

Balaguer, P. T., Térouanne, B., Boussioux, A.-M., and Nicolas, J.-C. (1989a). Use of bioluminescence in nucleic acid hybridization reactions. *J. Biolumin. Chemilumin.* **4**, 302–309.

Balaguer, P., Térouanne, B., Eliaou, J. F., Humbert, M., Boussioux, A. M., and Nicolas, J. C. (1989b). Use of glucose-6-phosphate dehydrogenase as a new label for nucleic acid hybridization reactions. *Anal. Biochem.* **180**, 50–54.

Balaguer, P., Térouanne, B., Boussioux, A., and Nicolas, J. (1991a). Quantification of DNA sequences obtained by polymerase chain reaction using a bioluminescent adsorbent. *Anal. Biochem.* **195**, 105–110.

Balaguer, P., Térouanne, B., Boussioux, A. M., and Nicolas, J. C. (1991b). Papillomavirus quantification using asymmetric amplification and a rapid bioluminescent assay. *In* "Bioluminescence & Chemiluminescence: Current Status" (P. E. Stanley and L. J. Kricka, eds.), pp. 143–146. John Wiley & Sons, Chichester.

Barker, R. H., Jr., Banchongaksorn, T., Courval, J. M., Suwonkerd, W., Rimwungtragoon, K., and Wirth, D. F. (1992). A simple method to detect *Plasmodium falciparum* directly from blood samples using the polymerase chain reactions. *Am. J. Trop. Med. Hyg.* **46**, 416–426.

Beck, S., and Köster, H. (1990). Applications of dioxetane chemiluminescent probes to molecular biology. *Anal. Chem.* **62**, 2258–2270.

Bettens, F., Pichler, W. J., and de Weck, A. L. (1991). Incorporation of biotinylated nucleotides for the quantification of PCR-amplified HIV-1 DNA by chemiluminescence. *Eur. J. Clin. Chem. Clin. Biochem.* **29**, 685–688.

Blackburn, G. F., Shah, H. P., Kenten, J. H., Leland, J., Kamin, R. A., Link, J., Peterman, J., Powell, M. J., Shah, A., Talley, D. B., Tyagi, S. K., Wilkins, E., Wu, T.-G., and Massey, R. J. (1991). Electrochemiluminescence detection for development of immunoassays and DNA probe assays for clinical diagnostics. *Clin. Chem.* **37**, 1534–1539.

Bouveresse, E., and Bourgeois, J. P. J. (1992). Chemiluminescent enzyme immunoassay for the detection of hepatitis B surface antigen (HBsAg). *Clin. Chem.* **38**, 1090.

Boxall, E. H. (1992). Enhanced luminescent assays for hepatitis markers: Assessment of post vaccine responses. *Arch. Virol. Suppl.* **4**, 156–159.

Bronstein, I., and Dimond, P. (1990). Chemiluminescent compounds for diagnostic tests. *Diagn. Clin. Test.* **28**, 36–39.

Bronstein, I., and Kricka, L. J. (1989). Clinical applications of luminescent assays for enzymes and enzyme labels. *J. Clin. Lab. Anal.* **3**, 316–322.

Bronstein, I., and Kricka, L. J. (1990). Instrumentation for luminescent assays. *Am. Clin. Lab.* **9(1)**, 33–37.

Bronstein, I., and Sparks, A. (1992). Sensitive enzyme immunoassays with chemiluminescent detection. *In* "Immunochemical Assays and Biosensor Technology for the 1990s" (R. M. Nakamura, Y. Kasahara, and G. A. Rechnitz, eds.), pp. 229–250. American Society for Microbiology, Washington, D.C.

Bronstein, I., and Voyta, J. C. (1989). Chemiluminescent detection of herpes simplex virus I DNA in blot and in-situ hybridization assays. *Clin. Chem.* **35**, 1856–1857.

Bronstein, I., Edwards, B., and Voyta, J. C. (1989a). 1,2-Dioxetanes: Novel chemiluminescent enzyme substrates. Applications to immunoassays. *J. Biolumin. Chemilumin.* **4**, 99–111.

Bronstein, I., Kerschner, J. H., Voyta, J. C., and Jablonski, E. G. (1989b). Chemiluminescent detection of HIV infected cells using in-situ hybridization assay technique. Presented at the V International Conference on AIDS, Montreal, Quebec, Canada, June 4-9.

Bronstein, I., Voyta, J. C., and Edwards B. (1989c). A comparison of chemiluminescent and colorimetric substrates in a hepatitis B virus DNA hybridization assay. *Anal. Biochem.* **180**, 95–98.

Bronstein, I., Voyta, J. C., Lazzari, K. G., Murphy, O., Edwards, B., and Kricka, L. J. (1990). Rapid and sensitive detection of DNA in Southern blots with chemiluminescence. *Bio Techniques* **8**, 310–313.

Bronstein, I., Juo, R.-R., Voyta, J. C., and Edwards, B. (1991). Novel chemiluminescent adamantyl 1,2-dioxetane enzyme substrates. *In* "Bioluminescence and Chemiluminescence: Current Status" (P. Stanley and L. J. Kricka, eds.), pp. 73–82. John Wiley, Chichester.

Bull, T., and Shanson, D. (1992). Evaluation of a commercial chemiluminescent gene probe system "AccuProbe" for the rapid differentiation of mycobacteria, including "MAIC X," isolated from blood and other sites, from patients with AIDS. *J. Hosp. Infect.* **21**, 143–149.

Carter, T. J. N., and Kricka, L. J. (1982). Analytical applications of chemiluminescence. *In* "Clinical and Biochemical Luminescence" (L. J. Kricka and T. J. N. Carter, eds.), pp. 135–151. Marcel Dekker, New York.

Cate, R. L., Ehrenfels, C. W., Wysk, M., Tizard, R., Voyta, J. C., Murphy, O. J., and Bronstein, I. (1991). Genomic Southern analysis with alkaline-phosphatase-conjugated oligonucleotide probes and the chemiluminescent substrate AMPPD. *Genetic Analysis, Techniques Applications* **8**, 102–106.

Chanteloup, E., Clement, A., and Payne, J. (1992). Chemiluminescent enzyme immunoassay for the evaluation of the immune status to rubella virus. *Clin. Chem.* **38**, 1085.

Clyne, J. M., Running, J. A., Stempien, M., Stephens, R. S., Akhavan-Tafti, H., Schaap, A. P., and Urdea, M. S. (1989). A rapid chemiluminescent DNA hybridization assay for the detection of *Chlamydia trachomatis*. **4**, 357–366.

Collomb, J., Finance, C., Alabouch, S., and Laporte, J. (1992). Radioactive and enzymatic cloned cDNA probes for bovine enteric coronavirus detection by molecular hybridization. *Arch. Virol.* **125**, 25–37.

Conway, B., Adler, K. E., Bechtel, L. J., Kaplan, J. C., and Hirsch, M. S. (1990). Detection of HIV-1 DNA in crude cell lysates of peripheral blood mononuclear cells by the polymerase chain reaction and nonradioactive oligonucleotide probes. *J. Acq. Immune Def. Syndr.* **3**, 1059–1064.

Cook, R. F., Cook, S. J., and Issel, C. J. (1992). A nonradioactive micro-assay for released reverse transcriptase activity of a lentivirus. *Bio Techniques* **13**, 380–386.

Dalessio, J., and Ashley, R. (1992). Highly sensitive enhanced chemiluminescence immunodetection methods for herpes simplex virus type-2 Western immunoblot. *J. Clin. Microbiol.* **30**, 1005–1010.

Daly, J., Clifton, N., Seshkin, K., and Gooch, W. (1991). Use of rapid, nonradioactive DNA probes in culture confirmation tests to detect *Streptococcus agalactiae, Haemophilus influenzae,* and *Enterococcus* spp. from pediatric patients with significant infections. *J. Clin. Microbiol.* **29,** 80–82.

DiCesare, J., Grossman, B., Katz, E., Picozza, Ragusa, R., and Woudenberg, T. (1993). A high-sensitivity electrochemiluminescence-based detection system for automated PCR product quantification. *Bio Techniques.* **15,** 152–157.

Donahue, C., Jurgensen, S., Nycz, C., Schram, J. L., Shank, D., Vonk, G. P., and Walker, G. T. (1993). Detection of mycobacterial DNA using strand displacement amplification and a chemiluminescent microwell assay. *J. Biolumin. Chemilumin.* **8,** 77.

Dumornay, W., Roblin, P., Gelling, M., Hammerschlag, M., and Worku, M. (1992). Comparison of a chemiluminometric immunoassay with culture for diagnosis of chlamydial infections in infants. *J. Clin. Microbiol.* **30,** 1867–1869.

Durrant, I. (1992). Detection of horseradish peroxidase by enhanced chemiluminescence. *In* "Nonisotopic DNA Probe Techniques" (L. J. Kricka, ed.), pp. 167–183. Academic Press, San Diego.

Durrant, I., Benge, L. C. A., Sturrock, C., Devenish, A. T., Howe, R., Roe, S., Moore, M., Scozzafava, G., Proudfoot, L. M. F., Richardson, T. C., and McFarthing, K. G. (1990). The application of enhanced chemiluminescence to membrane-based nucleic acid detection. *Bio Techniques* **8,** 564–570.

Eis-Hubinger, A. M., Kaiser, R., Kleim, J. P., Dlugosch, D., Estor, A., Kleeman, E., Lange, C. E., and Schneweis, K. E. (1992). Detection of varicella zoster virus infections using polymerase chain reaction. *Hautarzt* **43,** 767–771.

Escarceller, M., Rodriguez-Frias, F., Jardi, R., San-Segundo, B., and Eritja, R. (1992). Detection of hepatitis B virus DNA in human serum samples: Use of digoxigenin-labeled oligonucleotides as modified primers for the polymerase chain reaction. *Anal. Biochem.* **206,** 36–42.

Farmar, J. G., and Castaneda, M. (1991). An improved preparation and purification of oligonucleotide-alkaline phosphatase conjugates. *Bio Techniques* **11,** 588–589.

Faulkner, L. R., and Glass, R. S. (1982). Electrochemiluminescence. *In* "Chemical and Biological Generation of Excited States" (W. Adam, and G. Cilento, eds.), pp. 191–227. Academic Press, New York.

Fouly, H. M., Domier, L. L., and D'Arcy, C. J. (1992). A rapid chemiluminescent detection method for barley yellow dwarf virus. *J. Virol. Meth.* **39,** 291–298.

Fuchs, F., Leparc, I., Kopecka, H., Garin, D., and Aymard, M. (1993). Use of cRNA digoxigenin-labelled probes for detection of enteroviruses in humans and in the environment. *J. Virol. Meth.* **42,** 217–226.

Geiger, C. P., and Caselmann, W. H. (1992). Non-radioactive hybridization with hepatitis C virus-specific probes created during polymerase chain reaction: A fast and simple procedure to verify hepatitis C virus infection. *J. Hepatol.* **15,** 387–390.

Geiger, R. E. (1992). Detection of alkaline phosphatase by bioluminescence. *In* "Nonisotopic DNA Probe Techniques" (L. J. Kricka, ed.), pp. 113–126. Academic Press, San Diego.

Gratton, C. A., Lim-Fong, R., Prasad, E., and Kibsey, P. C. (1990). Comparison of a DNA probe with culture for detecting *Chlamydia trachomatis* directly from genital specimens. *Mol. Cell. Probes* **4,** 25–31.

Gudibande, S. R., Kenten, J. H., Link, J., Friedman, K., and Massey, R. J. (1992). Rapid non-separation electrochemiluminescent DNA hybridization assays for PCR products, using 3′-labelled oligonucleotide probes. *Mol. Cell. Probes* **6,** 495–503.

Gustaferro, C. A., and Persing, D. H. (1992). Chemiluminescent universal probe for bacterial ribotyping. *J. Clin. Microbiol.* **30,** 1039–1041.

Harber, M. J. (1982). Applications of luminescence in medical microbiology and hematology.

In "Clinical and Biochemical Luminescence" (L. J. Kricka and T. J. N. Carter, eds.), pp. 189–218. Marcel Dekker, New York.

Hauber, R., and Geiger, R. (1987). A new, very sensitive, bioluminescence-enhanced detection system for protein blotting. Ultrasensitive detection systems for protein blotting and DNA hybridization, I. *J. Clin. Chem. Clin. Biochem.* **25,** 511–514.

Hauber, R., and Geiger, R. (1988). A sensitive, bioluminescence-enhanced detection method for DNA dot-hybridization. *Nucleic Acid Res.* **16,** 1213.

Hauber, R., Miska, W., Schleinkofer, L., and Geiger, R. (1988). The application of a photon-counting camera in very sensitive, bioluminescence-enhanced detection systems for protein blotting. Ultrasensitive detection systems for protein blotting and DNA hybridization, II. *J. Clin. Chem. Clin. Biochem.* **26,** 147–148.

Hauber, R., Miska, W., Schleinkofer, L., and Geiger, R. (1989). New, sensitive, radioactive-free bioluminescence-enhanced detection system in protein blotting and nucleic acid hybridization. *J. Biolumin. Chemilumin.* **4,** 367–372.

Hawkins, E., and Cumming, R. (1990). Enhanced chemiluminescence for tissue antigen and cellular viral DNA detection. *J. Histochem. Cytochem.* **38,** 415–419.

Henchal, E. A., Polo, S. L., Vorndam, V., Yaemsiri, C., Innis, B. L., and Hoke, C. H. (1991). Sensitivity and specificity of a universal primer set for the rapid diagnosis of dengue virus infections by polymerase chain reaction and nucleic acid hybridization. *Am. J. Trop. Med. Hyg.* **45,** 418–428.

Hornsleth, A., Aaen, K., and Gundestrup, M. (1988). Detection of respiratory syncytial virus and rotavirus by enhanced chemiluminescence enzyme-linked immunosorbent assay. *J. Clin. Microbiol.* **26,** 630–635.

Ireland, D., and Samuel, D. (1989). Enhanced chemiluminescence ELISA for the detection of antibody to hepatitis B virus surface antigen. *J. Biolumin. Chemilumin.* **4,** 159–163.

Ishii, J. K., and Ghosh, S. S. (1993). Bead-based sandwich hybridization characteristics of oligonucleotide-alkaline phosphatase conjugates and their potential for quantitating target RNA sequences. *Bioconjugate Chem.* **4,** 34–41.

Iwen, P. C., Blair, T. M., and Woods, G. L. (1991). Comparison of the Gen-Probe PACE 2 system, direct fluorescent-antibody, and cell culture for detecting *Chlamydia trachomatis* in cervical specimens. *Am. J. Clin. Pathol.* **95,** 578–582.

Jablonski, E., Moomaw, E. W., Tullis, R. H., and Ruth, J. L. (1986). Preparation of oligodeoxynucleotide-alkaline phosphatase conjugates and their use as hybridization probes. *Nucleic Acids Res.* **14,** 6115–6128.

Jacobs, R. M., Smith, H. E., Gregory, B., Valli, V. E., and Whetstone, C. A. (1992). Detection of multiple retroviral infections in cattle and cross-reactivity of bovine immunodeficiency-like virus and human immunodeficiency virus type 1 proteins using bovine and human sera in a western blot assay. *Can. J. Vet. Res.* **56,** 353–359.

Jang, D., Sellors, J. W., Mahony, J. B., Pickard, L., and Chernesky, M. A. (1992). Effects of broadening the gold standard on the performance of a chemiluminometric immunoassay to detect *Chlamydia trachomatis* antigens in centrifuged first void urine and urethral swab samples from men. *Sex. Trans. Dis.* **19,** 315–319.

Kenten, J. H., Casadei, J., Link, J., Lupolid, S., Willey, J., Powell, M., Rees, A., and Massey, R. (1991). Rapid electrochemiluminescence assays of polymerase chain reaction products. *Clin. Chem.* **37,** 1626–1632.

Kenten, J. H., Gudibande, S., Link, J., Willey, J. J., Curfman, B., Major, E. O., and Massey, R. J. (1992). Improved electrochemiluminescent label for DNA probe assays: Rapid quantitative assays of HIV-1 polymerase chain reaction products. *Clin. Chem.* **38,** 873–879.

Khalil, O. S., Hanna, C. F., Huff, D., Zurek, T. F., Murphy, B., Pepe, C., and Genger, K. (1991a). Reaction tray and noncontract transfer method for heterogeneous chemiluminescence immunoassays. *Clin. Chem.* **37,** 1612–1617.

Khalil, O. S., Zurek, T. F., Tryba, J., Hanna, C. F., Hollar, R., Pepe, C., Genger, K., Brentz, C., Murphy, B., Abunimeh, N., Carver, R., Harder, P., Coleman, C., Robertson, E., and Wolf-Rogers, J. (1991b). Abbott Prism: A multichannel heterogeneous chemiluminescence immunoassay analyzer. *Clin. Chem.* **37**, 1540–1547.

Kricka, L. J. (1991). Chemiluminescent and bioluminescent techniques. *Clin. Chem.* **37**, 1472–1481.

Kricka, L. J. (1992). Nucleic acid hybridization test formats: Strategies and applications. *In* "Nonisotopic DNA Probe Techniques" (L. J. Kricka, ed.), pp. 3–28. Academic Press, San Diego.

Kricka, L. J., and Carter, T. J. N. (1982). Luminescent immunoassays, *In* "Clinical and Biochemical Luminescence" (L. J. Kricka and T. J. N. Carter, eds.), pp. 153–178. Marcel Dekker, New York.

Kricka, L. J., Thorpe, G. H. G., and Stott, R. A. W. (1987). Enhanced chemiluminescence enzyme immunoassay. *Pure Appl. Chem.* **59**, 651–654.

Kuroda, N., Nakashima, K., Akiyama, S., Shiraki, H., and Maeda, Y. (1992). Photographic chemiluminescent ELISA for detection of anti-human T-cell leukemia virus type-I antibodies by using synthetic peptides as antigens. *Clin. Chim. Acta* **211**, 113–119.

Martin, C. S., and Bronstein, I. (1993). Imaging of chemiluminescent signals with cooled CCD camera systems. Presented at the 2nd European Seminars on Low Light Imaging, Florence, Italy, September 1-4.

Martin, C. S., and Bronstein, I. (1994). Imaging of chemiluminescent signals with cooled CCD camera systems. *J. Biolumin. Chemilumin.* **9**, 145–153.

Martinez, M. L., and Weiss, R. C. (1993). Detection of feline infectious peritonitis virus infection in cell cultures and peripheral blood mononuclear leukocytes of experimentally infected cats using a biotinylated cDNA probe. *Vet. Microbiol.* **34**, 259–271.

Matthews, J. A., Batki, A., Hynds, C., and Kricka, L. J. (1985). Enhanced chemiluminescent method for the detection of DNA dot-hybridization assays. *Anal. Biochem.* **151**, 205–209.

McCapra, F. (1966). The chemiluminescence of organic compounds. *Q. Rev. Chem. Soc.* **20**, 485–510.

McCartney, R. A., Harbour, J., Roome, A. P. C. H., and Caul, E. O. (1993). Comparison of enhanced chemiluminescence and microparticle enzyme immunoassay for the measurement of hepatitis-B surface antibody. *Vaccine* **11**, 941–945.

McKimm-Breschkin, J. L. (1992). Rapid treatment of whole cells and RNA viruses for analysis of RNA slot blot hybridization. *Virus Res.* **22**, 199–206.

Mercer, L. J., Robinson, D. C., Sahm, D. F., Lawrie, M. J., and Hajj, S. N. (1990). Comparison of chemiluminescent DNA probe to cell culture for the screening of *Chlamydia trachomatis* in a gynecology clinic population. *Obstet. Gynecol.* **76**, 114–117.

Miliukiene, V., Dikiniene, N., Vidziunaite, R., Mikalauskiene, G., Veleckaite, A., and Mikulskis, P. (1991). The comparative evaluation of spectrophotometric and chemiluminescent detection in an immunoenzyme test of antibodies to the antigens of the bovine leukosis virus. *Zh. Mikrobiol. Epidemiol. Immunobiol.* **7**, 64–66.

Musiani, M., Zerbini, M., Gibellini, D., Gentilomi, G., La Placa, M., Ferri, E., and Girotti, S. (1991a). Chemiluminescent assay for the detection of viral and plasmid DNA using dogoxigenin-labeled probes. *Anal. Biochem.* **194**, 394–398.

Musiani, M., Zerbini, M., Gibellini, D., Gentilomi, G., Venturoli, S., Gallinella, G., Ferri, E., and Girotti, S. (1991b). Chemiluminescence dot blot hybridization assay for detection of B19 parvovirus DNA in human sera. *J. Clin. Microbiol.* **29**, 2047–2050.

Musiani, M., Zerbini, M., Gentilomi, G., Gibellini, D., Gallinella, G., Venturoli, S., and La Placa, M. (1992). Detection of CMV DNA in clinical samples of AIDS patients by chemiluminescence by hybridization. *J. Virol. Meth.* **38**, 1–9.

Nelson, N. C., and Kacian, D. L. (1990). Chemiluminescent DNA probes: A comparison of the acridinium ester and dioxetane detection system and their use in clinical diagnostic assays. *Clin. Chim. Acta* **194,** 73–90.

Nelson, N. C., Reynolds, M. A., and Arnold, L. J. Jr. (1992). Detection of acridinium esters by chemiluminescence. *In* "Nonisotopic DNA Probe Techniques" (L. J. Kricka, ed.), pp. 275–310. Academic Press, San Diego.

Neman-Simha, V., Delmas-Beauvieux, M. C., Geniaux, M., and Bebear, C. (1991). Evaluation of a chemiluminometric immunoassay and a direct immunofluorescence test for detecting *Chlamydia trachomatis* in urogenital specimens. *Eur. J. Clin. Microbiol. Infect. Dis.* **10,** 662–665.

Nicolas, J. C., Térouanne, B., Balaguer, P., Boussioux, A. M., and Crastes de Paulet, A. (1990). A bioluminescent solid phase for immunoassays and DNA probes. *Ann. Biol. Clin. (Paris)* **48,** 573–579.

Nicolas, J.-C., Balaguer, P., Térouanne, B., Villébrun, M. A., and Boussioux, A.-M. (1992). Detection of glucose 6-phosphate dehydrogenase by bioluminescence. *In* "Nonisotopic DNA Probe Techniques" (L. J. Kricka, ed.), pp. 203–225. Academic Press, San Diego.

O'Neill, H. J., and Coyle, P. V. (1992). Two anti-parvovirus B19 IgM capture assays incorporating a mouse monoclonal antibody specific for B19 viral capsid proteins Vp1 and Vp2. *Arch. Virol.* **123,** 125–134.

Ou, C.-Y., McDonough, S. H., Cabanas, D., Ryder, T. B., Harper, M., Moore, J., and Schochetman, G. (1990). Rapid and quantitative detection of enzymatically amplified HIV-1 DNA using chemiluminescent oligonucleotide probes. *AIDS Res. Hum. Retroviruses* **6,** 1323–1329.

Papsidero, L. D., Dittmer, R. P., Vaickus, L., and Poiesz, B. J. (1992). Monoclonal antibodies and chemiluminescence immunoassay for detection of the surface protein of human T-cell lumphotropic virus. *J. Clin. Microbiol.* **30,** 351–358.

Podleckis, E. V., Hammond, R. W., Hurtt, S. S., and Hadidi, A. (1993). Chemiluminescent detection of potato and pome fruit viroids by dogoxigenin-labeled dot blot and tissue blot hybridization. *J. Virol. Meth.* **43,** 147–158.

Pollard-Knight, D., Read, C. A., Downes, M. J., Howard, L. A., Leadbetter, M. R., Pheby, S. A., McNaughton, E., Syms, A., and Brady, M. A. W. (1990a). Nonradioactive nucleic acid detection by enhanced chemiluminescence using probes directly labeled with horseradish peroxidase. *Anal. Biochem.* **185,** 84–89.

Pollard-Knight, D., Simmonds, A. C., Schaap, A. P., Akhavan, H., and Brady, M. A. W. (1990b). Nonradioactive DNA detection on Southern blots by enzymatically triggered chemiluminescence. *Anal. Biochem.* **185,** 353–358.

Pollini, C. P., Giunchedi, L., and Credi, R. (1993). A chemiluminescent immunoassay for the diagnosis of grapevine closteroviruses on nitrocellulose membrane. *J. Virol. Meth.* **42,** 107–116.

Pronovost, A. D., Baumgarten, A., and Hsiung, G. D. (1981). Sensitive chemiluminescent enzyme-linked immunosorbent assay for quantification of human immunoglobulin G and detection of herpes simplex virus. *J. Clin. Microbiol.* **13,** 97–101.

Puchhammerstoeckl, E., Heinz, F. X., and Kunz, C. (1993). Evaluation of 3 nonradioactive DNA detection systems for identification of herpes simplex DNA amplified from cerebrospinal fluid. *J. Virol. Meth.* **43,** 257–266.

Radziszewski, B. (1877). Untersuchungen über Hydrobenzamid, Amarin und Lophin. *Chem. Berlin* **10,** 70–75.

Rapier, J. M., Villamarzo, Y., Schochetman, G., Ou, C. Y., Brakel, C. L., Donegan, J., Maltzman, W., Lee, S., Kirtikar, D., and Gatica, D. (1993). Nonradioactive colorimetric microplate hybridization assay for detecting amplified human immunodeficiency virus DNA. *Clin. Chem.* **39,** 244–247.

Robertson, E. F., Weare, J. A., Randell, R., Holland, P. V., Madsen, G., and Decker, R. H. (1991). Characterization of a reduction-sensitive factor from human plasma responsible for apparent false activity in competitive assays for antibody to hepatitis B core antigen. *J. Clin. Microbiol.* **29**, 605–610.

Sarkar, F. H., Sakr, W. A., Li, Y. W., Sreepathi, P., and Crissman, J. D. (1993). Detection of human papillomavirus (HPV) DNA in human prostatic tissues by polymerase chain reaction (PCR). *Prostate* **22**, 171–180.

Schmidt, B. L. (1991). A rapid chemiluminescence detection method for PCR-amplified HIV-1 DNA. *J. Virol. Meth.* **32**, 233–244.

Schmidt, B. L., and Gschnait, F. (1991). Detection of causal agents of AIDS using polymerase chain reaction and chemiluminescence measurement. *Hautarzt* **42**, 754–758.

Scieux, C., Bianchi, A., Henry, S., Brunat, N., Abdennader, S., Vexiau, D., Janier, M., Morel, P., and Lagrange, P. H. (1992a). Evaluation of a chemiluminometric immunoassay for detection of *Chlamydia trachomatis* in the urine of male and female patients. *Eur. J. Clin. Microbiol. Infect. Dis.* **11**, 704–708.

Scieux, C., Bianchi, A., Vassias, I., Meouchy, R., Felten, A., Morel, P., and Perol, Y. (1992b). Evaluation of a new chemiluminometric immunoassay, Magic Lite *Chlamydia*, for detecting *Chlamydia trachomatis* antigen from urogenital specimens. *Sex. Transm. Dis.* **19**, 161–164.

Septak, M. (1989). Acridinium ester-labelled DNA oligonucleotide probes. *J. Biolumin. Chemilumin.* **4**, 351–356.

Simmonds, A. C., Cunningham, M., Durrant, I., Fowler, S. J., and Evans, M. R. (1991). Enhanced chemiluminescence in filter-based DNA detection. *Clin. Chem.* **37**, 1527–1528.

Sritharan, V., and Barker, R. H., Jr. (1991). A simple method for diagnosing *M. tuberculosis* infection in clinical samples using PCR. *Mol. Cell. Probes* **5**, 385–395.

Stanley, P. E. (1992a). A survey of more than 90 commercially available luminometers and imaging devices for low light measurements of chemiluminescence and bioluminescence, including instruments for manual, automatic and specialized operation, for HPLC, LC, GLC and microtitre plates. Part 1: Descriptions. *J. Biolumin. Chemilumin.* **7**, 77–108.

Stanley, P. E. (1992b). A survey of more than 90 commercially available luminometers and imaging devices for low light measurements of chemiluminescence and bioluminescence, including instruments for manual, automatic and specialized operation, for HPLC, LC, GLC and microtitre plates. Part 2: Photographs, *J. Biolumin. Chemilumin.* **7**, 157–169.

Stanley, P. E. (1993a). A survey of some commercially available kits and reagents which include bioluminescence or chemiluminescence for their operation—including immunoassays, hybridization, labels, probes, blots and ATP-based rapid microbiology: Products from more than 40 companies. *J. Biolumin. Chemilumin.* **8**, 51–63.

Stanley, P. E. (1993b). Commercially available luminometers and imaging devices for low-light measurements and kits and reagents utilizing bioluminescence or chemiluminescence: Survey update I. *J. Biolumin. Chemilumin.* **8**, 237–240.

Stauber, S., Siegel, G., Janitschke, K., Schulze-Forster, K., and Simon, D. (1991). Detection of polymerase chain reaction (PCR) products of *Toxoplasma gondii* by enhanced chemiluminescence (ECL). *J. Biolumin. Chemilumin.* **6**, 283.

Suzuki, K., Okamoto, N., Watanabe, S., and Kano, T. (1992). Chemiluminescent microtiter method for detecting PCR amplified HIV-1 DNA. *J. Virol. Meth.* **38**, 113–122.

Suzuki, K., Craddock, B. P., Kano, T., and Steigbigel, R. T. (1993). Chemiluminescent enzyme-linked immunoassay for reverse transcriptase, illustrated by detection of HIV reverse transcriptase. *Anal. Biochem.* **210**, 277–281.

Tham, K. M., and Stanislawek, W. L. (1992a). Detection of chicken anaemia agent DNA sequences by the polymerase chain reaction. *Arch. Virol.* **127**, 245–255.

Tham, K. M., and Stanislawek, W. L. (1992b). Polymerase chain reaction amplification for direct detection of chicken anemia virus DNA in tissues and sera. *Avian Dis.* **36**, 1000–1006.

Thorpe, G. H. G., Kricka, L. J., Moseley, S. B., and Whitehead, T. P. (1985). Phenols as enhancers of the chemiluminescent horseradish peroxidase-luminol-hydrogen peroxide reaction: Application in luminescence-monitored enzyme immunoassays. *Clin. Chem.* **31,** 1335–1341.

Tumolo, A., Nguyen, Q., Witney, F., Murphy, O. J., Voyta, J. C., and Bronstein, I. (1992). Detection of DNA on membranes with alkaline phosphatase-labeled probes and chemiluminescent AMPPD substrate. *In* "Nonisotopic DNA Probe Techniques" (L. J. Kricka, ed.), pp. 127–145. Academic Press, San Diego.

Urdea, M. S., Running, J. A., Horn, T., Clyne, J., Ku, L., and Warner, B. D. (1987) A novel method for the rapid detection of specific nucleotide sequences in crude biological samples without blotting or radioactivity; application to the analysis of hepatitis B virus in human serum. *Gene* **61,** 253–264.

Urdea, M. S., Kolberg, J., Clyne, J., Running, J. A., Besemer, D., Warner, B., and Sanchez-Pescador, R. (1989). Application of a rapid non-radioisotopic nucleic acid analysis system to the detection of sexually transmitted disease-causing organisms and their associated antimicrobial resistances. *Clin. Chem.* **35,** 1571–1575.

Urdea, M. S., Kolberg, J., Warner, B. D., Horn, T., Clyne, J., Ku, L., and Running, J. A. (1990). A novel method for the rapid detection of hepatitis B virus in human serum samples without blotting or radioactivity. *In* "Luminescence Immunoassay and Molecular Applications" (K. van Dyke and R. van Dyke, eds.), pp. 275–292. CRC Press, Boca Raton, Florida.

Vlaspolder, F., Matsaers, J. A., Blog, F., and Notowicz, A. (1993). Value of a DNA probe assay (Gen-Probe) compared with that of culture for diagnosis of gonococcal infection. *J. Clin. Microbiol.* **31,** 107–110.

Vlieger, A. M., Medenblik, A. M., van-Gijlswijk, R. P., Tanke, H. J., van-der-Ploeg, M., Gratama, J. W., and Raap, A. K. (1992). Quantitation of polymerase chain reaction products by hybridization-based assays with fluorescent, colorimetric, or chemiluminescent detection. *Anal. Biochem.* **205,** 1–7.

Wages, J. M., Dolenga, L., and Fowler, A. K. (1993). Electrochemiluminescence detection and quantitation of PCR-amplified DNA. *Amplifications* **10,** 1–3.

Weare, J. A., Robertson, E. F., Madsen, G., Hu, R., and Decker, R. H. (1991). Improvement in the specificity of assays for detection of antibody to hepatitis B core antigen. *J. Clin. Microbiol.* **29,** 600–604.

Weeks, I., Behesti, I., McCapra, F., Campbell, A. K., and Woodhead, J. S. (1983). Acridinium esters as high-specific-activity labels in immunoassay. *Clin. Chem.* **29,** 1474–1479.

Welnicki, M., and Hiruki, C. (1993). Chemiluminescent assay for the detection of viral and viroid RNAs using digoxigenin labelled probes. *J. Biolumin. Chemilumin.* **8,** 127.

Whitehead, T. P., Thorpe, G. H. G., Carter, T. J. N., Groucutt, C., and Kricka, L. J. (1983). Enhanced luminescence procedure for sensitive determination of peroxidase-labelled conjugates in immunoassay. *Nature* **305,** 158–159.

Yang, J. Q., Tata, P. V., Park-Turkel, H. S., and Waksal, H. W. (1991). The application of AmpliProbe in diagnostics. *BioTechniques.* **11,** 392–397.

Zachar, V., Mayer, V., Aboagye-Mathiesen, G., Norskov-Laruitsen, N., and Ebbesen, P. (1991). Enhanced chemiluminescence-based hybridization analysis for PCR-mediated HIV-1 DNA detection offers an alternative to ^{32}P-labelled probes. *J. Virol. Meth.* **33,** 391–395.

8

Detection of Viral Pathogens Using PCR Amplification

Bruce J. McCreedy

Roche Biomedical Laboratories
Research Triangle Park, North Carolina 27709

I. Overview of PCR Amplification

The polymerase chain reaction (PCR) was introduced to the scientific community in 1985 by scientists from Cetus Corporation and has since become an integral part of the molecular biology laboratory (Saiki *et al.*, 1985; Mullis and Faloona, 1987). Several improvements in the technology and associated equipment have significantly changed the way PCR is performed today relative to the earlier PCR protocols. Well-characterized user-friendly PCR assays designed for diagnostic applications in clinical laboratories have been described for the detection of a variety of human viral pathogens.

PCR is a relatively simple procedure consisting of repetitive cycles of oligonucleotide primer-directed DNA replication (for reviews see Persing and Landry, 1989; Gibbs, 1990). A selected genetic target is copied in a three-step reaction consisting of (1) denaturing the target DNA at elevated temperature (90–95°C), (2) cooling the reaction to promote annealing of oligonucleotide primers that are complementary to either strand of the target DNA, and (3) extending the bound primers by DNA polymerase action, resulting in a replication of the target sequence (Fig. 1). The reaction is repeated in successive cycles of heating and cooling, referred to as thermocycles; each cycle doubles the input target sequence. Because the newly synthesized DNA copies can also serve as templates in subsequent replication

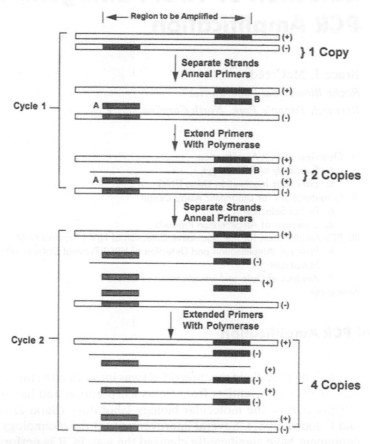

Figure 1   Cycles of polymerase chain reaction. A specific target region of DNA is amplified during a thermocycle consisting of (1) heating to separate the double-stranded DNA target; (2) cooling to promote binding of target-specific oligonucleotide primers [hatched (A) and stippled (B) bars]; and (3) extending the bound primers using a thermostable DNA polymerase. Two copies of the target region DNA are produced from the first cycle of amplification. The original template DNA and the copies (5′ ends terminate with the incorporated primer sequence) serve as templates during the second cycle of amplification. During subsequent rounds of amplification, the shorter DNA copies, with both ends defined by the primer sequences, are preferentially amplified over the original full-length template DNA. The "short product" DNA accumulates geometrically during successive rounds of amplification, generating the specific product DNA that serves as the substrate for detection protocols such as those employing labeled DNA probes.

cycles, a 2^n accumulation of product copies is possible, where n is the number of thermocycles. For this reason, PCR is often referred to as an exponential amplification technology. In practice, less PCR product than is theoretically predicted is actually synthesized because of lower reaction efficiency after many cycles (>25) of PCR amplification.

Several improvements that have enhanced the analytical sensitivity and specificity of PCR procedures have been introduced since 1985. The discovery of thermostable DNA polymerases such as *Taq* DNA polymerase (Saiki *et al.*, 1988) greatly simplified the PCR procedure by eliminating the need to add fresh enzyme after each cycle of denaturation because of inactivation of the thermolabile DNA polymerase enzyme used in earlier PCR protocols. The use of *Taq* DNA polymerase also resulted in increased specificity of PCR amplification reactions by permitting higher primer annealing and polymerase extension temperatures to be used, decreasing the amount of nonspecific target amplification. Implementation of target-specific PCR protocols with optimized reaction buffers and target-specific thermocycling profiles, and incorporation of innovative primer designs capable of accommodating mismatched bases due to sequence heterogeneity (Kwok *et al.*, 1990) furthered the potential usefulness of PCR-based diagnostic procedures. A partial list of PCR-based assays for detection of various viral pathogens is given in Table 1. Development of new equipment such as high sample capacity thermocycling machines (microprocessor-controlled heating–cooling blocks), capable of rapidly and accurately attaining target temperatures, increased assay sample throughput and precision. Better liquid handling equipment, including

TABLE I
Viral Pathogens Detected by PCR Amplification

Pathogen[a]	Reference
HIV-1	Ou *et al.* (1988), Rogers *et al.* (1988), Protocol 1
HIV-2	Rayfield *et al.* (1988)
HTLV-I and HTLV-II	Kwok *et al.* (1988), Palumbo *et al.* (1992)
HSV Type I and Type 2	Rosenberg and Lebon (1991), Aslanzadeh *et al.* (1992)
Hepatitis B Virus	Larzul *et al.* (1988), Kaneko *et al.* (1989)
Hepatitis C Virus	Young *et al.* (1993)
Enteroviruses	Rotbart (1990)
Cytomegalovirus	Demmler *et al.* (1988), Buffone *et al.* (1991), Einsele *et al.* (1991)
Human papillomavirus	Yi and Manos (1990), Schiffman *et al.* (1991)
Human parvovirus B19	Koch and Aeller (1990)
Human adenovirus	Allard *et al.* (1990)

[a] Abbreviations: HIV, human immunodeficiency virus; HTLV, human T-cell lymphotropic virus; HSV, herpes simplex virus.

robotic work stations that eliminate the need for individual pipetting of PCR reaction mixes, has improved the cost effectiveness of routine PCR testing by reducing the hands-on time required to perform PCR assays.

A. Specificity of PCR Assays

The high analytical sensitivity of PCR-based procedures facilitates the detection of minute amounts of pathogens in biological specimens that are often undetectable using more conventional technology. However, the accumulation of high copy numbers of specific target nucleic acid sequences presented problems with controlling the specificity of PCR amplifications. False positive results due to re-amplification of PCR products carried over from previous amplifications of positive specimens and controls plagued early attempts to implement routine clinical PCR testing procedures. Table 2 illustrates the power and potential pitfalls of routine use of an optimized PCR protocol for the detection of human immunodeficiency virus type 1 (HIV-1) proviral DNA. A 100-μl PCR assay can generate up to 10^9 copies of PCR product (amplicons) from only 10 copies of input HIV-1 DNA. Carryover contamination of subsequent reactions with as little as 10^{-6} μl (aerosol) amplification product can result in virtually all false positive reactions.

TABLE 2
Carryover Contamination during PCR Amplification[a]

	Volume carryover (μl)	Copies of PCR product (0 copy, 85%, 30 cycles)	Positive reactions (%)
Spill	100	10^9	100
	10	10^8	100
	1	10^7	100
Smudge	0.1	10^6	100
	0.01	10^5	100
	10^{-3}	10^4	100
	10^{-4}	10^3	100
Aerosol	10^{-5}	10^2	100
	10^{-6}	10	100
	10^{-7}	1	63
	10^{-8}	0.1	9
	10^{-9}	0.01	1

[a] After 30 cycles of amplification, an optimized PCR protocol for the amplification of HIV-1 proviral DNA can generate up to 10^9 copies of the target DNA from a 100 μl reaction containing 10 copies of HIV-1 input DNA with reaction efficiency of 85%. Carryover contamination and subsequent PCR amplification, with as little as 10^{-6} μl of previously amplified product DNA, can provide sufficient input target DNA to create false positive results from a clinically negative specimen.

Physical separation of pre- and post-PCR areas within the laboratory, dispensing of reagents into single-use aliquots, and use of positive displacement pipettors or aerosol-resistant pipette tips helped reduce the possibility of carryover contamination (Kwok and Higuchi, 1989). However, given the stability of unmodified DNA products and the exceedingly low level of carryover contamination that can result in false positive results, many laboratories still experienced some problems controlling reaction specificity because of low-level PCR product contamination. More recently, enzymatic, photochemical, and chemical methods to modify PCR amplification products so they cannot be re-amplified have been described that significantly reduce the possibility of false positive PCR results due to carryover contamination (Longo *et al.*, 1990; Cimino *et al.*, 1991; Aslanzadeh, 1992; see also Chapter 2).

B. Enzymatic Inactivation Using UNG

Incorporation of dUTP and the enzyme uracil-*N*-gylcosylase (UNG) into PCR mixes is a simple and effective means of carryover contamination control. UNG enzyme was first recognized as a component of a DNA excision and repair system observed among many bacterial species (Lindahl *et al.*, 1977). The function of the enzyme is to remove spontaneously arising uracil residues, the result of deamination of cytosine residues, from the DNA chain prior to incorporation of a new cytosine residue. By substituting dUTP for dTTP in PCR mixes, U is substituted for T by *Taq* polymerase during the primer extension (elongation) step of PCR amplification. The resulting product amplicons will contain deoxyuracil residues in contrast to the naturally occurring deoxythymidine residues present in template DNA derived from specimens. As shown in Fig. 2, carryover contamination of target nucleic acids (specimens) with deoxyuracil-containing amplicons *prior to* PCR amplification can be controlled by the inclusion of UNG in the amplification reaction mix.

UNG cleaves the uracil residues from the contaminating DNA amplicons, resulting in abasic sites in the carryover DNA. The abasic sites are hydrolyzed under the buffer conditions and elevated temperatures that are optimal for *Taq* polymerase during PCR amplification and thus do not serve as efficient templates for re-amplification. However, newly synthesized deoxyuracil-containing product DNA can still be used for detection by ethidium bromide staining of agarose gels and probe hybridization methods, or as templates for cloning (UNG⁻ hosts only) and sequencing experiments. Therefore, inclusion of dUTP and UNG enzyme in PCR mixes inactivates any contaminating PCR product DNA and modifies newly synthesized products so they, in turn, can be inactivated rather than serve as templates in

Specimen Target DNA

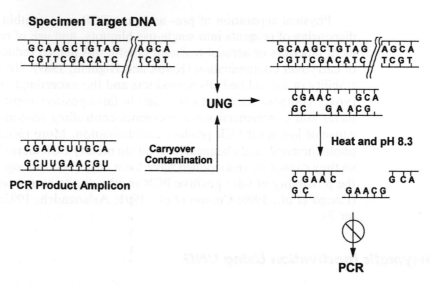

Figure 2 UNG restriction of uracil-containing amplification products. Incorporation of dUTP into PCR reaction mixes results in substitution of uracil for thymine in the amplified product DNA. When UNG enzyme is added prior to PCR amplification of specimen DNA, previously amplified product DNA, the source of carryover contamination, is modified by the removal of uracil residues whereas specimen DNA (which contains only thymine residues paired with adenine) is unaffected by UNG. Subsequent PCR amplification of the specimen DNA proceeds as usual. However, reamplification of the modified carryover DNA does not occur due to strand scission at the de-uracilated sites in the carryover DNA, which occurs when the temperature is raised during the denaturation step of the first amplification cycle.

subsequent PCR amplifications. When incorporating UNG restriction procedures into PCR protocols, note that UNG denatures at temperatures above 55°C; therefore, it will not cleave newly synthesized products. However, UNG will renature on cooling of PCR assays. To prevent UNG from degrading newly synthesized amplicons and compromising assay sensitivity, reactions must be kept at 55°C prior to product analysis or UNG activity should be destroyed by the addition of NaOH (~0.2 N final concentration) to establish basic conditions. Inactivation of 10^6–10^7 uracil-containing amplicons is readily achievable using UNG restriction methods. However, some GC-rich products may not be efficiently inactivated. One must optimize PCR mixes that incorporate UNG and empirically determine the efficiency of product inactivation.

The use of UNG in PCR protocols can also improve the sensitivity of PCR amplifications by degrading nonspecific extension products that result from low-stringency primer binding that occurs below the optimal primer annealing temperature. Once the reaction temperature is raised to around 55°C, UNG activity is lost, permitting the accumulation of specific amplified target sequences that are produced following high-stringency primer annealing. In this capacity, and in similar "hot-start" PCR protocols (Chou *et al.*, 1992), reaction sensitivity is enhanced because of less production and amplification of nonspecific sequences during the initial PCR cycles. Another benefit of UNG carryover control is that RNA templates can be used with this method because UNG does not cleave uracil residues in ribonucleic acid chains. This feature is particularly advantageous when RNA-PCR procedures are employed for the detection of viral pathogens with RNA genomes or when targeting mRNA for expression studies provided that the reverse transcriptase/DNA polymerase enzyme(s) employed can use either thymidine or uracil during transcription and amplification.

Using combinations of physical separation of sample preparation and amplification/product detection areas, aerosol-resistant pipettors and tips, and amplification product inactivation methods, successful PCR testing programs have been implemented in both small and large volume clinical laboratories (for reviews see White *et al.*, 1992; McCreedy and Callaway, 1993; see also Chapter 2).

II. CONSIDERATIONS FOR DIAGNOSTIC ASSAY DESIGN

When applying PCR amplification methods for diagnostic use, one must carefully consider the variables that may influence the potential utility of the assay. The selection of the nucleic acid target for amplification, the design of oligonucleotide primers and probes, the optimization of specimen preparation and PCR amplification conditions, and the method of amplified product detection and inactivation must all be optimized for the intended pathogen. The following discussion provides an overview of areas that must be considered when designing amplification reactions. Readers should keep in mind that each system must be individually optimized (for more comprehensive reviews, see Innis and Gelfand, 1990; Persing, 1993).

The selection of an appropriate amplification target for diagnostic applications should begin with a review of the published sequence information available for the organism. Nucleic acid sequence databases such as GenBank (Los Alamos, NM) are useful in the search for sequence information for isolates of a particular organism or a related group of organisms. Computer disks containing volumes of sequence information and programs to analyze the information are available from several commercial sources. The target selected can be DNA or RNA and should be a genetically stable (conserved) region. When RNA is targeted, a separate reverse transcription step must be included in the assay design to convert RNA to cDNA, which then serves as template DNA for PCR amplification. For diagnostic applications, the target should be between 100 and 600 bp in size, have approximately 50% GC content, be devoid of stretches of the same nucleotide, and not possess significant regions of self-complementarity or kinked (hairpin) secondary structures.

A. Primer Selection

The design and selection of primers are influenced driven by the target sequence. In general, primers should be between 25 and 35 bases in length, avoid contiguous stretches of the same nucleotide, possess melting temperatures (T_m) between 55 and 65°C, and avoid complementarity within a primer and between primers, especially near the 3′ ends. Shorter primer sequences (<20 bases) are less capable of accommodating mismatched bases with the target DNA or RNA template and are not recommended for diagnostic applications. A thymine residue can be included at the 3′ end of each primer to ensure DNA polymerase extension of a bound primer that is mismatched with the template base at the 3′ end of the primer sequence (Kwok *et al.*, 1990). Therefore, both the overall length and the base composition of primer sequences contribute to the ability to accommodate sequence mismatches and avoid potential false negative amplification results. The size and composition of the resulting PCR product should also be considered during primer selection when amplification product inactivation methods (described earlier) are intended as part of the PCR assay protocol. Selection of primers for PCR amplification protocols has been greatly simplified by the availability of computer programs that accurately predict T_m, secondary structures, and primer–primer interaction (Rychlik and Rhoades, 1989; Lucas *et al.*, 1991).

B. Detection of Amplification Products

For diagnostic applications, the use of DNA probes for detection of amplified products is recommended over non-probe-based detection strategies such as ethidium bromide staining of DNA following agarose gel electrophoresis (Manitatis *et al.*, 1982) or nested amplification using a second set of primers that recognize the initial amplification product. Because specificity of amplification varies widely among amplification protocols, visualization of the expected size DNA fragment following amplification does not guarantee that the specific target amplicon is present. Nested PCR protocols involve the opening and transfer of amplification product tubes, which greatly increases the possibility of carryover contamination. In addition, product inactivation protocols cannot be used in nested PCR systems because the first PCR product must serve as a template for the second (nested) amplification reaction. A single-tube nested PCR method has been described that may circumvent some of the problems associated with diagnostic use of two-tube nested PCR protocols (Yourno, 1992). In general, DNA probe-based detection protocols are rapid and simple, and add another level of specificity to diagnostic PCR protocols for clinical laboratory use. The abundance of the PCR product generated after amplification facilitates detection by a DNA probe by allowing the product concentration, not the amount of labeled probe added to a detection reaction, to drive the detection reaction sensitivity. The introduction of reporter molecules, or ligands for subsequent detection protocols, into primer or probe sequences has greatly simplified the detection of PCR amplification products. Several rapid and sensitive nonradioactive DNA probe detection protocols have been described for routine clinical laboratory use, including the use of acridinium ester derivatized DNA probes (Ou *et al.*, 1990) and microtiter plate based probe detection of biotinylated PCR products (Fig. 3; Butcher and Spadoro, 1992).

III. PCR AMPLIFICATION FOR THE QUALITATIVE DETECTION OF HIV-1 PROVIRAL DNA

The related human retroviruses, human immunodeficiency virus types I and II (HIV-1, HIV-2) are the etiologic agents of acquired immunodeficiency syndrome (AIDS). In North America, HIV-1 is responsible for the majority of AIDS cases whereas infection with HIV-1 or HIV-2 is associated with AIDS in Africa. Infection with HIV-1 and progression to AIDS-related complex (ARC) and AIDS occurs in successive stages of disease. Acute infection is characterized by a short period, usually between 2 and 10 wk, when circulating virus may be detectable (viremia) but no serological evidence of infection (no detectable antibodies) is present. Recently infected individuals in this "seronegative window" period are likely to be highly infectious because of the presence of high titers of circulating viral particles prior to the stimulation of an immune response and the production of anti-viral antibodies (seroconversion). After seroconversion, most circulating viruses are complexed and removed from the circulation. During this stage, which may last for years, most HIV-infected individuals can be reliably detected using standard enzyme immunoassay (EIA) antibody detection methods. Over time, HIV infection leads to a progressive depletion of $CD4^+$ T lymphocytes with associated deterioration of immune system function. Progression to clinical AIDS is characterized by the occurrence of opportunistic infections and other potentially life-threatening illnesses associated with immune system dysfunction. Direct detection of viral particles, or viral nucleic acid, offers the potential to detect early HIV infection (Wolinsky *et al.*, 1989) and

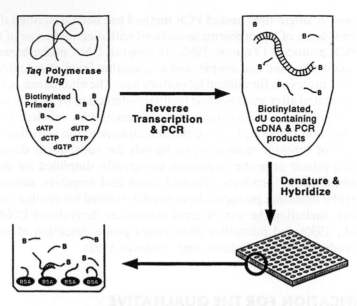

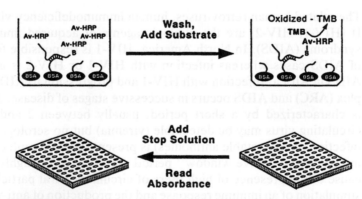

Figure 3 Nonisotopic DNA probe detection format. In this scheme, biotin moieties are incorporated into the PCR primers, which in turn become part of the amplified PCR product. A specific DNA capture probe conjugated to a bovine serum albumin (BSA) molecule is coated onto the wells of a microtiter plate. Capture of the amplified PCR product by the immobilized DNA probe is visualized in an enzyme immunoassay format. Added avidin–horseradish peroxidase conjugate attaches to the biotin moieties present in the probe-bound PCR product DNA. A color reaction is produced in positive wells after addition of a chromogen/substrate [tetramethylbenzidine (TMB)/ H_2O_2], which is measured by determination of optical density at 450 nm. Schematic reproduced with permission of Roche Molecular Systems, Inc., Branchburg, NJ.

to resolve inconclusive HIV antibody test results that may occur due to early seroconversion or nonspecific cross-reactivity in the case of uninfected individuals (Jackson *et al.*, 1990; Celum *et al.*, 1991). Direct detection of viral particles or nucleic acid is the most reliable method for diagnosis of HIV infection in newborns of HIV-infected women because of the presence of circulating maternal antibodies in the infant for up to 15 mo (Rogers *et al.*, 1989). PCR is ideally suited to these situations because of its ability to amplify HIV-specific nucleic acid sequences from small volumes of blood [e.g., HIV proviral DNA in peripheral blood mononuclear cells (PBMC)] or plasma (virion-associated HIV RNA). The following protocol describes a method for the amplification and detection of HIV-1 proviral DNA from whole blood specimens.

A. Protocol: Amplification and Detection of HIV-1 Proviral DNA in Whole Blood Specimens

1. Materials and Reagents

Leukoprep blood collection tubes or ACD Vacutainer® tubes (Becton Dickinson)

Sterile Tris-buffered saline: 10 mM Tris, pH 7.5, 0.15 M NaCl

Sample extraction buffer: 50 mM KCl, 7.5 mM MgCl$_2$, 0.5% Tween 20, 0.5% NP-40, 10 mM Tris pH 8.3, 100 mg/ml proteinase K

2× PCR master mix: 20 mM Tris, pH 8.5, 100 mM KCl, 400 μM each of dATP, dCTP, dGTP, and dUTP, 25 pmol SK462 (upstream primer), 25 pmol SK431 (downstream primer), 0.05 U/μl *Taq* DNA polymerase (2.5 U/50 μl); store 2× master mix at 4°C; immediately prior to use, add uracil-N-glycosylase (UNG) enzyme to 0.04 U/μl (2.0 U/50 μl).

NOTE: *2× PCR master mix is delivered at 50 μl per PCR assay tube. Master mix is diluted to 1× by addition of 50 μl sample in sample extraction buffer; final reaction volume is 100 μl.*

SK462: 5'-AGTTGGAGGACATCAAGCAGCCATGCAAAT-3'
SK431: 5'-TGCTATGTCAGTTCCCCTTGGTTCTCT-3'

Positive control DNA: HIV-1 plasmid DNA (Perkin-Elmer, Norwalk, CT); dilute to 20 copies/50 μl in dH$_2$O containing 1 μg human placental DNA

Specimen negative control: HIV negative donor PBMC or "mock" specimen tube (buffers only)

SK102 capture probe: 5'-GAGACCATCAATGAGGAAGCTGCA-GAATGGGAT-3' end-labeled at the 5' end with 32p.

2. Specimen Collection and Preparation

Collect 5–7 ml whole blood into a Leukoprep or ACD Vacutainer® tube.

NOTE: *All specimens should be considered capable of transmitting an infectious agent. Universal precautions should be observed when handling specimens. Laboratory coats, gloves, and safety goggles should be worn at all times. Specimens should be processed in a biological containment hood (BL-2 containment).*

1. Centrifuge whole blood in leukoprep tubes to separate PBMC. Alternatively, centrifuge whole blood collected into ACD anticoagulant and layered onto Ficoll–Hypaque for 30 min at 400 *g* to separate PBMC.
2. Collect the PBMC (buffy coat), transfer to a sterile plastic 2-ml screw-cap tube, and wash with 1.5 ml Tris-buffered saline, pH 7.5. Centrifuge at 200 *g* for 10 min in a microfuge to pellet the cells.
3. Remove the supernatant. Resuspend the pellet in 1.5 ml Tris-buffered saline and centrifuge at 200 *g* for 10 min in a microfuge.
4. Remove the supernatant and resuspend the pellet in 400 μl sample extraction buffer (final concentration of cells should be set at ~3 × 10^6 cells/ml).
5. Vortex the resuspended pellet and incubate at 60°C for 30 min (water bath or heat block).
6. Transfer to a 95–98°C water bath/heat block for 30 min to inactivate proteinase K.
7. Remove tubes; let cool to room temperature. Store cell lysates in refrigerator.

NOTE: *A specimen preparation negative control should be introduced at Step 3 and processed together with specimens through Step 9.*

3. PCR Amplification

NOTE: *This protocol is described for amplifications performed using the GeneAmp 9600 Thermocycler (Perkin-Elmer) with thin wall "microamp" tubes (Perkin-Elmer).*

1. Dispense 50 μl complete 2× PCR master mix (UNG enzyme added) into each reaction tube according to the number of samples to be amplified, plus a positive (HIV-1 DNA plasmid) and specimen negative control tube. In addition, an amplification control (master mix + 50 μl dH$_2$O) should be included. All specimens and controls should be amplified in duplicate reactions.

2. Using a positive displacement pipettor or micropipettor and aerosol-resistant tips, add 50 μl specimen lysate to the appropriately labeled reaction tubes (less specimen can be added and the volume made up to 50 μl with glass-distilled water if necessary; target cell number per PCR reaction is ~150,000 cells or ~1 μg genomic DNA). Pipette positive, negative, and amplification controls last.

3. Cap each tube after addition of sample. After each row of 8 tubes is completed, securely cap the row of tubes using the roller tool provided with the GeneAmp thermocycler. Single reaction tubes with individual caps are available and are recommended for small volume uses.

4. Place the tubes in the thermocycler and run the following thermocycling profile:

Step	Conditions	Comments
Step 1	50°C, hold for 2 min	UNG restriction incubation
Step 2	5 cycles of 95°C, 10 sec; 55°C, 10 sec; 72°C, 10 sec	Initial rounds of HIV-specific thermocycling
Step 3	30 cycles of 90°C, 10 sec; 60°C, 10 sec; 72°C, 10 sec	Thermocycles copying mostly specific primer-defined "short" product DNA
Step 4	1 cycle of 72°C, 10 min	
Step 5	72°C, hold until specimens are removed	

5. Ramp the thermocycler down to 20°C and remove the reaction tray. Denature the UNG enzyme by adding 100 μl 0.4N NAOH, 80 mM EDTA or freeze at −70°C.

6. Detect the amplified HIV-1 DNA using 32P-end labeled SK-102 probe as described by Kwok and Sninsky (1993).

NOTE: *Nonradioactive detection protocols using chemiluminescent (Whetsell et al., 1992) and colorimetric methods (Butcher and Spadoro, 1992; Jackson et al., 1991) have also been described and may be preferred by some users.*

B. Analysis of Data and Interpretation of Results

1. Assay Controls

Analysis of the data begins with interpretation of positive and negative controls. Both duplicate amplifications of the 20-copy HIV-1 DNA positive control should yield unambiguously positive signals. Neither the specimen negative control nor the amplification control reaction should produce any positive reaction product. If either or both of the positive control amplifications do not give positive signals, assay sensitivity cannot be guaranteed and the run should be repeated. If either or both of the duplicate amplifications of the specimen negative and amplification controls give signals above a predetermined background level, assay specificity cannot be ensured (potential carryover contamination or specimen mix-up has occurred) and the run should be repeated. In some instances, repeat analysis of the amplified products can be performed (without re-amplification of the specimen) to rule out pipetting errors during product analysis when negative control values are too high. Assay positive control signal levels must be determined empirically by analysis of known positive and negative specimens, as well as with dilutions of positive control plasmid DNA, to determine the reproducible analytical sensitivity of the system.

2. Specimen Results

When the assay control values are within range, the signals observed for the specimens can be analyzed to determine results. Specimens with clearly positive signals from each of the duplicate amplification reactions are interpreted as positive for HIV-1 proviral DNA. Alternatively, if neither value for a specimen yields a positive signal, the specimen is interpreted as negative for HIV-1 proviral DNA. If discordant results are obtained for the duplicate reactions, the specimen lysate should be re-amplified and detected in duplicate or re-isolated from the original blood specimen tube. If results remain discordant, the result is interpreted as inconclusive and a second specimen should be obtained for repeat analysis.

Several factors can lead to discrepant or false negative results including PCR inhibitors contained within the specimen, sample bias due to low copy numbers of the HIV target within the specimen, and poorly optimized preparation, amplification, or product analysis procedures. A separate control amplification of a resident human gene sequence (e.g., HLA DQ alpha or beta-globin) can be performed to check for the presence of gross inhibitors. Amplification of a prepared specimen in duplicate reactions helps avoid false negative results due to sample bias with low copy number specimens. Depending on volume and cost constraints, single or duplicate amplification reactions from two different preparations of the same specimen further decrease the possibility of not detecting low copy number positive specimens. The specimen preparation and PCR amplification parameters for the assay described here have been optimized and the results obtained from clinical specimens have been described (Jackson *et al.*, 1991; Butcher and Spadoro, 1992; Lynch *et al.*, 1992).

With proper consideration given to assay design and optimization and implementation of proven PCR quality control procedures, most laboratories should be capable of generating consistent, high-quality results using PCR amplification and DNA probe detection methods.

REFERENCES

Allard, A., Gierones, R., Juto, P., and Wadell, G. (1990). Polymerase chain reaction for detection of adenoviruses in stool samples. *J. Clin. Microbiol.* **28**, 2659–2667.

Aslanzadeh, J. (1992). Application of hydroxylamine hydrochloride for post-PCR sterilization. *Ann. Clin. Lab. Sci.* **22**, 280.

Aslanzadeh, J., Osman, D. R., Wilhelm, M. D., Espy, M. J., and Smith, T. F. (1992) A prospective study of the polymerase chain reaction for detection of herpes simplex virus in cerebrospinal fluid submitted to the clinical virology laboratory. *Mol. Cell. Probes* **6**, 367–373.

Buffone, G. J., Demmler, G. J., Schimbor, C. M., and Greer, J. (1991). Improved amplification of cytomegalovirus DNA from urine after purification of DNA with glass beads. *Clin. Chem.* **37**, 1945–1949.

Butcher, A., and Spadoro, J. (1992). Using PCR for detection of HIV-1 infection. *Clin. Immunol. News.* **12**, 73–76.

Celum, C. L., Coombs, R. W., Lafferty, W., Inci, T. S., Louie, P. H., Gates, C. A., McCreedy, B. J., Egan, R., Grove, T., Alexander, S., Koepsell, T., Weiss, N., Fisher, L., Casey, L., and Holmes, K. K. (1991). Indeterminate human immunodeficiency virus type 1 Western blots: Seroconversion risk, specificity of supplemental tests, and an algorithm for evaluation. *J. Infect. Dis.* **164**, 656–664.

Chou, Q., Russell, M., Birch, D. E., Raymond, J., and Bloch, W. (1992). Prevention of pre-PCR mis-priming and primer dimerization improves low-copy-number amplifications. *Nucleic Acids Res.* **29**, 1717–1723.

Cimino, G. D., Metchette, K. C., Tessman, J. W., Hearst, J. E., and Isaacs, S. T., (1991). Post-PCR sterilization: A method to control carryover contamination for the polymerase chain reaction. *Nucleic Acids Res.* **19**, 99–107.

Demmler, G. J., Buffane, G. J., Schimbor, C. M., and May, R. A. (1988). Detection of cytomegalovirus in urine from newborns by using polymerase chain reaction DNA amplification. *J. Infect. Dis.* **158**, 1177–1184.

Einsele, H., Erhinger, G., Steidle, M., Vallbaracht, A., Miller, M., Schmidt, H., Saal, J. G., Waller, H. D., and Muller, G. A. (1991). Polymerase chain reaction to evlauate antiviral therapy for cytomegalovirus disease. *Lancet* **338**, 1170–1172.

Gibbs, R. A. (1990). DNA amplification by the polymerase chain reaction. *Anal. Chem.* **62**, 1202–1214.

Innis, M. A., and Gelfand, D. H. (1990) Optimization of PCR's *In* "PCR Protocols: A Guide to Methods and Applications" (D. H. Gelfand, J. S. Sninsky, and T. White, eds.), pp. 3–12 Academic Press, San Diego.

Jackson, J. B., MacDonald, K. C., Cadwell, J., Sullivan, C., Klein, W. E., Hanson, M., Sannerud, K. J., Stramer, S. L., Fildes, N. J., Kwok, S. Y., Sninsky, J. J., Bowman, R. J., Polesky, H. F., Balfour, H. H., and Osterholm, M. T. (1990). Absence of human immunodeficiency virus infection in blood donors with indeterminate Western blot tests for antibody to HIV-1. *N. Engl. J. Med.* **312**, 217–222.

Jackson, J. B., Ndugwa, C., Mmiro, F., Kataaha, P., Guay, E., Dragon, E. A., Goldfarb, J., and Olness, K. (1991). Non-isotopic polymerase chain reaction methods for the detection of HIV-1 in Ugandan mothers and infants. *AIDS* **5,** 1463–1467.

Kaneko, S., Miller, R. H., Feinstone, S. M. Unoura, M., Kobayashi, K., Hattori, N., and Purcell, R. H. (1989). Detection of serum hepatitis B virus DNA in patients with chronic hepatitis using the polymerase chain reaction assay. *Proc. Natl. Acad. Sci. USA* **86,** 312–316.

Koch, W. C., and Adler, S. P. (1990). Detection of human parvovirus B19 by using the polymerase chain reaction. *J. Clin. Microbiol.* **28,** 65–69.

Kwok, S., and Higuchi, R. (1989). Avoiding false positives with PCR. *Nature (London)* **399,** 237–238.

Kwok, S., and Sninsky, J. J. (1993). PCR detection of human immunodeficiency virus type I proviral DNA sequences. *In* "Diagnostic Molecular Microbiology: Principles and Applications" (D. H. Persing, T. Smith, F. Tenover, and T. White eds.), pp. 309–321. American Society of Microbiology, Washington, DC.

Kwok, S., Kellogg, D., Ehrlich, G., Poiesz, B., Bhagavati, S., and Sninsky, J. J. (1988). Characterization of a sequence of human T cell leukemia virus type I from a patient with chronic progressive myelopathy. *J. Infect. Dis.* **158,** 1193–1197.

Kwok, S., Kellog, D. E., McKinney, N., Spasic, D., Gods, L., Levinson, C., and Sninsky, J. J. (1990). Effects of primer template mismatches on the polymerase chain reaction: Human immunodeficiency virus type I model studies. *Nucleic Acids Res.* **18,** 999–1005.

Larzul, D., Guigue, F., Sninsky, J. J., Mack, D. H., Brechot, C., and Guesdon, J .L. (1988). Detection of hepatitis B virus sequences in serum by using in vitro enzymatic amplification. *J. Virol. Meth.* **20,** 227–237.

Lindahl, T., Ljungquist, S., Siegart, W., Nyberg, B., and Sperens, B. (1977). DNA N-glycosidases. Properties of uracil-DNA glycosidase from *Escherichia coli. J. Biol. Chem.* **252,** 3286–3294.

Longo, M. C., Berninger, M. S., and Hartley, J. L. (1990). Use of uracil DNA glycosylase to control carryover contamination in polymerase chain reactions. *Gene* **93,** 125–128.

Lucus, K., Busch, M., Mossinger, S., and Thompson, J. A. (1991). An improved microcomputer program for finding gene- or gene family-specific oligonucleotides suitable as primers for polymerase chain reactions or as probes. *Comput. Appl. Biosci.* **7,** 525–529.

Lynch, C. E., Madej, R., Louie, P., and Rodgers, G. (1992). Detection of HIV-1 DNA by PCR: Evaluation of primer pair concordance and sensitivity of a single primer pair. *J. Acq. Immune Def. Syndr.* **5,** 433–440.

Maniatis, T., Fritsch, E. F., and Sambrook, T., (eds.) (1982). "Molecular Cloning: A Laboratory Manual." Cold Spring Harbor Laboratory Press, Cold Spring Harbor, NY.

McCreedy, B. J., and Callaway, T. H. (1993). Laboratory design and workflow. *In* "Diagnostic Molecular Microbiology: Principles and Applications" (D. H. Persing, T. Smith, F. Tenover, and T. White, eds.), pp. 149–159. American Society for Microbiology, Washington, DC.

Mullis, K. B., and Faloona, F. (1987). Specific synthesis of DNA *in vitro* via a polymerase-catalyzed chain reaction. *Meth. Enzymol.* **155,** 335–350.

Ou, C.-Y., Kush, S., Mitchell, S. W., Mack, D. H., Sninsky, J. J., Krebs, J. W., Fedrino, P., Warfield, D., and Schochetman, G. (1988). DNA amplification for direct detection of HIV-1 in DNA of peripheral blood mononuclear cells. *Science* **239,** 295–297.

Ou, C.-Y., McDonough, S. H., Cabanas, D., Ryder, T. B., Harper, M., Moore, J., and Schochetman, G. (1990). Rapid and quantitative detection of enzymatically amplified HIV-1 DNA using chemiluminescent oligonucleotide probes. *AIDS Res. Hum. Retroviruses* **6,** 1323–1329.

Palumbo, P. E., Weiss, S. H., McCreedy, B. J., Alexander, S. S., Denny, T. N., Klein, C. W., and Altman, R. (1992) Evaluation of HTLV infection in a cohort of injecting drug users. *J. Infect. Dis.* **176,** 896–902.

Persing, D. H. (1993). Target selection and optimization of amplification reactions. *In* "Diagnostic Molecular Microbiology: Principles and Applications" (D. H. Persing, T. Smith, F. Tenover, and T. White, eds.), pp. 88–104. American Society for Microbiology, Washington, DC.

Persing, D. H., and Landry, M. L. (1989). *In vitro* amplification techniques for the detection of nucleic acids: New tools for the diagnostic laboratory. *Yale J. Biol. Med.* **62,** 159–171.

Rayfield, M., DeCock, K., Hayward, W., Goldstein, L., Krebs, J., Kwok, S., Lee, S., McCormik, J., Moreau, J. M., Odehouri, K., Schochetman, G., Sninsky, J., and Ou, C.-Y. (1988). Mixed human immunodeficiency virus (HIV) infection in an individual: Demonstration of both HIV type 1 and type 2 proviral sequences by using polymerase chain reactions. *J. Infect. Dis.* **158,** 1170–1176.

Rogers, M. T., Ou, C.-Y., Rayfield, M., Thomas, P. A., Schoenbaum, E. E., Abrams, E., Kraginski Moore, J., and Keel, A. (1989). Use of the polymerase chain reaction for early detection of proviral sequences of human immunodeficiency virus in infants born to seropositive mothers. New York City Collaborative Study of Maternal Transmission Study Group. *N. Engl. J. Med.* **320,** 1644–1654.

Rosenberg, F., and Lebon, P. (1991). Amplification and characterization of herpes virus DNA in cerebropsinal fluid from patients with acute encephalitis. *J. Clin. Microbiol.* **29,** 2412–2417.

Rotbart, H. A. (1990). Diagnosis of enteroviral meningitis with polymerase chain reaction. *J. Pediatr.* **117,** 85–89.

Rychlik, W., and Rhoades, R. E. (1989). A computer program for choosing optimal oligonucleotides for filter hybridization, sequencing and *in vitro* amplification of DNA. *Nucleic Acids Res.* **17,** 8543–8551.

Saiki, R. K., Scharf, S., Faloona, F., Mullis, K. B., Horn, G. T., Erlich, H. A., and Arnheim, N. (1985). Enzymatic amplification of β-globin genomic sequences and restriction site analysis for the diagnosis of sickle cell anemia. *Science* **230,** 1350–1354.

Saiki, R. K., Gelfand, D. H., Stoffel, S., Scharf, S. J., Higuchi, R., Horn, G. T., Mullis, K. B., and Erlich, H. A. (1988). Primer-directed enzymatic amplification of DNA with a thermostable DNA polymerase. *Science* **239,** 487–491.

Schiffman, M. H., Bauer, H. M., Lorincz, A. T., Manos, M. M., Byrne, J. C., Glass, A. G., Cadell, D. M., and Howley, P. M. (1991). Comparison of Southern blot hybridization and polymerase chain reaction methods for the detection of human papillomavirus DNA. *J. Clin. Microbiol.* **29,** 1183–1187.

Whetsell, A. J., Drew, J. B., and Milman, G. (1992). Comparison of three nonradiosotopic polymerase chain reaction-based methods for detection of human immunodeficiency virus type 1. *J. Clin. Microbiol.* **30,** 845–853.

White, T. J., Madej, R., and Persing, D. H. (1992). The polymerase chain reaction: Clinical applications. *Adv. Clin. Chem.* **29,** 161–196.

Wolinsky, S. M., Rinaldo, C. R., Kwok, S., Sninsky, J. J., Gupta, P., Imagawa, G., Farzadegan, H., Jacobson, L. P., Grovit, K. S., Lee, M. H., Chmiel, J. S., Ginzburg, H., Kaslow, R. A., and Phair, J. P. (1989). Human immunodeficiency virus type I (HIV-I) infection a median of 18 months before a diagnostic Western blot. *Ann. Intern. Med.* **111,** 961–972.

Yi, T., and Manos, M. M. (1990). Detection and typing of genital human papillomaviruses. *In* "PCR Protocols: A Guide to Methods and Applications." (M. A. Innis, D. H., Gelfand, J. J. Sninsky, and T. J. White, eds.), pp. 356–367. Academic Press, San Diego.

Young, K. Y., Resnick, R. M., and Myers, T. W. (1993). Detection of hepatitis C virus RNA by a combined reverse transcription polymerase chain reaction assay. *J. Clin. Microbiol.* **31,** 882–886.

Yourno, J. (1992). A method for nested PCR with single closed reaction tubes. *PCR Meth. Appl.* **2,** 60–65.

Quantification of RNA Targets Using the Polymerase Chain Reaction

Francois Ferre, Patrick Pezzoli, Eric Buxton, Chris Duffy, Annie Marchese, and Anne Daigle

The Immune Response Corporation
Carlsbad, California 92008

I. INTRODUCTION

To the mathematician, the polymerase chain reaction (PCR) is merely a challenging mathematical model that may be described as a "branching process" (Nedelman *et al.*, 1991). To the molecular archeologist, PCR is nothing short of a revolution. But what is PCR to the virologist? To paraphrase Kary Mullis, its inventor (Mullis and Faloona, 1987), "PCR is to the molecular biologist (virologist) what a screwdriver is to the handyman, an indispensable tool." PCR, first applied to the diagnosis of sickle cell anemia (Saiki *et al.*, 1985), rapidly found its way into virology laboratories where it was used to detect a number of human viruses such as the human immunodeficiency virus (HIV-1; Kwok *et al.*, 1987), cytomegalovirus (CMV: Seto and Yen, 1987), and human papillomavirus (HPV; Shibata *et al.*, 1988).

Currently, PCR is the method of choice for the detection of viral nucleic acids present in small amounts in biological samples. In recent years, major improvements in the sensitivity and specificity of PCR methodology have permitted routine diagnosis of viral infections with a high degree of accuracy in clinical laboratories (Jackson, 1993; Ferre, 1994). In 1988, coinciding with its maturation as a diagnostic tool, PCR emerged as a potentially powerful quantitative technology (Abbott *et al.*, 1988; Chelly *et al.*, 1988; Davidson *et al.*, 1988; Rappolee *et al.*, 1988).

In virology, quantitative evaluation of viral target sequences is necessary in two major applications: (1) to decipher complex pathogenesis problems and (2) to monitor the efficacy of antiviral therapies. In pathogenesis, the goal is to obtain an estimate of viral burden that is as close as possible to the real value, also referred to as absolute quantification. Our current struggle to understand the role of viral load in the complex pathogenesis of HIV-1 infection provides a vivid example of the dimensions of the challenge. An accurate estimation of the amount of HIV-1 that is present in an infected individual would definitely shed light on the disease process. Furthermore, quantitative information on the level of viral gene expression in a variety of tissues is crucial to demonstrating the direct role of HIV-1 in tissue pathogenicity. Although achieving absolute quantification while monitoring therapy is the ultimate goal, precise assessment of viral load over the course of a clinical trial is usually adequate.

Most current knowledge about the formulation of quantitative PCR assays to address these challenges is derived from HIV studies. Furthermore, since our own experience is also in the quantitative analysis of HIV targets, a great deal of our discussion will focus on this specific area. Even more specifically, we will limit the scope of this chapter to RNA target quantification by PCR (for recent reviews on DNA quantification in virology using PCR, see Ferre, 1992,1994; Ferre *et al.*, 1994). We first review the different PCR methods for viral RNA quantification. A discussion follows on the ability of these technologies to provide absolute quantification for pathogenicity studies and/or precise quantification for therapeutic monitoring. Finally, we present our own work consisting of the development and validation of a quantitative HIV-1 RNA PCR assay for the monitoring of anti-HIV-1 immunotherapy.

II. PCR METHODS FOR VIRAL RNA QUANTIFICATION

One of the first attempts to quantify viral RNA using PCR was published in 1989 by Arrigo and co-workers, who attempted to evaluate the level of expression of a number of HIV-1 RNA species (Arrigo *et al.*, 1989). The first hurdle these investigators had to overcome was the need to increase PCR sensitivity so that low abundance HIV-1 RNA species could be detected without increasing the cycle number. Indeed, to remain in the linear range of amplification in which the amount of amplified target is proportional to the initial concentration of target molecules, it is imperative to limit the number of cycles. To accomplish this goal, Arrigo et al. pioneered the use of end-labeled oligonucleotide primers during PCR. End-labeling the primers, and thereby directly labeling the PCR products, eliminated the need for nucleic acid hybridization and reduced the number of PCR cycles. A number of primer pairs also needed to be tested to ensure low background amplification levels. Since no standard was used in these early attempts, the quantitative ability of PCR was judged by the linear relationship between the initial amount of HIV-1 RNA target input and the amount of amplified product. Arrigo et al. reported that the signal produced by RNA PCR were quantitative across a 2000-fold range of RNA concentration. These data demonstrated that low abundance HIV-1 RNA species could be quantified by PCR. In addition, an important legacy of this approach was the direct use of labeled primer in PCR, which has become a central part of many subsequent quantitative PCR RNA methods.

Currently, however, this type of quantitative analysis would be considered rather crude because it carries very little quantitative information. For instance, no information was available on the level of precision and/or accuracy of the PCR evaluation. Nonetheless, major differences (10-fold and over) in RNA content can be demonstrated using this or closely related methodologies. Using a method containing a probing step, striking differences in the levels of HIV-1 *gag* RNA were consistently observed between peripheral blood mononuclear cells (PBMCs) and lymphoid tissue mononuclear cells (LTMCs) in all patients analyzed, regardless of the stage of disease (Pantaleo *et al.*, 1993). As presented here, subsequent attempts to refine these basic technologies have been made by numereous groups, to address the questions of precision and accuracy.

A. Quantification of Viral RNA Using an Internal Control

For the vast majority of current methods, quantification of RNA by PCR requires two enzymes: a reverse transcriptase (RT) to convert RNA into cDNA and *Taq* polymerase to amplify the cDNA. However, a thermostable DNA polymerase (*Tth*) from the thermophilic bacterium *Thermus thermophilus* has been isolated that is capable of performing both reverse transcription and DNA amplification (Myers and Gelfand, 1991). The impact of this development for the future of RNA quantification is discussed subsequently.

Quantifying RNA from reactions that use RT has been a real challenge. RT enzymes are notorious for their poor processivity, which in turn provides poor reproducibility for cDNA synthesis. Specifically, researchers have reported that the efficiency of the cDNA synthesis step is variable, ranging from 5% to 90% (Noonan *et al.*, 1990; Simmonds *et al.*, 1990; Henrard *et al.*, 1992). Furthermore, Gilliland *et al.* (1990) have shown that the amplification yield from a cDNA fragment can vary as much as 6-fold among duplicate reactions. The lack of reproducibility in the RT step prompted the use of an internal RNA control that was co-reverse-transcribed and co-amplified with the target RNA.

Endogenous RNAs such as β_2-microglobulin or ribosomal RNA have been used extensively for the relative quantification of cellular RNA but, to our knowledge, not for the quantification of viral RNA. One advantage of this type of RNA control is that it provides an indication of the integrity as well as the quantity of the extracted total RNA. However, a major disadvantage of this approach is that it can only be used for the quantification of RNA in cells and cannot be used when testing plasma or serum specimens, explaining why this approach was not suitable for a number of virological investigations in which plasma and serum are important sources of virus.

The concept of using an exogenous internal control for the quantification of RNA was introduced by Chelly and co-workers (1988) for the evaluation of the level of transcription of the dystrophin gene in human muscle and non-muscle tissues. The exogenous RNA control, which was different in nature from the target, was added to the reaction, co-reverse-transcribed, and co-amplified with the target RNA. One advantage of this PCR format over the endogenous control is that body fluids such as plasma can be used for viral RNA studies. This approach was adopted by Aoki-Sei *et al.* (1992) for the quantification of plasma HIV-1 viremia using a known amount of rabbit globin mRNA that was co-amplified with the HIV-1 RNA *gag* sequences. In addition, the amount of control RNA can be adjusted to the amount of target RNA to avoid any competition between the two reactants. Finally, the exogenous control represents a more consistent source of standard because, in contrast to endogenous RNA, it is not subject to temporal variation in expression. However, this method does not provide an internal control for the quantity and integrity of the RNA material.

Further refinements of the "co-amplification of exogenous RNA" approach have been reported. To eliminate potential differences in the amplification efficiencies of the target and internal controls, researchers have proposed that the same set of primers should be used and that both RNA sequences should be as similar as possible. Internal standards that differ from the target by the substitution of a single base have been utilized for this purpose (Becker-Andre and Hahlbrock, 1989). Efficient methods for the generation of synthetic control RNA (cRNA), which is very close in composition to the target sequence, have been published (Simon *et al.*, 1992; Vanden Heuvel *et al.*, 1993). Most of the recently developed quantitative PCR RNA procedures have used a cRNA molecule as a control and/or a standard. The cRNA concept has been used in two different co-amplification PCR formats.

The first co-amplification format or so-called competitive PCR approach (cRT-PCR) is considered by many investigators to be the ultimate tool for viral RNA quantification (see review by Clementi *et al.*, 1993). In this method, the target RNA is co-amplified with serial dilutions of the synthetic cRNA control and both amplicons are separated by gel electrophoresis. The competitive cRNA can be of different length than the target sequence (Scadden *et al.*, 1992; Piatak *et al.*, 1993a) or it can have a mutation that creates a new restriction site (Becker-Andre and Hahlbrock, 1989; Stieger *et al.*, 1991). However, the variable efficiencies of enzyme digestions make it preferable to use sequences that can be separated directly by gel electrophoresis. To quantify the amount of RNA target, at least four or five cRNA dilutions are made that encompass the expected range of target RNA concentration. A target RNA:cRNA ratio of 1 obtained from the amplified products reflects the initial concentration of both template species, regardless of the total amounts produced by the amplification reaction. Thus, the method does not require that the quantification be done in the exponential phase of the co-amplification reaction. This result implies that the sensitivity of this method can be boosted by simply increasing the number of cycles (up to 40 cycles) and that simple detection/quantification methods such as densitometric evaluation of ethidium bromide-stained gels can be used.

However, one major drawback of this procedure is that it usually requires at least six amplification reactions per sample and therefore lacks practicality. The use of a single competitive cRNA dilution results in an imprecise value when the two RNA species are present in different amounts. Serial dilutions to generate a target RNA:cRNA ratio close to 1 are needed for precise analysis. Finally, note that the formation of heteroduplexes between the cRNA and target amplicons has been reported (Becker-Andre and Hahlbrock, 1989; Gilliland *et al.*, 1990; Piatak *et al.*, 1993a); in reactions using endonuclease digestons, this amplification artifact can affect the calculation of transcript concentration by a factor of 3. The cRT-PCR format is a promising tool for absolute quantification of viral RNA molecules and may play a major role in understanding the pathogenesis of HIV-1 infections. However, this approach is cumbersome and has limited value when monitoring the effectiveness of antiviral therapies in hundreds of specimens. Furthermore, only scarce data are available on the precision and reproducibility of this method (Becker-Andre and Hahlbrock, 1989).

Despite these limitations, cRT-PCR has been used successfully for the evaluation of viral load in research and clinical settings. Kaneko and co-workers (1992) used cRT-PCR to monitor hepatitis C virus (HCV)-infected patients receiving interferon-α therapy. Cross-sectional studies involving HIV-1-infected individuals at different disease stages have confirmed the correlation between increases in HIV-1 RNA load and disease progression (Bagnarelli *et al.*, 1992; Piatak *et al.*, 1993b).

The second cRNA co-amplification format, developed by Wang *et al.* (1989) for the quantification of interleukins, capitalizes on the fact that the PCR primers have the same efficiency when amplifying both the target and the control sequences during the exponential phase of the reaction. When the starting concentration of cRNA is known, the concentration of target mRNA can be determined by extrapolating against a standard curve generated from the co-amplification of a serial dilution of cRNAs. This procedure also requires the amplification of four to five control dilutions per sample. As does cRT-PCR, this approach seems to be overly cumbersome and has limited applicability, particularly in a clinical setting.

However, once the equivalency of the amplification efficiencies for the target and control nucleic acids has been established, a single time point per sample is sufficient for quantitative analysis (see Section III; Powell and Kroon, 1992; Ferre *et al.*, 1994; Mulder *et al.*, 1994). Because quantification is performed during the exponential phase or just after a plateau is reached (typically 25 to 33 cycles), a range (2 to 4 orders of magnitude depending on the assay) of RNA target concentrations exists in which the target RNA:cRNA ratio precisely reflects the concentration of target RNA in the sample (Ferre *et al.*, 1994; Mulder *et al.*, 1994). One disadvantage of this approach is that very sensitive detection methods are needed to identify the smaller amounts of PCR product generated from the fewer cycles required to ensure linear amplification. This PCR format has been used to quantify HIV-1 *tat* mRNA in PBMCs (Furtado *et al.*, 1993) and to study acute HIV-1 infection in plasma (Mulder *et al.*, 1994). The assay developed by Mulder and co-workers is depicted in Fig. 1. Finally, we have used a similar method to quantify HIV-1 *gag* mRNA in PBMCs and to monitor the effectiveness of immunotherapeutic treatments. A detailed protocol and validation for this approach are presented later in this chapter.

B. Quantification of Viral RNA Using an External Control

In 1991, Thomas Merigan's group developed a quantitative HIV-1 RNA PCR assay for plasma and serum that used an HIV-1 cRNA external standard (Holodniy *et al.*, 1991a). This method was a radical departure from accepted protocols because, at the time, an internal control was mandatory. A refined and validated version of this approach has been published (Winters *et al.*,

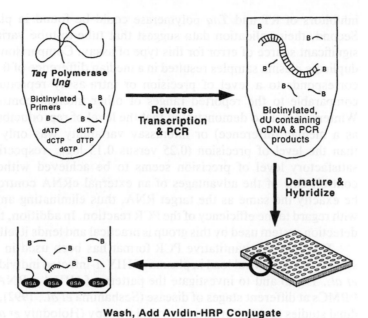

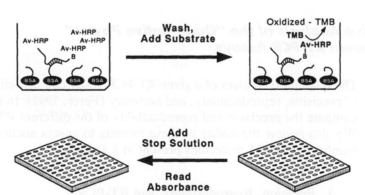

Figure 1 The Roche polymerase chain reaction test format. Tth, *Thermus thermophilus*; Ung, uracil-*N*-glycosylase; dU, deoxyuridine; BSA, bovine serum albumin; HRP, horseradish peroxidase; AV, avidin; TMB, tetramethylbenzidine.

1993). As previously mentioned, the rationale for using an internal control was linked to the belief that RT-PCR was inherently variable on a tube-to-tube basis and that an internal standard would ultimately control for the presence of PCR inhibitors. Interestingly, Winters *et al.* addressed those two concerns and found solutions to them. First, these investigators reported that no data have been presented that indicate that other than heparin, inhibitors of RT and *Taq* polymerase could be found in plasma samples. Second, their validation data suggest that tube-to-tube variability is not a significant source of error for this type of assay. Comparisons of 363 sets of duplicate plasma samples resulted in a median difference of $0.15 \log_{10}$, which corresponds to a level of precision or intra-assay reproducibility that is comparable to the reported ranges of other internal control approaches. Winters *et al.* also demonstrated that the level of reproducibility (expressed as a median difference) or interassay variability was only slightly lower than the level of precision (0.25 versus $0.15 \log_{10}$, respectively). Thus, a satisfactory level of precision seems to be achieved without an internal control. One of the advantages of an external cRNA control is that it can be exactly the same as the target RNA, thus eliminating any possible bias with regard to the efficiency of the PCR reaction. In addition, the colorimetric detection system used by this group is practical and lends itself to automation.

This type of quantitative PCR format has been used in cross-sectional studies to assess viral load in plasma of HIV-1-infected individuals (Holodniy *et al.*, 1991a) and to investigate the pattern of HIV-1 mRNA expression in PBMCs at different stages of disease (Seshamma *et al.*, 1992), and in longitudinal studies to monitor anti-retroviral therapy (Holodniy *et al.*, 1991b; Winters *et al.*, 1992).

C. Comparative Analysis of the "Quantitative Power" of the Different RT-PCR Assays

The quantitative power of a given RT-PCR assay can be defined by its level of precision, reproducibility, and accuracy (Ferre, 1992). In this section, we compare the precision and reproducibility of the different RT-PCR formats. We also review the ability of these formats to assess accurately the actual numbers of HIV-1 molecules present in a sample.

1. Precision, Reproducibility, and RT-PCR

One question represents an interesting point of contention in quantitative RT-PCR circles: Is an internal control mandatory for maximum precision and/or reproducibility?

In 1992, Masters *et al.* reported that RNA quantity measurements [usually based on optical density (OD) at 260 and 280 nm] are major sources of error when quantifying the amount of RNA target. These investigators and others (Murphy *et al.*, 1990) advocated using an endogenous control that is amplified in a separate reaction vessel, instead of an internal control, to normalize the amount of amplified target to the total RNA content. Other groups emphasized that the two amplicons could potentially interfere with each other in the co-amplification format (Noonan *et al.*, 1990; Horikoshi *et al.*, 1992). Finally, as previously discussed in this chapter, HIV-1 RNA quantification using a cRNA external standard produced an adequate level of precision for monitoring anti-retroviral therapy. Nonetheless, for reasons discussed next, using an internal control still might increase the precision. Because no control for the quantity and integrity of the RNA is possible when quantifying in plasma or serum, the addition of an internal control to the plasma sample prior to extraction makes it possible to boost the precision of the test by normalizing the differences in extraction efficiencies between samples (S. Kwok, personal communication). We have also attempted to control for differences in extraction efficiencies of HIV-1 RNA in plasma samples by adding a known amount of ribosomal RNA to the sample and then assessing the postextraction yield by OD readings. In this way, a fixed amount (1 μg) of extracted ribosomal RNA from each sample can be added to the reaction. However, this procedure is rather cumbersome and a gain in precision has not been proven.

The second and stronger point in favor of using a cRNA internal control is that, in our hands, the co-amplification format does yield more precision. We have shown in seven independent experiments that, after normalization with the co-amplified control, there is a significant gain in the precision of the assay relative to normalization without co-amplification (Table 1). However, as shown in Fig. 2, the variance in the amount of target amplicon is equivalent with or without co-amplification. Not only is the variability between replicates in each PCR generally lower in the co-amplification format, but the level of precision achieved for each PCR is also more reproducible (Table 1). The internal control might eliminate the high variability between replicate samples seen on occasion in the external standard format (Winters *et al.*, 1993). Thus, the internal standard assay format should not only be more precise but also more robust. As reported in detail in Section III, we have shown that a 2.2-fold increase in HIV-1 RNA copy number between two consecutive samples run in the same PCR assay was significantly greater than the variation due to experimental error, and therefore represented a significant change in copy number.

Currently, two enzymes, RT and *Taq* polymerase, are required to perform PCR on RNA substrates. The use of an enzyme such as *Tth* DNA polymerase, which is capable of performing both reverse transcription and amplification in one tube, might add precision to the assay. Since this enzyme is thermostable and thermoactive, reverse transcription can be performed at elevated temperatures, thereby increasing the specificity of the assay (Mulder *et al.*, 1994). In addition, reverse transcription through GC-rich regions with stable secondary structures would be facilitated at higher temperatures, thereby increasing the reproducibility of this step.

TABLE 1
Impact of the HIR Internal Control on Assay Precision

	HIV adjustment			
	Separate amplification		Co-amplification	
PCR assay number	Mean RNA copy number (SD)	Mean fold difference (SD)	Mean RNA copy number (SD)	Mean fold difference (SD)
1	1270 (181)	1.3 (0.2)	1266 (44)	1 (0.1)
2	1608 (586)	2 (0.2)	1538 170	1.3 (0.1)
3	2038 (420)	1.5 (0.1)	1932 (220)	1.3 (0.1)
4	827 (17)	1 (0)	862 (17)	1 (0)
5	1699 (449)	1.4 (0.4)	1395 (412)	1.6 (0.4)
6	1390 (248)	1.3 (0.2)	1379 (25)	1 (0)
7	621 (265)	2 (0.5)	565 (107)	1.3 (0.2)
Mean		1.5 (0.4)		1.2 (0.2)
Range		1–2		1–1.6

To date, the high interassay variability and poor reproducibility of RT-PCR assays have hampered real-time RNA quantification. With current improvements that are leading to higher assay robustness, it might be possible to perform credible real-time quantification. The assay developed by Kwok and presented throughout this chapter is certainly a prime candidate for such a task. Winters *et al.* (1993) also presented evidence that real-time analysis can be conducted using the external control format. Nonetheless, to perform real-time quantitative analysis in situations in which it will be tremendously valuable, such as monitoring therapy over several years, more data must be accumulated. Additionally, logistical concerns arise for commercial applications that must be addressed to ensure the consistency of all reagents over a long period of time.

Finally, Nedelman and co-workers (1992) developed a mathematical model of PCR, the multiple branching-process model, that was used to compare the level of precision of the three PCR assay formats discussed in this chapter (i.e., no control, cRNA external control, and cRNA internal control methods). These investigators reported that when the three methods are compared under assumptions that are maximally optimistic for each, the

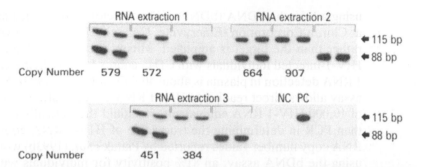

Figure 2 Assessment of the precision of the HIV-1 RNA PCR assay. Three HIV-1 RNA extractions were performed (HIR controls were from the same source); each PCR was assayed in duplicate as follows: co-amplification of HIV-1/HIR (*left*), HIV-1 RNA amplification alone (*middle*), HIR amplification alone (*right*). The copy number without co-amplification, calculated from the middle duplicate, was obtained by dividing the cpm from the duplicated HIV-1 target bands by the mean of the cpm from the three duplicated HIR control bands. PC, HIV-1 DNA positive control; NC, negative control.

cRNA internal control method was superior to the other two methods because of its insensitivity to nonexponential growth (in the competitive cRNA format) and interexperimental variability.

2. Absolute Quantification and RT-PCR

To achieve absolute quantification with the RT-PCR format, access to a quantifiable RNA standard is imperative. At this stage the cRNA molecule becomes handy. When synthesized *in vitro,* the cRNA molecule can be sized by gel electrophoresis and the purity and quantity of the RNA can be determined by OD readings. The accurate measurement of the OD of the cRNA control is obviously essential. Thus, in theory, any method using this type of RNA control should be capable of absolute quantification. Because several groups have used cRNA controls for HIV-1 quantification, one might expect that a consensus opinion would emerge for determining the true value of HIV-1 RNA load in different tissues and at different disease stages. At this time, however, no consensus opinion has emerged. Piatak *et al.* (1993b), using a competitive RT-PCR format, were the first to report absolute quantification of HIV-1 RNA in plasma. Their estimation, which was orders of magnitude higher than previously reported values (Holodniy *et al.* 1991b; Aoki-Sei *et al.,* 1992), caused a significant amount of controversy. In light of other studies using different RT-PCR methods or other techniques of HIV-1 RNA quantification, there still seems to be a great deal of discrepancy over the actual values of HIV-1 RNA in plasma and/or PBMCs. For instance, using a branched DNA (bDNA) assay developed by Urdea and co-workers at Chiron Corporation (Emeryville, CA; see Chapter 6), in which the signal rather than the target is amplified, Mitsuya and others have shown that in HIV-1-infected individuals with $CD4^+$ counts (>500, the percentage of HIV-1 RNA detection in plasma is about 40% (Todd et al., 1993). Since the bDNA assay allows direct reading of HIV-1 RNA concentration with a cutoff value of 10,000 HIV-1 RNA equivalents, it should theoretically be more accurate than PCR in determining the true value of HIV-1 RNA copy number. The RNA copy number values reported by Piatak et al. (1993b) would have given, using the bDNA assay, an 81% positivity for individuals with $CD4^+$ count >500 (CDC Stage II or III). Winters et al. (1993), using the RT-PCR external standard format, have reported that in patients with $CD4^+$ counts >500, only about 25% had RNA values above 10,000 copies. We have found similar results in plasma of asymptomatic patients; 4 of 16 patients (25%) demonstrated >10,000 RNA copies per milliliter of plasma. Finally, Mulder and co-workers (1994) reported that the RNA values obtained from samples of seroconverting patients were up to 100-fold less than those reported by Piatak et al. (1993b). Is this result sufficient to infer that Piatak et al. overestimated the RNA load in plasma? Not exactly, since none of the other assays fully demonstrated that they were capable of absolute quantification. Mulder et al. (1994) mentioned that by optimizing the sample extraction protocol and recalibrating their standards, they achieved a 20- to 50-fold increase in RNA copy number. The point is that no one knows the true level of HIV-1 RNA in plasma, and claims of absolute quantification should be better substantiated.

What strategies must be implemented to put this important issue to rest (see also Ferre et al., 1994)? As a first step, comparison studies using a number of quantitative methods with well-defined sensitivities (e.g., different RT-PCR formats, the bDNA assay, an RNase protection assay, and Northern blotting)should be performed in parallel by different laboratories in an effort to assess the true HIV-1 RNA copy number in plasma. This type of analysis will hopefully curtail the inflationary trend in HIV-1 RNA load, which has already passed the 10^9 HIV-1 RNA copy/ml mark (Baumberger et al., 1993).

An important point to consider when using cRT-PCR for absolute quantification is that the competitor cRNA can be amplified more efficiently than the target RNA (Horikoshi et al., 1992; Volkenandt et al., 1992). For instance, differences in secondary structures between the RNAs can potentially lead to significant differences in the efficiency of the RT step. Therefore, a target RNA:cRNA ratio of 1 may not actually represent equal concentrations of the competing RNA segments. Therefore, because of this potential flaw, one could argue that the ultimate PCR tool for absolute quantification is the cRNA external control approach. Indeed, since the control cRNA molecules can be engineered to be exact copies of the target molecule, they should amplify with the same efficiency. Thus, using the latest technical refinements, such as *Tth* DNA polymerase, and the right cRNA PCR format, it may soon be possible to achieve absolute quantification.

III. DEVELOPMENT AND VALIDATION OF AN RT-PCR ASSAY FOR THE PRECISE QUANTIFICATION OF HIV-1 RNA IN THE PERIPHERAL BLOOD MONONUCLEAR CELLS OF INFECTED INDIVIDUALS

In this section, we describe a PCR method for the quantification of HIV-1 RNA in PBMCs of infected individuals. This quantitative method is based on a co-amplification format using a synthetic cRNA internal control (HIR) that contains several HIV primer sequences. Typically, known amounts of HIR cRNA (1000 copies) are added to the sample containing unknown amounts of HIV-1 RNA, co-reverse-transcribed, co-amplified, and then used to calculate the amount of HIV-1 RNA in the sample. To optimize the amplification reaction, we also modified the acid guanidinium thiocyanate/phenol/chloroform RNA extraction method (Chomczynski and Sacchi, 1987) by replacing the ethanol precipitation steps with an RNA capture on glass beads. This modification increases the speed of the extraction method and the purity of the RNA. The quantitative ability of this assay was tested on PBMCs infected with serial dilutions of infectious HIV-1 particles. A good correlation between the number of RNA molecules as ascertained by PCR and the levels of HIV p24 antigen in the supernatant of the culture was established. This quantitative PCR method is linear over a 2 $\log_{10}$ range of input template (50–2 $\times$ 10^4 copies). To address the intra-assay variability or precision, 30 sets of triplicate RNA isolations and RT-PCRs were run. We found that an increase in copy number of 120% or a decrease of 54% was a significant change. The interassay variability or reproducibility was also assessed; a minimum increase of 297% or a minimum decrease of 75% was significant. This assay was able to detect changes in HIV-1 RNA load in HIV-1-infected asymptomatic individuals enrolled in a double-blind adjuvant-controlled HIV-1 vaccine trial.

A. Materials and Methods

1. Construction of *gag* cRNA Standard

Two oligonucleotides—HIR1 (89-mer) and HIR2 (90-mer)—with a 30-nucleotide overlap were annealed at high stringency (72°C) and extended with *Taq* polymerase to generate a double-stranded piece of DNA of 149 bp long. The synthetic HIR DNA was inserted into the *Bam*HI/*Hind*III sites of the SP65(poly A) transcription vector (Promega, Madison, WI). *Eco*RI was used to linearize the plasmid, which then served as a template for transcription by SP6 polymerase according to the manufacturer's transcription protocol. The HIR RNA product of 209 bases [149 bases from the insert + 60 bases from the plasmid including poly(A) tail] was purified by agarose gel electrophoresis and extracted from the gel using RNaid glass beads (Bio 101, San Diego). The RNA was then treated with RNase-free DNaase (Promega), phenol/chloroform extracted, ethanol precipitated, and suspended in water. Careful 260/280 OD readings (triplicate measurements on different dilutions) were performed for quantification and the integrity of the pruified HIR RNA was ascertained on a 2% agarose gel. Yeast tRNA was added as a carrier to the control RNA (HIR) transcript, which was then stored as an ethanol precipitate.

2. Purification of RNA

PBMCs were isolated by Ficoll–Hypaque and washed. The chronically HIV-1-infected Hut-78 cell line HB2 was obtained from Hillcrest Biologicals (Cypress, CA). The RNA purification protocol was a modified version of the technique developed by Chomczynski and Sacchi (1987). Typically, 6×10^6 PBMCs were lysed in 0.7 ml 4 M guanidinium thiocyanate denaturing solution (solution D). At this point, samples may be kept frozen in solution D at $-70°C$ for at least 1 yr. Extraction was conducted as previously described (Chomczynski and Sacchi, 1987) with the following modifications: two extra chloroform extractions were performed and on recovery of 0.65–0.7 ml original aqueous phase, 7 μl RNaid glass beads were added and then mixed. On average, 1 μl beads captures 1 μg RNA. Binding was allowed at room temperature for 10 min with continuous mixing. The beads were lightly spun down to avoid difficult resuspension and washed twice with 200 μl RNA wash solution (Bio 101), followed by resuspension of the beads each time. The beads were then dried, resuspended in 30 μl diethylpyrocarbonate (DEPC)-treated water, eluted at 56°C for 3 min, and spun down full speed in a microcentrifuge for 10 min.

From the first spin 25 μl RNA solution was recovered that was spun again for 10 min to ensure that the beads would not interfere with OD readings. Eluted total cellular RNA was quantified by OD readings at 260/280. All RNA preparations were treated with 1 U RQ1 DNase (Promega) for 30 min at 37°C. The DNase was inactivated by heat denaturation at 95°C for 5 min. The sample was cooled on ice and spun in a microcentrifuge briefly to recover the condensate. The RNA was then adjusted to a final concentration of 1 μg /10 μl with DEPC-treated water.

3. *In Vitro* Infection of PBMCs

In this procedure, 6×10^7 phytohemagglutinin (PHA) stimulated PBMCs were infected with serial dilutions of cell-free supernatant fluid from chronically HIV-1-infected HB2 cells containing 5,000, 500, 50, and 5 infectious units/ml. Cells were washed with phosphate-buffered saline (PBS) and transferred to T-75 flasks with 30 ml growth media. Each day for 5 days, 6×10^6 cells were pelleted and lysed in solution D for RNA purification. Uninfected PBMCs were maintained as a negative control. Supernatant fluid was assayed for p24 antigen concentration using the Coulter p 24 antigen capture enzyme-linked immunosorbent assay (ELISA).

4. Amplification Method

a. cDNA Synthesis   HIR RNA and total cellular RNA were reverse transcribed in the same tube. A 30-μl reverse transcription reaction mixture containing 1 μg total cellular RNA, 1 μg HIR3 (1000 copies of HIR RNA), $1 \times$ RT buffer (10 mM Tris-HCl, pH 8.3, 50 mM KCl, 2.5 mm$M$ MgCl$_2$), 40 U RNasin (Promega), 0.5 mM each dNTP, 5 pmol SK39 *gag* primer, and 24 U avian myeloblastosis virus (AMV) reverse transcriptase (8 U/μl; Promega) was incubated at 42°C for 30 min. The reaction was terminated and the RNA–cDNA hybrid was separated by incubating the tubes at 98°C for 3 min and then quickly chilling on ice.

b. Amplification   A 50-μl amplification reaction mixture containing 5 μl cDNA, $1 \times$ PCR buffer, 0.25 mM each dNTP, 10 pmol ^{32}P-labeled SK38, and 50 pmol SK39 was covered with mineral oil and heated to 95°C in a dry bath under a laminar flow hood (all samples were run in duplicate). Then 5 μl diluted *Taq* polymerase (5 U) was added to each tube and the heating block was transferred to a Perkin-Elmer (Norwalk, CT) thermocycler preheated to 95°C. Amplification was initiated with cycles consisting of 1 min at 95°C and 2 min at 65°C for up to 33 cycles. Then 20 μl of the reaction, with loading dye, was loaded onto a 10% acrylamide gel. Gels were run until sufficient separation of target and control bands was achieved (about 1.5 hr at 280 V). Gels were dried and analyzed with the AMBIS Radioanalytic Imaging System.

c. Data Analysis and Quantification   Two identical rectangles are drawn on the AMBIS screen, one for the HIR amplicon (88 bp) and one for the HIV-1 target amplicon (115 bp). A separate large rectangle encompassing the full width of the gel below the HIR rectangle is also drawn for background reading (the AMBIS system simultaneously images the radioactive bands and quantifies the cpm inside the different rectangles). The HIV-1 RNA *gag* copy number is obtained with the following simple equation: [(cpm HIV target − cpm background)/(cpm HIR3 control − cpm background)] × 10^3 = no. *gag* HIV-1 RNA molecules/μg total cellular RNA. The calculated copy number should be considered a relative value (relative to the HIR3 control) rather than an absolute value. Further studies are necessary to ascertain that the HIR3 RNA copy number is effectively 1000 copies in absolute terms. Positive and negative controls must be run in each experiment. A seronegative sample and a water sample serve as negative controls. An HIV RNA positive sample and an HIV-1 positive DNA sample serve as positive controls. RNA samples are also amplified without reverse transcription to check for HIV-1 DNA contamination of the RNA extraction.

B. Results

1. Synthesis of the HIR RNA

We used the approach developed by Wang *et al.* (1989) to design the HIR internal standard. The HIR RNA served the dual purpose of being an internal RNA control in the reverse transcription step and allowing quantification in the subsequent PCR step. The HIR DNA was cloned into the SP6 transcription vector (pSP6-HIR) and was used to produce run-off HIR transcripts. The PCR product from the HIR RNA using the SK38/SK39 primer set is 88 bp and is designed to differ significantly in size from the 115-bp PCR product of the HIV-1 target RNA. The same primer set is used in the PCR amplification of both templates. As shown in Fig. 3B, no significant differences exist between the efficiencies of amplification of the RNA control and the target RNA.

Micrograms of highly purified HIR RNA were obtained and carefully quantified by OD readings (see Section III,A). The quantified HIR RNA was then diluted in yeast tRNA to a final concentration of 10^3 HIR RNA molecules per μg yeast tRNA and labeled HIR3. Subsequently 10-μg aliquots of HIR3 RNA were precipitated in ethanol and kept at −70°C. Unlimited quantities of control RNA can be generated using this method. When needed, precipitated HIR3 RNA can be pelleted, dried, resuspended in DEPC-treated water, and adjusted to 1 μg/10 μl after performing 260/280 OD readings.

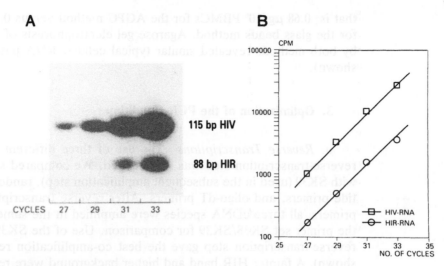

Figure 3 (A) AMBIS scan of a kinetic analysis of HIV-1/HIR co-amplification. (B) Plot of the co-amplification's kinetics.

2. RNA Isolation from PBMCs

The method of RNA isolation by acid guanidinium thiocyanate-phenol-chloroform extraction (AGPC) has been widely used for the recovery of total cellular RNA from small quantities of tissue or cells (Chomczynski and Sacchi, 1987). We developed a modified version of the AGPC method that is faster (1.5 hr instead of 4 hr), and easier to perform when dealing with small amounts of RNA. The two extra chloroform extractions that were added increased RNA purity 2-fold; the 260/280 ratios are typically 1.8 to 1.9. After the organic extractions, the aqueous phase containing the RNA is incubated with RNaid glass beads, for purification rather than being ethanol precipitated. Substitution of the precipitation step with the glass beads gives the greatest reduction in preparation time. The beads are then washed and the RNA is eluted in water at 56°C for 3 min (see Section III,A). Total RNA was isolated from 6×10^6 unstimulated PBMCs by the glass beads method and compared with RNA isolated, in parallel, by the AGPC method. RNA isolated by the glass beads method contained fewer contaminants, as judged by the 260/280 ratio, than RNA isolated by the AGPC method. In four separate experiments, the mean 260/280 ratio obtained by the AGPC method was 1.76 ± 0.03; the ratio did not change even after an additional wash with 75% ethanol. In contrast, when using the glass beads procedure, the 260/280 ratio was reproducibly in the range of 1.8–2.0 with a mean value of 1.92 ± 0.04. This method should allow for more accurate quantification of the total RNA. RNA yields obtained with both techniques were equivalent, that is, 0.68 μg/10^6 PBMCs for the AGPC method versus 0.60 μg/10^6 cells for the glass beads method. Agarose gel electrophoresis of RNAs isolated by both methods revealed similar typical cellular RNA patterns (data not shown).

3. Optimization of the PCR Condition

a. Reverse Transcriptions The use of three different primers in the reverse-transcription step was investigated. We compared specific priming with SK39 (used in the subsequent amplification step), random hexanucleo-tide primers, and oligo-dT primers. After reverse transcription with these primers, all three cDNA species were amplified in the same manner using the primer set SK38/SK39 for comparison. Use of the SK39 primer in the reverse-transcription step gave the best co-amplification results (data not shown). A fainter HIR band and higher background were reproducibly obtained by random priming whereas a weaker HIV amplified product was consistently obtained with the oligo-dT primers relative to SK39-driven reverse transcription. Different concentrations of SK39 ranging from 1 to 70 pmol were then assayed. In this range of primer concentrations, HIV/HIR RNA were amplified to comparable levels; lower primer concentrations (1–5 pmol) resulted in lower backgrounds (data not shown). Unless indicated otherwise, we subsequently used 5 pmol SK39 primer for the reverse-transcription step.

b. Amplification Mullis and co-workers previously showed that starting the PCR amplification at a high temperature (80°C) can increase the sensitivity of the reaction as well as reduce the background level quite dramatically (Faloona *et al.*, 1990; Mullis, 1991). We applied this principle to our quantitative RNA assay on blood samples and found that a lower background is clearly obtained by starting the amplification at 80°C instead of room temperature (data not shown). At 80°C, the stringency of the amplification greatly reduces the capability of creating nonspecific PCR fragments by limiting the occurrences of mismatch annealing and extension of the 5′ primer (SK38).

4. Quantitative Analysis

This quantitative approach was first tried on RNA isolated from HIV-1-infected HB2 cells diluted in PBMCs. In this reaction, 1 ng HB2 RNA/μg PBMC RNA was reverse transcribed with 10^3 molecules of HIR RNA per μg yeast tRNA (see Section III,A). Aliquots containing one-fifth of the cDNA mixture each were subjected to 24, 26, 28, 30, and 32 cycles of amplification. Reaction products were resolved by gel electrophoresis; the gels were dried and subjected to autoradiography (Fig. 3A). The amount of radioactivity from the HIV and HIR bands was quantified using the AMBIS System and was plotted as a function of the number of cycles (Fig. 3B). The rates of amplification were exponential between 24 and 30 cycles for both templates. The efficiencies of amplification were calculated from the slopes of these curves and were found to be 100% between 24 and 28 cycles. A 25% decrease in efficiency was observed for both templates between the 28th and 30th cycles. Between the 30th and 32nd cycles, the rates of amplification decreased significantly to below 50% efficiency. Interestingly, the amplification efficiency was the same for both targets throughout the reaction sequence, that is, the amount of HIV-1 RNA in 1 ng HB2 RNA can be calculated by comparison with the HIR3 RNA control even at 32 cycles. The calculated amount of HIV-1 RNA in 1 ng HB2 total RNA was 7.93×10^3 molecules. (Note: the RNA copy numbers reported throughout Section III of this chapter should be considered relative values.)

5. Quantitative Analysis of HIV-1 Infection *in Vitro*

To assess the quantitative ability of the PCR method, the level of RNA synthesized over the course of an HIV-1 infection was monitored and compared with the level of p24 antigen produced in the supernatant of infected cells. PHA-stimulated PBMCs were infected with serial dilutions of supernatant fluids of chronically HIV-1-infected HB2 cells. After infection, PBMCs were harvested daily, pelleted, and lysed in solution D for PCR RNA analysis. The supernatant fluids were assayed for p24 antigen concentration using the Coulter ELISA. As shown in Fig. 4, a good correlation was established between the amount of *gag* RNA in the infected PBMCs and the level of p24 antigen in the supernatant. Interestingly, on the day of infection, the approximately 10-fold differences observed in the amount of RNA (i.e., 9600, 1100, 60, and 6) parallel very closely the 10-fold differences in the input of virus into the culture (i.e., 5,000, 500, 50, and 5 IFU/ml). This result suggests that the quantitative range of this assay is close to 3 $\log_{10}$.

The fact that on the second day of infection the amount of *gag* RNA dropped sharply in all four dilutions of virus could be a reflection of the life cycle of the virus. On infection, HIV-1 RNA must be reverse transcribed to a double-stranded DNA that must integrate into the host genome before the synthesis of new genomic RNA molecules can be initiated. In the 5 IFU/ml titer culture, a sharp rise in *gag* RNA production preceded and then paralleled a sharp rise in p24 antigen production.

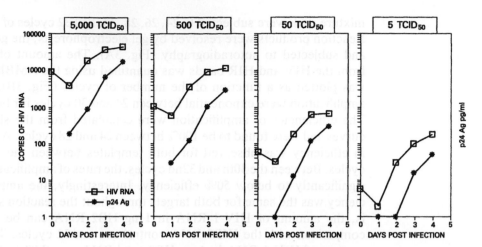

Figure 4   HIV-1 infection of peripheral blood mononuclear cells (PBMCs) in culture: comparison of HIV-1 *gag* RNA and p24 Ag expression. Day 0 indicates the first day of infection.

6. Validation of the HIV-1 RNA PCR Assay

To validate this quantitative PCR method, we have evaluated a number of analytical parameters such as sensitivity, specificity, precision, and reproducibility. We have previously defined this terminology as it applies to quantitative PCR methods (Ferre, 1992).

a. Sensitivity and Specificity   To evaluate the sensitivity of the assay, the best reference standard available was a chronically HIV-1-infected cell line (HB2) diluted into seronegative PBMCs. However, because the amount of HIV-1 RNA produced by this cell line is unknown, the sensitivity can only be described in terms of the number of HB2 cells necessary to identify a positive signal. The assay can precisely quantify to the equivalent of 0.6 HB2 cells diluted in 6×10^6 PBMCs. When translated into HIV-1 RNA molecules/μg total RNA, and relative to the HIR control, this result amounts to a mean of 136 ± 41 HIV-1 RNA molecules. This mean value was calculated from five independent PCR experiments in which samples were run in triplicate. The assay can also quantify with less proportionality to the equivalent of 0.06 HB2 cells in 6×10^6 PBMCs (44 ± 10 RNA copies; $n = 25$). The specificity of the assay was assessed by analyzing numerous seronegative samples from healthy individuals, of which none showed a positive HIV-1 RNA signal (data not shown).

b. Precision and within-Assay Variability   To determine the within-assay variability, which also examines the repeatability of replicates, HB2 dilutions in PBMCs and samples from HIV-1-infected individuals were assayed. In three duplicate runs, 30 sets of samples were prepared and analyzed to determine the within-assay variability (data not shown). A one-way analysis using sample as a factor was performed to estimate measurement (experimental) variability, which calculates the standard deviation ($\sigma = 0.236 \log_n$). A very conservative assessment of the assay's limits in terms of precision was obtained using 95% tolerance intervals, which are equivalent to $\sqrt{2} \times 2.3 \times \sigma$ (3.3σ). (Note: most people use confidence limits, which are less stringent.) This assessment of limits not only uses tolerance limits (2.3σ) but also takes into account the fact that the goal is to compare two consecutive samples. To take into consideration the experimental variability of assessing statistically significant changes over time between two clinical samples, the variability of each sample is accounted for by the factor $\sqrt{2}$. Using such a rigorous definition of the assay's limits, a 119% increase or a 54% decrease in copy number becomes necessary to record a significant change in HIV-1 RNA copy number between two samples.

   c. Reproducibility and between-Assay Variability   To evaluate the between-assay variability, aliquots from nine samples were run in five different PCR assays. Using the 95% tolerance interval, we showed that a 297% increase or a 75% decrease in RNA copy number represented a significant change in HIV-1 RNA values.

7. Application to the Monitoring of Therapy

   This validated assay has been used to assess changes in HIV-1 RNA copy numbers over time in HIV-1-infected asymptomatic subjects ($n = 103$) enrolled in a double-blind adjuvant-controlled HIV-1 vaccine trial. An example of the raw PCR data (autoradiogram) for a given patient is presented in Fig. 5. For each individual, numerous samples collected at baseline and over the course of a 1-yr study were analyzed in a single PCR assay (batch analysis; see Fig. 5). The fact that a concomitant and significant drop in the intensity of both HIV-1 and HIR RNA occurred in one of the duplicates of bleed number 28 is reflective of the role for which the HIR control was meant, that is, to compensate for the loss in HIV-1 RNA cpm due to changes in the PCR conditions. Each individual received three doses of inactivated HIV-1 immunogen in incomplete Freund's adjuvant (IFA) or IFA alone for the control group at 3-mo intervals. These individuals had a low HIV-1 RNA load (mean of 150 copies relative to cRNA control; range of 0–2500 copies) at baseline. A regression analysis of the slopes followed by a one-tailed weighted ANOVA of the HIV-1 RNA copy number, evaluated during the course of the study in both groups, revealed differences that favored the

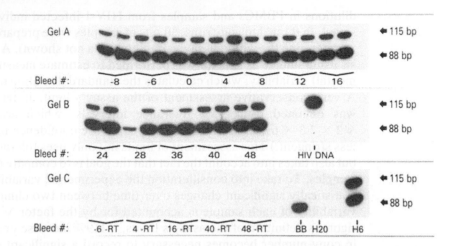

Figure 5 Example of an autoradiogram of the HIV-1 RNA PCR analysis obtained for a given patient in the trial. Bleed # −8, −6, and 0 represent baseline values. Five samples were assayed without reverse transcription to control for the presence of residual DNA (gel C: −6 −RT, 4 −RT, 16 −RT, 40 −RT, and 48 −RT). BB, Blood bank negative control RNA; H₂O, water control; H6, HB2 positive control RNA.

immunogen-treated group ($p = 0.04$) (Moss *et al.*, 1994). In other words, the rate of increase in HIV-1 RNA copy numbers over time was lower in the treated group than in the IFA control group.

IV. GENERAL CONCLUSION

Throughout this chapter we have discussed the merits of a number of PCR assays and their ability to quantify RNA targets. We have also mentioned a number of applications in research and clinical settings.

The more recently developed quantitative PCR assay based on the cRNA control and/or standard is potentially capable of evaluating the true value of HIV-1 RNA copy number. In addition to providing a powerful tool to decipher the role HIV-1 directly plays in AIDS pathogenesis, assessing the absolute number of HIV-1 RNA target should be very useful in monitoring progression of the disease. For instance, data suggest the existence of a threshold in HIV-1 RNA expression that is capable of predicting clinical outcome years before changes in CD4⁺ counts are recorded. Indeed, using an external standard cRNA PCR assay, Saksela *et al.* (1994) found that a temporal increase in the amount of HIV-1 RNA in PBMCs correlated with disease progression. More importantly, these investigators showed that the quantification of HIV-1 RNA in PBMCs has a prognostic value independent of CD4$^+$ count. Individuals with low initial HIV-1 RNA copies ($<10^3/\mu$g total RNA) had a normal CD4$^+$ count for at least 5 yr. In contrast, and irrespective of whether CD4$^+$ counts were normal at the time of sampling, individuals with high levels of HIV-1 expression (10^3 to $>10^5$ copres/μg total RNA) demonstrated accelerated disease progression within the next 2 yr. This type of information could potentially be very important for the clinician when deciding on appropriate therapeutic interventions. However, to be broadly useful, this information should be translatable in terms of absolute quantification, this is currently not possible. In other words, this hypothetical threshold is only relative to the cRNA standard described by Saksela and co-workers, so there is no guarantee that another group with another cRNA assay format would have found similar HIV-1 RNA copy numbers. Nonetheless, such studies reinforce the current thinking that HIV-1 viral load in PBMCs represents an important marker of disease progression, and therefore should prove helpful in assessing anti-HIV-1 therapies.

The increasing importance of viral RNA assessments for monitoring antiviral therapy is demonstrated by the rapid accrual of publications on the subject in recent years. This trend is also reflected by the continuous effort to refine assays to yield more precise and robust quantitative results. Finally, as anti-HIV-1 therapies become initiated in healthier individuals (Luque *et al.*, 1994, Trauger *et al.*, 1994); the need for PCR assays capable of quantifying low levels of HIV-1 RNA with adequate precision will increase. In Section III, we have described a validated assay that was designed to quantify low level expression of HIV-1 RNA in the PBMCs of healthy asymptomatic individuals receiving immunotherapeutic treatment. We have reported that this treatment, aimed at augmenting HIV-specific immunity, significantly affected the rate of increase in HIV-1 RNA load and favored the treated group (Moss *et al.*, 1994). Interestingly, a similar trend was observed when the HIV-1 DNA load in PBMCs was assessed in this control trial by a quantitative PCR assay (Trauger *et al.*, 1994). There has been no validation that treatment-induced changes in virus burden are associated with clinical benefit. However, as reviewed previously in this chapter, it is now apparent that virus burden, as estimated by quantitative PCR methodologies, should be valuable as a correlate of disease progression.

REFERENCES

Abbott, M. A., Poiesz, B. J., Byrne, B. C., Kwok S., Sninsky, J. J., and Ehrlich, G. D. (1988). Enzymatic gene amplification: Qualitative and quantitative methods for detecting proviral DNA amplified in vitro. *J. Infect. Dis.* **158,** 1158–1169.

Aoki-Sei, S., Yarchoan, R., Kageyama, S., Hoekzema, D. T., Pluda, J. M., Wyvill, K. M.,

Broder, S., and Mitsuya, H. (1992). Plasma HIV-1 viremia in HIV-1 infected individuals assessed by polymerase chain reaction. *AIDS Res. Hum. Retroviruses* **7**, 1263–1270.

Arrigo, S. J., Weitsman, S., Rosenblatt, J. D., and Chen, I. S. Y. (1989). Analysis of rev gene function on human immunodeficiency virus type 1 replication in lymphoid cells by using a quantitative polymerase chain reaction method. *J. Virol.* **63**, 4875–4881.

Bagnarelli, P., Menzo, S., Valenza, A., Manzin, A., Giacca, M., Ancarani, F., Scalise, G., Varaldo, P. E., and Clementi, M. (1992). Molecular profile of human immunodeficiency virus type 1 infection in symptomless patients and in patients with AIDS. *J. Virol.* **66**, 7328–7335.

Baumberger, C., Kinloch-de-Loes, S., Yerly, S., Hirschel, B., and Perrin, L. (1993). High levels of circulating RNA in patients with symptomatic HIV-1 infection. *AIDS* **7**, S59–S64.

Becker-Andre, M., and Hahlbrock, K. (1989). Absolute mRNA quantification using the polymerase chain reaction (PCR). A novel approach by a PCR aided transcript titration assay (PATTY). *Nucleic Acids Res.* **17**, 9437–9446.

Chelly, J., Kaplan, J. C., Maire, P., Gautron, S., and Kahn, A. (1988). Transcription of the dystrophin gene in human muscle and non-muscle tissues. *Nature* **333**, 858–860.

Chomcynski, P., and Sacchi, N. (1987). Single-step method of RNA isolation by acid guanidinium thiocyanate–phenol–chloroform extraction. *Anal. Biochem.* **162**, 156–159.

Clementi, M., Menzo, S., Bagnarelli, P., Manzin, A., Valenza, A., and Varaldo, P. E. (1993). Quantitative PCR and RT-PCR in virology. *PCR Meth. Appl.* **2**, 191–196.

Davidson, N. O., Powell, L. M., Wallis, S. C., and Scott, J. (1988). Thyroid hormone modulates the introduction of a stop codon in rat liver apolipoprotein B messenger RNA. *J. Biol. Chem.* **263**, 13482–13485.

Faloona, J., Weiss, S., Ferre, F., and Mullis, K. (1990). Direct detection of HIV sequences in blood: High-gain polymerase chain reaction. VIth International Conference on AIDS, San Francisco (Abstract # 1019).

Ferre, F. (1992). Quantitative or semi-quantitative PCR: Reality versus myth. *PCR Meth. Appl.* **2**, 1–9.

Ferre, F. (1994). Polymerase chain reaction and HIV. In "Clinics of Laboratory Medicine; HIV–AIDS Issue" (R. Pomerantz, ed), pp. 313–334. Saunders, Philadelphia.

Ferre, F., Marchese, A., Pezzoli, P., et al., (1994). Quantitative PCR: An overview. In "The Polymerase Chain Reaction" (K. B. Mullis, F. Ferre, and R. Gibbs, eds.), pp. 67–88. Birkhauser, Boston-Basel-Berlin.

Furtado, M. R., Balachandran, R., Gupta, P., and Wolinsky, S. M. (1993). Analysis of alternatively spliced human immunodeficiency virus type-1 mRNA species, one of which encodes a novel tat-env fusion protein. *Virology* **185**, 258–270.

Gilliland, G., Perrin, S., and Bunn, H. F. (1990). Competitive PCR for quantitation of mRNA. In "PCR Protocols" (M. A. Innis, D. H. Gelfand, J. J. Sninsky, and T. J. White, eds.), pp. 60–69. Academic Press, San Diego.

Henrard, D. R., Mehaffey, W. F., and Allain, J. P. (1992). A sensitive viral capture assay for detection of plasma viremia in HIV infected individuals. *AIDS Res. Hum. Retroviruses* **8**, 45–52.

Holodniy, M., Katzenstein, D. A., Sengupta, S., Wang, A. M., Casipit, C., Schwartz, D. H. Konrad, M., Groves, E., and Merigan, T. C. (1991a). Detection and quantification of human immunodeficiency virus RNA in patient serum by use of the polymerase chain reaction. *J. Infect. Dis.* **163**, 862–866.

Holodniy, M., Katzenstein, D. A., Israelski, D. M., and Merigan, T. C. (1991b). Reduction in plasma human immunodeficiency virus ribonucleic acid after dideoxynucleoside therapy as determined by the polymerase chain reaction. *J. Clin. Invest.* **88**, 1755–1759.

Horikoshi, T., Danenberg, K. D., Stadbauer, T. H. W., Volkenandt, M., Shea, L. C. C., Aigner, K., Gustavsson, B., Leichman, L., Frosing, R., Ray, M., Gibson, N. W., Spears, C. P., and Danenberg, P. V. (1992). Quantitation of thymidylate synthase, dihydrofolate reductase, and DT-diaphorase gene expression in human tumors using the polymerase chain reaction. *Cancer Res.* **52**, 108–116.

Jackson, J. B. (1993). Detection and quantitation of human immunodeficiency virus type 1 using molecular DNA/RNA technology. *Arch. Pathol. Lab. Med.* **117**, 473.

Kaneko, S., Murakami, S., Unoura, M., and Kobayashi, K. (1992). Quantitation of hepatitis C virus RNA by competitive polymerase chain reaction. *J. Med. Virol.* **37**, 278–282.

Kwok, S., Mack, D. H., Mullis, K. B., Poiesz, B., Ehrlich, G., Blair, D., Friedman-Kien, A., and Sninsky, J. J. (1987). Identification of human immunodeficiency virus sequences by using in vitro enzymatic amplification and oligomer cleavage detection. *J. Virol.* **61**, 1690–1694.

Luque, F., Caruz, A., Pineda, J. A., Torres, Y., Larder, B., and Leal, M. (1994). Provirus load changes in untreated and zidovudine-treated human immunodeficiency virus type 1-infected patients. *J. Infect. Dis.* **169**, 267–273.

Masters, D. B., Griggs, C. T., and Berde, C. B. (1992). High sensitivity quantification of RNA from gels and autoradiograms with affordable optical scanning. *BioTechniques* **13**, 902–911.

Moss, R. B., Ferre, F., Trauger, R., Jensen, F., Daigle, A., Richieri, S. P., and Carlo, D. J. (1994). Inactivated HIV-1 immunogen: Impact on markers of disease progression. *J. AIDS* **7**, 521–527.

Mulder, J., McKinney, N., Christopherson, C., Sninsky, J., Greenfield, L., and Kwok, S. (1994). Rapid and simple PCR assay for quantitation of human immunodeficiency virus type 1 RNA in plasma: Application of acute retroviral infection. *J. Clin. Microbiol.* **32**, 292–300.

Mullis, K. B. (1991). The polymerase chain reaction in an anemic mode: How to avoid cold oligodeoxyribonuclear fusion. *PCR Meth. Appl.* **1**, 1–4.

Mullis, K. B., and Faloona, F. (1987). Specific synthesis of DNA in vitro via a polymerase catalysed chain reaction. *Meth. Enzymol.* **155**, 335–350.

Murphy, L. D., Herzog, C. E., Rudick, J. B., Fojo, A. T., and Bates, S. E. (1990). Use of the polymerase chain reaction in the quantitation of mdr-1 gene expression. *Biochemistry* **29**, 10351–10356.

Myers, T. W., and Gelfand, D. H., (1991). Reverse transcription and DNA amplification by a *Thermus thermophilus* DNA polymerase. *Biochemistry* **30**, 7661–7666.

Nedelman, J., Heagerty, J. P., and Lawrence, C. (1992). Quantitative PCR: Procedures and precision. *Bull. Math. Biol.* **54**, 477–502.

Noonan, K. E., Beck, C., Holzmayer, T. A., Chin, J. E., Wunder, J. S., Andrulis, I. L., Gazdar, A. F., Willman, C. L., Griffith, B., Von Hoff, D. D., and Roninson, I. B. (1990). Quantitative analysis of MDR1 (multidrug resistance) gene expression in human tumors by polymerase chain reaction. *Proc. Natl. Acad. Sci. USA* **87**, 7160–7164.

Pantaleo, G., Graziosi, C., Demarest, J. F., Butini, L., Montroni, M., Fox, C. H., Orenstein, J. M., Koller, D. P., and Fauci, A. S. (1993). HIV infection is active and progressive in lymphoid tissue during the clinically latent stage of disease. *Nature* **362**, 355.

Piatak, M. Jr., Luk, K., Williams, B., and Lifson, J. D. (1993a). Quantitative competitive polymerase chain reaction for accurate quantitation of HIV DNA and RNA species. *BioTechniques* **14**, 70–80.

Piatak, M. Jr., Saag, M. S., Yang, L. C., Clark, S. J., Kappes, J. C., Luke, K.-C. Hahn, B. H., Shaw, G. M., and Lifson, J. D. (1993b). High levels of HIV-1 in plasma during all stages of infection determined by competitive PCR. *Science* **259**, 1749–1754.

Powell, E. E., and Kroon, P. A. (1992). Measurement of mRNA by quantitative PCR with nonradioactive label. *J. Lipid Res.* **33**, 609–614.

Rappolee, D. A., Mark, D., Banda, M. J., and Werb, Z. (1988). Wound macrophages express TGF-α and other growth factors in vivo: analysis by mRNA phenotyping. *Science* **241**, 708–712.

Saiki, R. K., Scharf, S., Faloona, F., Mullis, K. B., Horn, G. T., Erlich, H. A., and Arnheim,

N. (1985). Enzymatic amplification of β-globin genomic sequences and restriction site analysis for diagnosis of sickle cell anemia. *Science* **230,** 1350–1354.

Saksela, K., Stevens, C., Rubinstein, P., and Baltimore, D. (1994). Human immunodeficiency virus type 1 mRNA expression in peripheral blood cells predicts disease progression independently of the numbers of CD4$^+$ lymphocytes. *Proc. Natl. Acad. Sci. USA* **91,** 1104–1108.

Scadden, D. T., Wang, Z., and Groopman, J. E. (1992). Quantitation of plasma human immunodeficiency virus type 1 RNA by competitive polymerase chain reaction. *J. Infect. Dis.* **165,** 1119–1123.

Seshamma, T., Bagasra, O., Trono, D., Baltimore, D., and Pomerantz, R. J. (1992). Blocked early-stage latency in the peripheral blood cells of certain individuals infected with human immunodeficiency virus type 1. *Proc. Natl. Acad. Sci. USA* **89,** 10663–10667.

Seto, E., and Yen, T. S. B. (1987). Detection of cytomegalovirus infection by means of DNA isolated from paraffin-embedded tissues and dot hybridization. *Am. J. Pathol.* **127,** 409–413.

Shibata, D., Fu, Y. S., Gupta, J. W., Shah, K. V., Arnheim, N., and Martin, W. J. (1988). Detection of human papillomavirus in normal and dysplastic tissue by the polymerase chain reaction. *Lab. Invest.* **59,** 555–559.

Simmonds, P., Balfe, P., Peutherer, J. F., Ludlam, C. A., Bishop, J. O., and Leigh Brown, A. J. (1990). Human immunodeficiency virus infected individuals contain provirus in small numbers of peripheral mononuclear cells and at low copy numbers. *J. Virol.* **64,** 864–872.

Simon, L., Levesque, R. C., and Lalonde, M. (1992). Rapid Quantitation by PCR of endomycorrhizal fungi colonizing roots. *PCR Meth. Appl.* **2,** 76–80.

Stieger, M., Demolliere, C., Ahlborn-Laake, L., and Mous, J. (1991). Competitive polymerase chain reaction assay for quantitation of HIV-1 DNA and RNA. *J. Virol. Method.* **34,** 149–160.

Todd, J., Stempien, M., Kojima, E., Shirasaka, T., Mitsuya, H. and Urdea M. (1993). Quantitation of HIV plasma RNA using the branched DNA (bDNA) and reverse transcription-coupled polymerase chain reaction (RT-PCR) assays. IXth International Conference on AIDS, Berlin (Abstract #PO-B41-2482).

Trauger, R., Ferre, F., Daigle, A. E., Jensen, F. C., Moss, R. B., Mueller, S. H., Richieri, S. P., Slade, H. B., and Carlo, D. J. (1994). Effect of immunization with inactivated envelope-depleted human immunodeficiency virus type 1 (HIV-1) immunogen in incomplete Freund's adjuvant on HIV-1 immunity, viral DNA, and percentage of CD4 cells. *J. Infect. Dis.* **169,** 1256–1264.

Vanden Heuvel, J. P., Tyson, F. L., and Bell, D. A. (1993). Construction of recombinant RNA templates for use as internal standards in quantitative RT-PCR. *Bio Techniques* **14,** 395–398.

Volkenandt, M., Dicker, A. P., Banerjee, D., Fanin, R., Schweitzer, B., Horikoshi, T., Danenberg, K., Danenberg, P., and Bertino, J. R. (1992). Quantitation of gene copy number and mRNA using the polymerase chain reaction. *Proceedings of the Society for Experimental Biology and Medicine* **200,** 1–6.

Wang, M., Doyle, M. V., and Mark, D. F. (1989). Quantitation of mRNA by the polymerase chain reaction. *Proc. Natl. Acad. Sci. USA* **86,** 9717–9721.

Winters, M. A., Holodniy, M., Katzenstein, D. A., and Merigan, T. C. (1992). Quantitative RNA and DNA gene amplification can rapidly monitor HIV infection and antiviral activity in cell culture. *PCR Meth. Appl.* **1,** 257–262.

Winters, M. A., Tan, L. B., Katzenstein, D. A., and Merigan, T. C. (1993). Biological variation and quality control of plasma human immunodeficiency virus type 1 RNA quantitation by reverse transcriptase polymerase chain reaction. *J. Clin. Microbiol.* **31,** 2960–2966.

Multiplex Polymerase Chain Reaction

James B. Mahony and Max A. Chernesky
McMaster University Regional Virology and Chlamydiology Laboratory
St. Joseph's Hospital
Hamilton, Ontario, Canada L8N 4A6

I. INTRODUCTION

In a few short years, the polymerase chain reaction (PCR) has had a major impact on how clinical laboratories diagnose infectious diseases. PCR has been used for the diagnosis of infections caused by several noncultivatable viruses including human papillomavirus (HPV), parvovirus B19, and BK/JC papovaviruses. Arguably, the greatest impact of PCR has been in the field of retrovirology, where it has been used to detect infections in seronegative

219

individuals. Quantitative PCR (see Chapter 9), is also proving useful in evaluating the efficacy of new antiviral agents in clinical trials.

Multiplex PCR (M-PCR) is the simultaneous detection of more than one target sequence. The first reports appearing in the literature that used M-PCR involved the diagnosis of inherited genetic diseases. The majority of M-PCR studies have involved the detection of mutations in the dystrophin gene and the cystic fibrosis transmembrane regulatory gene (Chamberlain *et al.*, 1988,1990; Fortina *et al.*, 1992; Kilimann *et al.*, 1992; Ko *et al.*, 1992; Morrall and Estivill, 1992; Picci *et al.*, 1992; Richards *et al.*, 1993). A computer search of the medical literature (Medline: Mesh Headings, PCR and multiplex, September 1993) turned up 112 papers on M-PCR of which only 11 addressed the detection of infectious agents. The development of multitarget co-amplification, or M-PCR, has not yet had an impact on clinical laboratories, but as more virologists explore the tremendous potential of this technique, M-PCR will assume a major role in the diagnostic virology laboratory.

II. METHODOLOGY

A. Principle

M-PCR is the simultaneous amplification of more than one target sequence in a single reaction tube using more than one primer pair. This co-amplification of two or more targets in a single reaction is dependent on the compatibility of the PCR primers used in the reaction. All primers in the reaction must have similar melting temperatures (T_m) so they anneal to and dissociate from complementary DNA sequences at approximately the same temperatures, allowing each amplification to proceed at the selected temperature. This procedure could not be done if one primer set was annealing at the time that another primer set was dissociating from its target. Therefore, all primers must be selected so their T_ms are within a few degrees (°C) of each other. Each amplification proceeds independently of the others (as long as none of the reagents is present at rate-limiting concentrations) and each specific amplification product or amplicon is synthesized in an unencumbered way. Primers should also be chosen that define amplicons of approximately the same size range (100–500 bp), so each is synthesized efficiently and at equal rates. Each M-PCR assay must also have a detection step capable of identifying each amplicon. This can be done by gel electrophoresis with visual identification of separate amplicons of different size or by hybridization with specific DNA probes and detection using spectrophotometry, fluorometry, autoradiography, or chemiluminescence.

B. Applications

The first infectious disease application of M-PCR was for the detection of HPV using either degenerate primers or consensus primers. Gregoire *et al.* (1989) used degenerate primers containing deoxyinosine at the variable base locations and showed that these consensus primers could detect all the HPV genotypes tested. Manos *et al.* (1989) also used degenerate primers, targeted to the L1 gene instead of the E1 gene used by Gregoire *et al.*, to detect several HPV genotypes recovered from the cervix. More recently, Vandervelde *et al.* (1992) used six pairs of HPV primers targeted to the E7 region to detect dysplastic changes in cervix tissue samples from Belgian women. Jullian *et al.* (1993) used HPV 16- and HPV 18-specific primers in an M-PCR assay to detect HPV in cervix tissue samples of women with normal cytology. Human T-lymphotropic virus (HTLV) types I and II have been detected in peripheral blood using type I- and type II-specific primers for three different genes (*env, pol,* and *tax*) in an M-PCR assay described by Wattel *et al.* (1992). Two different bloodborne viruses, human immunodeficiency virus type 1 (HIV-1) and hepatitis C virus (HCV), have been detected by M-PCR using primers for the HIV-1 *gag* region and the cloned C-100 NS-3/4 region of HCV, respectively (Nedjar *et al.*, 1991). Cytomegalovirus (CMV) and Epstein–Barr virus (EBV) have been detected in preserved paraffin sections of lung tissue from immunocompromised patients (Burgart *et al.*, 1992); more recently, M-PCR has been used to test paraffin-embedded small bowel tissues from patients with celiac disease for adenovirus type 12, CMV, and herpes simplex virus (HSV) (Vesey *et al.*, 1993). Our laboratory has developed an M-PCR assay for the detection of *Chlamydia trachomatis* and *Neisseria gonorrhoeae* in genitourinary specimens collected from men with urethritis (Mahony *et al.*, 1993b). Table 1 summarizes the viruses and primers used in selected studies.

C. Strengths and Weaknesses

The obvious advantage of M-PCR is the ability to detect more than one agent in a single test. For specimens such as respiratory tract secretions, from which several different viruses can be recovered, this ability offers potential cost savings. The second advantage of M-PCR is its high degree of sensitivity and ability to detect both noncultivatable virus and neutralized virus present in antigen–antibody complexes. As technology advances, target quantification, which is now working its way into commercial PCR tests, will also be incorporated into M-PCR assays for several agents. Applications of quantitative PCR are still evolving, but the major ones will be in monitoring individual

TABLE 1
Selected Papers Using M-PCR for the Detection of Viruses and Bacteria[a]

Agents[b]	Target	Amplicon size (bp)	Primers	Sequence (5' to 3')	Reference
HPV (consensus)	L1	145	(GP5)	TTTGTTACTGTGGTAGATAC	Snijders et al. (1990)
			(GP6)	TGATTTACAGTTTATTTTC	
	L1		(GP11)	TTTATCACAGTGGTAGACAA	
			(GP12)	TGAAATTTCTTTATATTAC	
HIV-I	gag	114	(SK38)	ATAATCCACCTATCCCAGTAGGAGAGAAAT	Nedjar et al. (1991)
			(SK39)	TTTGGTCCTTGTCTTATGTCCAGAATGC	
HCV	C-100	660	(SN01)	TCAATGCCGTGGCCTACTAC	
			(SN04)	TTCCCTGTCAGGTATGATTG	
			(SN02)	TTGGAATTCAGGCAATACGGGGCCAAAGCA	
			(SN03)	TTGTCTTAGATCGATTTCAGCCTTGA	
CMV		159	IE Exon 4	GTTCGAGTGGACATGCTGCGGCATAG	Burgart et al. (1992)
				CATCTCCTGAAAGGCTCATGAACCT	
EBV		245	IR3 repeat	TGTCTGACGAGGGCAGGTACAGGAC	
				GCAGCCAATGCAACTTGGACGTTTTTC	
HTLV I	env	467	—	CTCCCTTCTAGTGACGCTCCAGG	Wattel et al. (1992)
HTLV II	env		—	GCCACCGGTACCGCTCGGCGGGGAG	
HTLV I	pol	142	(SK110)	CCCTACAATCCAACCAGCTCAG	
			(SK111)	GTGGTGAAGCTGCCATCGGGTTTT	
HTLV II	pol	142	(SK110)	CCATACAACCCCACCAGCTCAG	
			(Sk111)	GTGGTGGATTTGCCATCGGGTTTT	

Organism	Target	Size	Primer	Sequence	Reference
HTLV I	tax	91	(SK43)	CGGATACCCAGTCTACGTGT	
			(SK44)	GAGCCGATAAACGCGTCCATCG	
HTLV II	tax	91	(SK43)	TGGATACCCGTCTACGTGT	Vesey et al. (1993)
			(SK44)	GAGCTGACAACGCGTCCATCG	
Adenovirus 12	275		—	TGTGCTGATGGAAATTGTCAT	
			—	ACGATGTCGAGTATCATCATT	
CMV	435		—	CCAAGGGCCTCTGATAACCAAGCC	
			—	CAGCACCATCCTCCTCTTCCTCTGG	
HSV	92		—	CATCACCGACCCGGAGAGGGAC	
			—	GGGCCAGGCGCTTGTTGGTGTA	
HPV 16	E6	407	16-1	TTCAGGACCCACAGGAGCGA	Jullian et al. (1993)
			16-2	CATACAATGACCGGTCCACC	
HPV 18	E6	389	18-1	GCTACCTGATCTGTGCACGG	
			18-2	TGCAGCACGAATGGCACTGG	
C. trachomatis	Plasmid	241	KL1	TCCGGAGCGAGTTACGAAGA	Mahony et al. (1993b)
			KL2	AATCAATGCCCGGGATTGGT	
N. gonorrhoeae	Plasmid	390	HO1	GCTACGCATACCCGCGTTGC	
			HO3	CGAAGACCTTCGAGCAGACA	
HPV	L1	450	MY9	GCACCAGGGATCATAACTAATGG	
			MY11	CGTCCACAGAGAGGGAATACTGATC	

[a] See also Manos et al. (1989) for detection of multiple HPV genotypes using degenerate primers MY9/MY11 and Chamberlain et al. (1990) and Richards et al. (1993) for M-PCR assays for the detection of mutations within the Duchenne muscular dystrophy and cystic fibrosis transmembrane conductance regulator gene.

[b] Abbreviations: HPV, human papillomavirus; HIV, human immunodeficiency virus; HCV, hepatitis C virus; CMV, cytomegalovirus; EBV, Epstein–Barr virus; HTLV, human T-cell lymphotropic virus; HSV, herpes simplex virus.

patient response to treatment and in evaluation of new drugs and therapeutic regimes. The weaknesses of M-PCR are relatively minor in nature. Cost, which was once the major drawback due to expensive thermal cyclers, is now decreasing in importance as more laboratories acquire instruments and competition plays an expanding role in the market. Other weaknesses of M-PCR are similar to those of conventional PCR: the considerable time required to develop and evaluate new assays and the need for effective anticontamination measures. To some degree, these issues are related to the "learning curve," as each individual laboratory becomes proficient in PCR, these drawbacks will become less significant.

III. DEVELOPING A MULTIPLEX PCR ASSAY

A. Primers

The first step in developing an M-PCR assay is choosing primers that have similar conditions for optimal amplification. Optimizing the conditions for M-PCR (discussed later) involves setting conditions for each primer pair, combining the primers in a single tube, and re-evaluating the reaction conditions to obtain maximum sensitivity for the cocktail. When choosing primers, it would be foolish to choose pairs with widely different optimal $MgCl_2$ concentrations (e.g., 1.5 mM and 4.5 mM) since it would be difficult to find a concentration of $MgCl_2$ at which both amplifications would proceed optimally. More importantly, the annealing temperatures for the primer sets should be similar. Primers should be selected that have similar GC contents and melting temperatures (T_m) so a single annealing temperature can be found that works optimally for both primer sets. Obviously M-PCR would not work very well if the T_ms of the two primer sets were 45°C and 65°C, since no one temperature could be used for annealing without some nonspecific annealing of the second primer set (usually the one with the higher T_m, in this case, 65°C). The presence of four oligonucleotide primers in a reaction instead of the usual two presents a greater opportunity for primer dimer formation, so it is imperative that there be no base pair homology between any of the primers. Primers should not have more than three repeating bases and should have no sequence complementarity with other primers, especially at the 3' end. The GC content should be 40–60% and the T_m should be 50–60°C. If primers are rich in G content, 5°C may have to be added to the T_m for complete denaturation. We have had the best success with T_ms near 55°C. Often when an individual primer pair has a T_m of 55°C, inclusion of a second primer pair with the same T_m requires few if any changes in amplification conditions. Our M-PCR for HPV, *C. trachomatis,* and *N. gonorrhoeae* was designed using primers with individual T_ms of 55°C, so the annealing step of the M-PCR assay could be done at 55°C. This multiplex assay is capable of detecting 1 fg of each target sequence using 35 cycles of 1 min denaturation at 94°C, 1 min annealing at 55°C, and 2 min extension at 72°C.

B. Optimizing Conditions

Determining the optimal conditions for an M-PCR assay is similar to determining the optimal conditions for a solid-phase antibody enzyme-linked immunosorbent assay (ELISA) by checkerboard titration. Basically, optimal conditions for each individual reagent are determined in combination with other reagents. We typically utilize a five-member sensitivity panel to determine the optimal $MgCl_2$ concentration, then repeat the experiment using various primer and dNTP concentrations. When all other conditions have been optimized, we raise the annealing temperature stepwise until the maximum sensitivity and specificity (usually when there are minimal nonspecific amplification products) are obtained. If the individual PCRs are carefully constructed and the two primer pairs have a similar T_m (a difference of <5°C), there is a good chance that the first attempt at co-amplification will be successful. The testing of a five-member sensitivity panel containing serial 10-fold dilutions of both DNA targets by first M-PCR (using all primer pairs together) and then by individual PCR (using each primer set separately) is necessary to determine whether the conditions are near optimum and to determine the analytical sensitivity of the M-PCR assay. If the sensitivity of M-PCR is 1 fg for the first virus and 1 pg for the second virus, the ratio of the primer sets can be altered to try to bring the sensitivities closer to each other. We typically run the primers at molar ratios of 1:1, 2:1, 3:1, 1:2, and 1:3. If this manipulation does not improve the sensitivity for the second target, consider a new set of primers for this target. As a last resort, adding 5% (v/v) dimethylsulfoxide (DMSO) to the reaction may improve sensitivity, as has been done for M-PCR with more than five primer sets (Burgart *et al.,* 1992). We have found that DMSO has little or no effect on an M-PCR using two or three primer sets. If the decreased sensitivity for amplification of one target is due to "mispriming," the hot-start method alone or in conjunction with paraffin can be incorporated into M-PCR (Erlich *et al.,* 1991; Chou *et al.,* 1992). The primers can be diluted in an attempt to minimize mispriming in specimens that contain large amounts of DNA.

1. M-PCR Protocol for *C. trachomatis* and *N. gonorrhoeae*

First-void urine (20 ml) collected fresh or stored frozen can be tested by PCR. An aliquot (0.1 ml) is centrifuged in a microfuge at 12,000 g for 20 min at room temperature; the pellet is resuspended in 0.1 ml lysis buffer containing 50 mM KCl, 10 mM Tris-HCl, pH 8.3, 2.5 mM $MgCl_2$, 0.01% gelatin, 1% Tween 20, and 200 μg/ml proteinase K. The specimens are heated at 55°C for 1 hr, followed by 10 min at 94°C to destroy the proteinase. Swab specimens collected in culture transport media can be processed for PCR in the same manner as urine specimens. Nonspecific amplification occurs occasionally, in which case it may be necessary to add a phenol/chloroform extraction and ethanol precipitation after proteinase K inactivation. If the swab specimen has been collected in a collection tube containing detergent (e.g. Chlamydiazyme® or VIDAS® Specimen Collection Tubes), no further processing is required. KL1/KL2 primers for *C. trachomatis* (Mahony *et al.*, 1993a) and HO1/HO3 pairs for *N. gonorrhoeae* (Ho *et al.*, 1992) are used.

1. Prepare the following master mix (enough for 30 100-μl reactions) in a small polypropylene tube:

Component	Volume	Final concentration
Glass-distilled sterile H_2O	2280 μl	—
10× Reaction buffer [500 mM KCl, 100 mM Tris-HCl, pH 8.3, 25 mM $MgCl_2$, 0.1% (w/v) gelatin]	300 μl	1×
dNTP mix, 12.5 mM each	48 μl	200 μM each
Primer 1: KL1, 200 μM	15 μl	1.0 μM
Primer 2: KL2, 200 μM	15 μl	1.0 μM
Primer 3: HO1, 200 μM	15 μl	1.0 μM
Primer 4: HO3, 200 μM	15 μl	1.0 μM
Cetus AmpliTaq® (5 U/ml)	12 μl	2.0 U/100 μl
Total	2700 μl	

2. Aliquot 90 μl master mix into 0.5-ml microfuge tubes and cover with 75 μl light mineral oil.

3. Add 10 μl processed specimen (urine, endocervical or urethral swab). Set up the following controls:

    a. 10 μl lysis buffer (treated in the same manner as the urine specimens, i.e., heated at 55°C for 1 hr and 94°C for 10 min) as a negative control for the processing step

    b. 10 μl dH_2O as a negative control for the master mix

c. 10 μl 0.1 fg, 1 fg, 10 fg, 100 fg, or 1 pg *Chlamydia* DNA as positive control

4. Perform the PCR cycles on a thermocycler programmed for 40 cycles of 1 min at 94°C (denaturation), 1 min at 55°C (annealing), and 2 min at 72°C (extension), followed by 8 min at 72°C, and a hold at 15°C overnight if necessary.

5. Load 10 μl amplified products with 2 μl tracking dye onto a 2% agarose gel and run at 75 V for 2 hr with 0.5× TBE as the running buffer. (Tracking dye: 50% sucrose, 50 m*M* EDTA, 0.1% bromophenol blue, 4 *M* urea, pH 7.0.)

C. Amplicon Analysis

Gel electrophoresis followed by hybridization with labeled DNA probes has been used to verify specific amplification products. If primer pairs are chosen so amplicons differ in size by 50–100 bp, agarose gel electrophoresis can be used to resolve M-PCR amplicons. Figure 1 shows the separation and identification of a 390-bp amplicon for *N. gonorrhoeae* and a 241-bp amplicon of *C. trachomatis* produced by M-PCR. With a well-characterized M-PCR assay that gives few or no nonspecific amplification products, gel electropho-

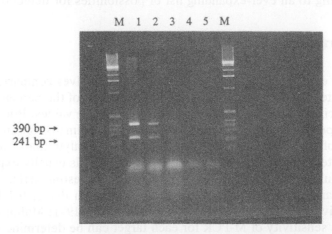

390 bp →
241 bp →

Figure 1 Assessment of the sensitivity of an M-PCR assay for the detection of *Chlamydia trachomatis* and *Neisseria gonorrhoeae* DNA. Serial 10-fold dilutions of *C. trachomatis* and *N. gonorrhoeae* DNA were tested by PCR using KL1/KL2 and HO1/HO3 primers, as described in the text. Amplification products were assessed by agarose gel electrophoresis. Outer lanes contain 1-kb marker DNA. Lane 1: 1 pg; Lane 2: 100 fg; Lane 3: 10 fg; Lane 4: 1 fg; Lane 5: 0.1 fg. Locations of 390-bp and 241-bp amplicons of *N. gonorrhoeae* and *C. trachomatis* are indicated by arrows.

resis alone is usually sufficient for detecting the presence or absence of specific sequences. During the development of the assay, hybridization with labeled probes should be included to verify specific amplification products. Southern blotting followed by hybridization is necessary for reactions that give many nonspecific products when the specific amplicon is obscured. Some PCR assays may never yield "clean reactions" with easily identifiable amplicons, making hybridization mandatory. Solid-phase hybridization using 96-well microtiter plates coated with avidin to capture biotin-containing amplicons, or complementary capture probes, has been used by many researchers and represents an easy way to assay PCR products for specific sequences. High-performance liquid chromatography (HPLC) has been used to detect specific amplicons and may be well suited to M-PCR for the detection of several amplicons differing in size by <50 bp. Simple M-PCRs with only two or three amplicons can be analyzed with oligonucleotide probes labeled with different enzymes and appropriate substrates giving absorbence at different wavelengths. The availability of dNTPs labeled with fluorescein isothiocyanate (FITC), rhodamine, or other fluorochromes presents the possibility of using fluorescence spectroscopy to detect several different amplicons in a single reading of a 96-well plate. Flow cytometry capable of distinguishing and quantifying several different fluorescent labels may also be useful for analyzing M-PCR products although, to our knowledge, this has not yet been reported. Chemiluminescence and other technologies will surely follow, adding to an ever-expanding list of possibilities for detection.

D. Evaluating Performance

The evaluation of any new diagnostic test involves comparison to an appropriate "gold standard" test and determination of the percentage sensitivity, specificity, and positive and negative predictive values. Prior to determining the performance characteristics on clinical specimens, the analytical sensitivity of a test, that is, the smallest amount of analyte that can be detected by the test, should be determined. For PCR, this is usually expressed as pg or fg purified DNA and is easily determined by testing serial 10-fold dilutions of target DNA. The most sensitive PCRs can detect 0.1 fg target or the equivalent of 1–5 target copies (Erlich *et al.*, 1991; Mahony *et al.*, 1993a). The sensitivity of M-PCR for each target can be determined by preparing a five-member sensitivity panel consisting of serial 10-fold dilutions of each target DNA pooled together, ranging from 1 pg to 0.1 fg (Fig. 1). Once the analytical sensitivity has been determined, the sensitivity and specificity can be determined using clinical specimens. A sample *Chlamydia–Neisseria* M-PCR result obtained with genitourinary specimens is shown in Fig. 2. An approximation of the sensitivity can be determined quickly by testing 10–20

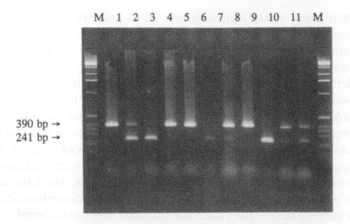

Figure 2 Detection of *Chlamydia trachomatis* and *Neisseria gonorrhoeae* DNA in genitourinary specimens by M-PCR. Outer lanes contain 1-kb marker DNA. Locations of 241-bp and 390-bp amplicons of *C. trachomatis* and *N. gonorrhoeae,* respectively, are indicated by arrows. Specimens in Lanes 3, 6, and 9 are positive for *C. trachomatis.* Specimens in Lanes 1, 4, 5, and 8 are positive for *N. gonorrhoeae.* Specimens in Lanes 2, 7, 10, and 11 are positive for both *C. trachomatis* and *N. gonorrhoeae.*

specimens that are positive by the gold standard test. This result can be used to determine the sensitivity of the test and, more importantly, whether any substance in the clinical specimen is interfering with the PCR. The sensitivity and specificity of the PCR can then be determined, first in a retrospective fashion using stored specimens, then prospectively using specimens collected sequentially from a clinically defined cohort of patients. Further evaluations of the test in various populations with different prevalences of infection (low, medium, and high) and different clinical presentations should be performed to assess the effect of disease prevalence (Sackett *et al.*, 1985) and spectrum bias (Lachs *et al.*, 1992) on test performance. For M-PCR tests, a sufficient number of specimens should be tested so a minimum of 15–20 positives is obtained for each virus or bacteria. If the numbers are small, 95% confidence intervals should be used for sensitivity, although this is rarely done.

E. Troubleshooting

Even the best designed PCR assays occasionally develop problems. Many problems are the results of changing reagents without properly testing them, as described in Section IV. Problems due to carryover contamination can be devastating and set a laboratory back 1–3 mo depending on the number of primers in use and the laboratory's knowledge of the problem. (See Section IV,A for ways to prevent carryover contamination.) Sometimes a particular

PCR assay works well with one set of primers, then suddenly gives many nonspecific products with a different set of primers or a new batch of the same pair of primers. The introduction of a new lot of previously used primers should follow an appropriate testing protocol to verify its performance and determine whether it is having any negative influence on sensitivity (see Section IV,B). Evaluation of a novel primer pair may reveal many nonspecific products that prevent visualization of the expected amplicon on ethidium bromide-stained gels and, more importantly, result in a marked decrease in analytical sensitivity because of incorporation of primers and nucleotides into nonspecific products. Nonspecific products can be eliminated in part or completely by raising the annealing temperature, which increases the fidelity of primer–target annealing and prevents primer dimer formation. If raising the annealing temperature does not eliminate nonspecific amplification products, raising or lowering the $MgCl_2$ concentration or decreasing the primer concentration, as described in Section III,B, may alleviate the problem. If none of these maneuvers improves the specificity of amplification, there may be too much target DNA in the sample, reducing the DNA concentration by making serial dilutions of the sample sometimes improves specificity. Using "hot-start" or uracil-N-glycosylase methods alone or together has been shown to increase the specificity of amplification (Erlich et al., 1991; Chou et al., 1992; Thornton et al., 1992). The hot-start involves raising the reaction temperature to a temperature above the T_m of the primers (usually 80°C), then adding the Taq polymerase or primers to the reaction tube (before cycling begins) to prevent the primers from annealing to noncomplementary DNA, thereby eliminating nonspecific products. Although useful in some PCRs, the hot-start method does not circumvent problems attributable to poorly designed primers that may contain short stretches of bases that appear in nontarget DNA present in the specimens (i.e., host derived). The use of paraffin wax to separate primers and enzymes physically from DNA in reaction tubes uses the same principle of separating primers and DNA to improve the fidelity of primer annealing.

Inhibitors of Taq polymerase may be present in clinical specimens, preventing the detection of specific viral nucleic acids by PCR. Although these inhibitors have been poorly characterized, they are usually proteinaceous or nonproteinaceous substances that inhibit polymerases nonspecifically, presumably by altering the quaternary structure of the polypeptide or by interfering with the active site of the enzyme. Certain salts such as oxalates and urea, heparin, iron-containing compounds such as hemoglobin, and urine have been reported to inhibit Taq polymerase (Mercier et al., 1990; Holodniy et al., 1991; Khan et al., 1991). Hemoglobin can be particularly troublesome if plasma, serum, or blood is being tested by PCR. Traces of hemoglobin can be removed by deproteination with organic solvents such as phenol or chloroform. Inhibitors of PCR can be removed by selective adsorption of nucleic acids to hydroxyapatite, silica gel, or immobilized captive DNA.

We have had success testing some urine and semen specimens that contain inhibitory substances by boiling or by making serial dilutons, then retesting by PCR. For the latter method, the idea is to dilute out the inhibitor before diluting out the target sequence so amplification can be done. This simple approach has not, however, been rigorously analyzed to ensure that target sequences are also not lost to dilution. The possibility of selective inhibition of one primer pair over another also has not been investigated systematically (Coultee *et al.,* 1991). These concerns accentuate the need for appropriate quality assurance measures for both conventional PCR and M-PCR.

IV. QUALITY ASSURANCE

A. Anticontamination Measures

Pre- and poststerilization measures have been described for preventing PCR false positive results due to carryover contamination with pre-amplified sequences. Although these methods are highly effective, they are not absolutely necessary and can be replaced largely by specific laboratory practices designed to prevent carryover. One precautionary measure is the use of three designated areas for: (1) specimen handling and preparation, (2) PCR set-up, and (3) amplicon detection. If space in the laboratory permits, these three steps are best carried out in separate rooms; however, it is possible to have separate areas in one room by using a combination of laminar flow hoods and/or enclosed plexiglass working stations, thereby creating "PCR-clean" areas free of amplified DNA. These clean areas should never be visited by personnel, pipettes, reagents, or tubes that have been used in or exposed to "dirty" areas where PCR tubes have been opened. Since amplicons have been detected on benches, in hair, and on laboratory coats, designating laboratory coats for each area is prudent. Months of developmental experiments can be lost by contaminating set-up areas with amplified DNA, so "spread out your PCR" and take every possible precaution to prevent carryover contamination. The use of commercial reagents such as uracil-*N*-glycosylase (in conjunction with uracil in place of thymidine) helps prevent carryover contamination. Other pre-sterilization measures include UV irradiation or sodium hydroxide treatment to render nucleic acid nonamplifiable.

B. Controls

Procedural controls for M-PCR are similar to controls that should be used for conventional PCR. These consist of both positive and negative controls, which are designed to monitor the sensitivity and specificity of each amplification run, respectively. The preparation of a batch of clinical specimens to be tested by PCR should include up to six negative specimens, depending on the number of specimens being tested. Use two negative controls for a batch of six specimens and six for a batch of 20 specimens. Possible negative controls include extraction buffer alone or negative cells or tissues interspersed with the clinical specimens to assess specimen contamination at the set-up stage. We also include one or two positive controls to ensure that the specimen preparation step worked, especially if extraction of the nucleic acid from tissue with enzymes, detergents, or organic compounds is involved. Each PCR run should include a small sensitivity panel consisting of serial 10-fold dilutions of DNA (ranging from 1 pg to 0.1 fg) to assess day-to-day fluctuation in assay sensitivity. This test is also useful in comparing results obtained in different laboratories. For M-PCR, the sensitivity panel should consist of specimens with both amplification targets and specimens containing single targets as well as appropriate negative controls. When switching lots of any reagent, each new lot should be tested by substituting the new reagent (for example, *Taq* polymerase, PCR master mix, or dNTPs) into the existing assay and running the sensitivity panel alone or running the sensitivity panel with a batch of specimens, and comparing the sensitivity obtained with the new reagent to that of the existing assay. Only when each new reagent is tested in this fashion and each is shown not to affect the performance of the assay adversely should any reagents in the assay be changed. This precaution is especially important when switching to a new lot of enzyme or trying out a new set of primers (perhaps purchased from a new company). If this control procedure is followed religiously, there should be few surprises necessitating costly repeat testing of large runs.

C. Limitations

The use of good quality reagents, as well as good laboratory practices, should ensure that PCR testing generates valid results. The inclusion of quality control procedures (positive and negative controls, specimen preparation controls, sensitivity controls, monitoring for inhibitors of *Taq* polymerase) ensures that M-PCR results are interpreted correctly and minimizes the limitations on this technology. Although most M-PCR assays should give clearly distinguishable positive and negative results (especially when DNA probing is used to verify the presence of specific amplicons), some types of specimens can present problems. Specimens containing very high concentrations of DNA sometimes give spurious nonspecific bands or a false negative result. These specimens must be serially diluted and retested at two or three different dilutions, retested with lower concentrations of primers, or both to decrease mispriming. The effect of widely varying concentrations of target DNA on the relative amplification efficiency of each target in M-PCR has not yet been investigated systematically. This may not turn out to be a problem in the clinical microbiology laboratory where clinical specimens usually contain 10^3-10^6 viral genome equivalents and the difference in target numbers may be small. Experimentation with various amounts of each primer pair may help answer this question. Advances in specimen preparation protocols, that is, silica gel or bead adsorption of RNA for detection by RT-PCR, and the implementation of new enzymes such as *Thermus thermophilus* polymerase, which has both reverse transcriptase and DNA polymerase activity, have had major effects in decreasing the variability of results obtained by RT-PCR and have allowed more laboratories to detect RNA viruses reliably. Commercial tests for several RNA viruses are currently in clinical trials and will soon be available. Future technological advances will permit the detection of both RNA and DNA viruses in the same specimen by M-PCR.

D. Interpretation

If proper anticontamination measures are followed so PCR false positives are either prevented altogether or readily identified, there is little difficulty interpreting positive test results. Only when both positive and negative controls give expected results can the results of a run be accepted. M-PCR results should show clearly distinguishable bands of the predicted amplicon size by gel electrophoresis. In a well-designed and characterized M-PCR assay, gel electrophoresis may be sufficient to identify specimens as positive. Confirmatory DNA probing (either by dot blot hybridization or by solid-phase DNA capture with oligonucleotide probing) may be necessary when multiple nonspecific bands obscure specific amplicons. It is good practice for laboratories to confirm positives for reportable diseases and highly communicable viral infections before reporting results; PCR testing should be included in this practice. We have used a confirmatory PCR for our in-house PCR for *C. trachomatis,* targeting a second plasmid sequence or a chromosomal sequence to confirm our initial PCR results (Mahony *et al.,* 1992). The first commercially available amplification test for *C. trachomatis,* Amplicor™ *Chlamydia* from Roche Molecular Systems (Branchburg, NJ) has, however, not included a confirmatory test. Since no test is 100% specific, a small number of false positives will be reported unless a confirmatory test is employed. The presence of *Taq* polymerase inhibitors in clinical specimens, which results in false negative results for truly positive specimens, is worrisome but by most accounts, this is an infrequent occurrence. We and others have observed chlamydial PCR inhibitors in urine occurring at a frequency below 0.3% (Mahony *et al.*, 1992; S. Herman, personal communication). Blood or serum specimens containing heme and semen specimens containing the polyamines spermine and spermidine (Ahokas and Erkkila, 1993) may also produce false negative PCR results because of the presence of these interfering compounds. A number of internal PCR controls including β-globin, HLA DQα, actin, and myosin genes have been used by some laboratories to monitor the presence of inhibitors (Coultee *et al.*, 1991; Holodniy *et al.*, 1991; Jullian *et al.*, 1993; Richards *et al.*, 1993). Although this procedure doubles the number of reactions, it is an effective means of detecting inhibitors in clinical specimens that may yield false negative results.

ACKNOWLEDGMENTS

We gratefully acknowledge the excellent contributions of Kathy Luinstra in performing PCR experiments and Lisa Rizzo and Lucia Weatherley in typing the manuscript.

REFERENCES

Ahokas, H., and Erkkila, M. J. (1993). Interference of PCR amplification by the polyamines, spermine and spermidine. *PCR Meth. Appl.* **3**, 65–68.

Burgart, L. J., Robinson, R. A., Heller, M. J., Wilke, W. W., Iakoubova, O. K., and Cheville, J. C. (1992). Methods in pathology—Multiplex polymerase chain reaction. *Modern Pathol.* **5**, 320–323.

Chamberlain, J. S., Gibbs, R. A., Ranier, J. E., Nguyen, P. N., and Caskey, C. T. (1988). Deletion screening of the Duchenne muscular dystrophy locus via multiplex DNA amplification. *Nucleic Acids Res.* **16**, 11141–11156.

Chamberlain, J. S., Gibbs, R. A., Ranier, J. E., and Caskey, C. T. (1990). Multiplex PCR for the diagnosis of Duchenne muscular dystrophy. *In* "PCR Protocols: A Guide to Methods and Applications." (M. A. Innis, D. H. Gelfand, J. Sninsky, and T. J. White, eds.), pp. 272–281. Academic Press, San Diego.

Chou, O., Russell, M., Birch, D. E., Raymond, J., and Bloch W. (1992). Prevention of pre-PCR mis-priming and primer dimerization improves low-copy-number amplifications. *Nucleic Acids Res.* **20**, 1717–1723.

Coultee, F., Saint-Antoine, P., Olivier, C., Kessous-Elbaz, A., Voyer, H., Berrada, F., Begin, P., Grioux, L., and Viscidi, R. (1991). Discordance between primer pairs in the polymerase chain reaction for detection of human immunodeficiency virus type 1: A role for *Taq* polymerase inhibitors. *J. Infect. Dis.* **164**, 817–818.

Erlich, H., Gelfand, D., and Sninsky, J. (1991). Recent advances in the polymerase chain reaction. *Science* **252**, 1643–1651.

Fortina, P., Conant, R., Parrella, T., Rappaport, E., Scanlin, T., Schwartz, E., Robertson, J. M., and Surrey, S. (1992). Fluorescence-based, multiplex allele-specific PCR (MASPCR) detection of the delta F508 deletion in the cystic fibrosis transmembrane conductance regulator (CFTR) gene. *Mol. Cell. Probes* **6**, 353–356.

Gregoire, L., Arella, M., Capione-Piccardo, J., and Lancaster, W. D. (1989). Amplification of human papillomavirus DNA sequences by using conserved primers. *J. Clin. Microbiol.* **27,** 2660–2665.

Ho, B. S. W., Feng, W. G., Wong, B. K. C., and Egglestone, S. I. (1992). Polymerase chain reaction for the detection of *Neisseria gonorrhoeae* in clinical samples. *J. Clin. Pathol.* **45,** 439–442.

Holodniy, M., Kim, S., Katzenstein, D., Konrad, M., Groves, E., and Merigan, T. C. (1991). Inhibition of human immunodeficiency virus gene amplification by heparin. *J. Clin. Microbiol.* **29**(4), 676–679.

Jullian, E. H., Dhellemmes, C., Saglio, O., Chavinie, J., and Pompidou, A. (1993). Methods in laboratory investigation—Improved detection of human papillomavirus types 16 and 18 in cervical scrapes by a multiplex polymerase chain reaction: A 4% prevalence among 120 French women with normal cytology. *Lab. Invest.* **68,** 242–247.

Khan, G., Kangro, H. O., Coates, P. J., and Heath, R. B. (1991). Inhibitory effects of urine on the polymerase chain reaction for cytomegalovirus DNA. *J. Clin. Pathol.* **44,** 360–365.

Kilimann, M. W., Pizzuti, A., Grompe, M., and Caskey, C. T. (1992). Point mutations and polymorphisms in the human dystrophin gene identified in genomic DNA sequences amplified by multiplex PCR. *Hum. Genet.* **89,** 253–258.

Ko, T. M., Tseng, L. H., Chiu, H. C., Hsieh, F. J., and Lee, T. Y. (1992). Dystrophin gene deletion in Chinese Duchenne/Becker muscular dystrophy patients via multiplex DNA amplification. *J. Formos. Med. Assoc.* **91,** 951–954.

Lachs, M. S., Nachamkin, I., Edelstein, P. H., Goldman, J., Feinstein, A. R., and Stanford Schwartz, J. (1992). Spectrum bias in the evaluation of diagnostic tests: Lessons from the rapid dipstick test for urinary tract infection. *Ann. Int. Med.* **117,** 135–140.

Mahony, J., Luinstra, K., Sellors, J., Jang, D., and Chernesky, M. (1992). Confirmatory polymerase chain reaction testing for *Chlamydia trachomatis* in first-void urine from asymptomatic and symptomatic men. *J. Clin. Microbiol.* **9,** 2241–2245.

Mahony, J., Luinstra, K., Sellors, J., and Chernesky, M. (1993a). Comparison of plasmid- and chromosome-based polymerase chain reaction assays for detecting *Chlamydia trachomatis* nucleic acids. *J. Clin. Microbiol.* **7,** 1753–1758.

Mahony, J., Luinstra, K., Tyndall, M., Krepel, J., Sellors, J., and Chernesky, M. (1993b). Detection of *Chlamydia trachomatis* and *Neisseria gonorrhoeae* in first void urine by multiplex PCR. 10th I.S.S.T.D.R. Meeting, Helsinki 1993 (Abstracts), p. 79.

Manos, M. M., Ting, Y., Wright, D. K., Lewis, A. J., Broker, T. R., and Wolinsky, S. M. (1989). Use of polymerase chain reaction amplification for the detection of genital human papillomaviruses. *Cancer Cells* **7,** 209–214.

Mercier, B., Gaucher, C., Feugeas, O., and Mazurier C. (1990). Direct PCR from whole blood, without DNA extraction. *Nucleic Acids Res.* **18,** 5908.

Morrall, N., and Estivill, X. (1992). Multiplex PCR amplification of three microsatellites within the CFTR gene. *Genomics* **13,** 1362–1364.

Nedjar, S., Biswas, R. M., and Hewlett, I. K. (1991). Co-amplification of specific sequences of HCV and HIV-1 genomes by using the polymerase chain reaction assay: A potential tool for the simultaneous detection of HCV and HIV-1. *J. Virol. Meth.* **35,** 297–304.

Picci, L., Anglani, F., Scarpa, M., and Zacchello, F. (1992). Screening for cystic fibrosis gene mutations by multiplex DNA amplification. *Hum. Genet.* **88,** 552–556.

Richards, B., Skoletsky, J., Shuber, A. P., Balfour, R., Stern, R. C., Dorkin, H. L., Parad, R. B., Witt, D., and Klinger, K. W. (1993). Multiplex PCR amplification from the CFTR gene using DNA prepared from buccal brushes/swabs. *Hum. Mol. Genet.* **2,** 159–163.

Sackett, D. L., Haynes, R. B., and Tugwell, P. (1985). "Clinical Epidemiology: A Basic Science for Clinical Medicine." Little, Brown, Boston.

Snijders, P. J. F., von den Brule, A. J. C., Schrijnemakers, H. F. J., Snow, G., Meijer,

C. J. L. M., and Walboomers, J. M. M. (1990). The use of general primers in the polymerase chain reaction permits the detection of a broad spectrum of human papillomavirus genotypes. *J. Gen. Virol.* **71,** 173–181.

Thornton *et al.* (1992). Utilizing uracil DNA glycosylase to control carryover contamination in PCR: Characterization of residual UDG activity following thermal cycling. *BioTechniques* **13,** 180–184.

Vandenvelde, C., Scheen, R., Van Pachterbeke, C., Loriaux, C., Decelle, J., Hubert, T., Delhaye C., Cattoor, J. P., Duys, M., and Van Beers, D. (1992). Prevalence of high risk genital papillomaviruses in the Belgian female population determined by fast multiplex polymerase chain reaction. *J. Med. Virol.* **36,** 279–282.

Vesy, C. J., Greenson, J. K., Papp, A. C., Snyder, P. J., Qualman, S. J., and Prior, T. W. (1993). Evaluation of celiac disease biopsies for adenovirus 12 DNA using a multiplex polymerase chain reaction. *Modern Pathol.* **6,** 61–64.

Wattel, E., Mariotti, M., Agis, F., Gordien, E., Prou, O., Courouce, A. M., Rouger, P., Wain-Hobson, S., Chen, I. S., and Lefriere, J. J. (1992). Human T lymphotropic virus (HTLV) type I and II DNA amplification in HTLV-I/II-seropositive blood donors of the French West Indies. *J. Infect. Dis.* **165,** 369–372.

Wilson, S. M., McNerney, R., Nye, P. M., Godfrey-Faussett, P. D., and Stoker, N. G., and Voller, A. (1993). Progress toward a simplified polymerase chain reaction and its application to diagnosis of tuberculosis. *J. Clin. Microbiol.* **31,** 776–782.

11

PCR *in Situ* Hybridization

Gerard J. Nuovo
Department of Pathology
State University of New York at Stoney Brook
Stony Brook, New York 11794

I. Introduction
   A. Review of the Methodology
   B. Strengths and Weaknesses of PCR *in Situ* Hybridization
II. Methods
   A. Protocols
   B. Troubleshooting
III. Applications
   A. The Equivocal Penile or Vulvar Biopsy
   B. *In Situ* Detection of PCR-Amplified HIV-1 Nucleic Acids and Tumor Necrosis
      Factor cDNA
   C. Localization of PCR-Amplified Hepatitis C cDNA
References

I. INTRODUCTION

A. Review of the Methodology

Haase *et al.* (1990) were the first investigators to describe the *in situ* detection of polymerase chain reaction (PCR)-amplified DNA in intact cells. This landmark work used cell suspensions in tubes. Clearly, it would be advantageous to apply this analysis to tissue sections on glass slides. This method was first reported by Nuovo *et al.* (1991a), who demonstrated the distribution of PCR-amplified human papillomavirus (HPV) DNA in cervical squamous intraepithelial lesions (SILs). Since this time, many groups have published protocols for PCR *in situ* hybridization, primarily for the detection of human immunodeficiency virus type 1 (HIV-1) DNA (Bagasra *et al.*, 1992; Chiu *et al.*, 1992; Nuovo, 1992; Nuovo *et al.*, 1992a; Embretson *et al.*, 1993a,b;

Patterson *et al.*, 1993). In 1992, Nuovo *et al.* reported the detection of a variety of human mRNAs in cultured cells using reverse transcriptase (RT) *in situ* PCR and later, in tissue sections (Nuovo *et al.*, 1992b,d; 1993a,b). Several other groups have described the detection of mRNAs and HIV-1 RNA by RT *in situ* PCR; similarly, PCR-amplified hepatitis C virus (HCV) RNA (cDNA) has been detected in tissue sections (Nuovo *et al.*, 1993a; Patterson *et al.*, 1993).

Perhaps the greatest impact of PCR *in situ* hybridization is in the area of HIV-1-related pathogenesis. One of the most perplexing issues regarding acquired immunodeficiency syndrome (AIDS) is that early studies stated that few of the T helper cells actually contain the virus, especially early in the disease process. This fact is important because the reported detection rate of <1 per 1000 CD4 cells is not consistent with the severe immunosuppression that is characteristic of AIDS (Harper *et al.*, 1986; Shapshak *et al.*, 1990). Using different variations of the PCR *in situ* hybridization technique, Nuovo *et al.* (1992c), Bagasra *et al.* (1992), and Patterson *et al.* (1993) independently demonstrated that up to 10% of circulating peripheral blood CD4 cells are actually infected by the virus early in the disease process, and over 80% are infected in advanced AIDS. Using PCR *in situ* hybridization, Nuovo *et al.* (unpublished data) and Embreston *et al.* (1993a,b) showed that in early HIV infection up to 25% of the dendritic cells and over 50% of the CD4 lymphocytes in the lymph nodes contain HIV-1 DNA. At end-stage AIDS, however, virtually all CD4 lymphocytes in the lymph nodes and in the peripheral blood are infected by the virus. Clearly, these numbers are consistent with the profound immunosuppression that is the central feature in patients with AIDS.

A great deal of variation is seen in the different published protocols for PCR *in situ* hybridization. Much of this variation is based on methods that attempt to inhibit diffusion of the amplified product from the site of origin and to prevent drying of the amplifying solution. The use of multiple primer pairs that produce DNA segments with 40–60 bp overhangs was offered as a way to limit diffusion of the amplified product (Haase *et al.*, 1990; Chiu *et al.*, 1992). Although multiple primer pairs were able to increase the intensity of the signal in PCR *in situ* hybridization, a single primer pair could function as well if not better when the hot-start modification was used (Nuovo *et al.*, 1991b). This finding suggested that the degree of target-specific amplification, not the size of the product, determined the success of PCR *in situ* hybridization. Similar conclusions about solution-phase PCR (i.e., that inhibition of the nonspecific pathways of DNA synthesis could enhance the detection of the target) were made subsequently (Nuovo *et al.*, 1991b; Chou *et al.*, 1992; see Erlich *et al.*, 1991, for a review of hot-start PCR).

Much attention has focused on the ability to label the product directly in *in situ* PCR, which would eliminate the need for a hybridization step.

Clearly, assurance is required that incorporation of the reporter molecule is completely target specific. As will be further explained later, it appears that such assurance is possible for RNA detection in tissue sections with RT *in situ* PCR but not for *in situ* PCR methods for DNA in tissue sections (Nuovo, 1992,1993).

B. Strengths and Weaknesses of PCR in Situ *Hybridization*

PCR *in situ* hybridization has three major strengths. First and foremost, PCR *in situ* hybridization combines the high sensitivity of PCR with the cell localizing ability of *in situ* hybridization. Why does one need to amplify the product to detect it by *in situ* hybridization? Researchers generally agreed that *in situ* hybridization is less sensitive than filter hybridization techniques and PCR (Lorincz *et al.,* 1989; Nuovo, 1989,1992,1993). Many studies have compared the sensitivities of these three techniques, often with respect to the detection of viruses associated with infectious diseases, particularly HPV and HIV-1. These studies have strongly suggested that (1) PCR, filter hybridization, and *in situ* hybridization are of equivalent sensitivity for productive viral infections, such as HPV in low-grade SILs; and (2) latent infection, as routinely noted in HIV-1 infection, and occult or subclinical infection can be detected by PCR or filter hybridization but not by *in situ* hybridization (Lorincz *et al.,* 1989; Nuovo, 1989,1992,1993). This result reflects the lower copy number of virus associated with these specific conditions. The reported detection thresholds of the three techniques vary considerably, in part reflecting methodological variations. For example, the hot-start modification of PCR increases the detection threshold from 1000–2000 copies to 1–10 copies, assuming 1 μg background nontarget DNA (Nuovo *et al.,* 1991b; Chou *et al.,* 1992). The detection threshold for filter hybridization is generally reported to be about 1 copy per 100 cells (Lorincz, 1989; Nuovo, 1989,1992,1993). Most groups report detection thresholds for *in situ* hybridization of about 10 copies per cell (Crum *et al.,* 1988,1991; Walboomers *et al.,* 1988; Nuovo *et al.,* 1991a). Claims that 1 copy per cell can be detected by *in situ* hybridization have been made (Lawrence *et al.,* 1990). In our experience, such reports are in conflict with the general inability to detect latent HIV-1 infection by *in situ* hybridization, which is associated with 1 to a few integrated copies of DNA (Bagasra *et al.,* 1992; Nuovo *et al.,* 1992c,1993b; Embretson *et al.,* 1993a,b; Patterson *et al.,* 1993). Interest in PCR *in situ* hybridization has grown out of the experience of virtually all investigators that the relatively high detection threshold of *in situ* hybridization is a major limiting factor for its usefulness.

A second strength of PCR *in situ* hybridization relates to the issue of sample contamination in solution-phase PCR. Sample contamination, which can lead to false positive results in PCR, has limited its usefulness as a diagnostic test. However, this problem has not been encountered in PCR *in situ* hybridization because the positive signal is localized to specific areas within the cell. Indeed, we have added purified viral DNA to cell and tissue specimens known to be negative for the virus and shown that the PCR product remains in the amplifying solution and does not enter the fixed cell.

The third strength of PCR *in situ* hybridization is the enormous amount of information it provides relative to the histological distribution of the amplified PCR product. For example, the observation that PCR-amplified HIV-1 DNA was restricted to the endocervical aspect of the transformation zone in the cervix suggested that this area may be its portal of entry (Nuovo *et al.*, 1993). Similarly, the demonstration that HPV RNA, as evident from RT *in situ* PCR, was found in most cancer cells but in none of the adjacent normal cells suggests that viral transcription is essential for the malignant phenotype (J. Chumas and G. J. Nuovo, unpublished observations).

The weakness of PCR *in situ* hybridization is related to the weaknesses of both PCR and *in situ* hybridization. Competing pathways in PCR can limit its specificity and sensitivity. Further, background signal from complexing of the probe and nontarget molecules in *in situ* hybridization can also lead to false positive results (Nuovo, 1992,1993). The need for controls for every slide will be stressed in this manuscript as a way to ensure that the conditions are optimized for PCR and that background is not causing problems in interpreting the hybridization signal.

II. METHODS

As noted earlier, methods for the *in situ* detection of PCR-amplified DNA have been reported by several groups. This chapter describes a manual "hot-start" modification for PCR *in situ* hybridization that allows the detection of one target copy per cell with a single primer pair (Nuovo *et al.*, 1991b, 1992a,e; Nuovo, 1992). This modification reduces much of the unwanted DNA synthesis due to mispriming and primer oligomerization and, under certain specified conditions, permits direct incorporation of target-specific labeled nucleotides in a process termed *in situ* PCR. This process has been expanded to use for RNA detection by preceding PCR with an RT step.

A. Protocols

The protocols listed here for the *in situ* detection of PCR-amplified DNA and cDNA are derived from two textbooks (Nuovo, 1992,1993).

1. Preparation of Tissue Sections and Cell Samples

1. Fix tissue samples in 10% buffered formalin, preferably 8–15 hr.
2. For cell cultures, wash the cells directly in the plates. Add 10% buffered formalin and let stand overnight. Place 2000–5000 cells per 1-cm area using a cytospin centrifuge.
3. Place several 4-μm tissue sections or 2 cytospins on silane-coated slides (Oncor, Gaithersburg, MD)
4. To remove the paraffin from tissue samples, place slides in fresh xylene for 5 min and then in 100% ethanol for 5 min. Then air dry.

2. Protease Digestion

Pepsin, trypsin, or proteinase K may be used. Proper protease concentration and time needed for digestion are based on tissue type and length of fixation and are established by trial and error. Insufficient protease treatment can result in lack of hybridization signal, whereas excessive protease treatment will cause destruction of tissue morphology.

1. Begin with pepsin at 2 mg/ml (stock solution is 20 mg pepsin, 9.5 ml water, 0.5 ml 2 N HCl). Treat the cells or tissue for 12–30 min.
2. For RT *in situ* PCR, inactivate using sterile diethylpyrocarbonate (DEPC)-treated water for 1 min followed by 100% ethanol for 1 min.
3. For PCR *in situ* hybridization, inactivate using a solution of Tris (0.1 M pH 7.5) and 0.1 M NaCl for 1 min followed by 100% ethanol for 1 min.

3. Reverse Transcription

1. Treat the RNase-free DNase (Boehringer Mannheim, Indianapolis, IN) overnight. Use 10 U/tissue section. Cover the solution with a polypropylene cover slip and incubate at 37°C.
2. Inactivate the DNase by washing the slides in DEPC-treated water for 1 min, followed by 100% ethanol for 1 min. Then air dry.

3. To a tube, add:

2 μl RT buffer
2 μl each of dATP, dCTP, dGTP, and dTTP (stock solution: 10 mM)
1 μl of 3' downstream primer (stock solution: 20 μM)
3 μl DEPC-treated water
1 μl RNase inhibitor
4 μl MgCl$_2$ (stock solution: 25 mM)
1 μl (2.5 U) RT

This is enough for 2 tissue sections. (Note: This solution is prepared as described in the RT PCR GeneAmp kit; Perkin-Elmer Corporation, Norwalk, CT.)

4. Add 10 μl to each tissue section and incubate at 42°C for 30 min.

4. PCR

1. For amplifying solution add to one tube:

2.5 μl PCR Buffer II (GeneAmp kit)
4.5 μl MgCl$_2$ (stock solution: 25 mM)
4 μl dNTPs (stock solution: 2.5 mM; dilution as per GeneAmp kit)
1 μl primer 1 (5' primer; stock solution: 20 μM)
1 μl primer 2 (3' primer; stock solution: 20 μM)
1 μl bovine serum albumin (BSA; stock solution: 2% w/v)
6.2 μl water

2. To a separate tube, add 4.0 μl water and 0.8 μl *Taq* polymerase (Ampli*Taq*, Perkin Elmer).
3. Cover each of two tissue sections with 10 μl amplifying solution. Cut plastic cover slips to size and cover the amplifying solution. Then anchor the cover slips with 2 small drops of nail polish.
4. Place slides in an aluminum foil boat, and then on the block of an automated thermal cycler (Perkin Elmer).
5. When the block has reached 55°C, add 2.4 μl *Taq* polymerase solution to each section by gently lifting the cover slip. Immediately overlay the section with about 1 ml preheated (82°C) mineral oil.
6. Bring block to 82°C, then to 94°C for 3 min (denaturing).
7. Incubate at 55°C, 2 min and 94°C, 1 min for 30 cycles; add more heated mineral oil as needed.
8. At conclusion, soak at 4°C.
9. Remove cover slip and oil by soaking in xylene for 5 min followed by 100% ethanol for 5 min. Then air dry slides.

5. *In Situ* Hybridization

1. Label oligoprobe using the Genius 5 oligoprobe 3' tailing kit (Boehringer Mannheim, Indianapolis, IN).
2. Make oligoprobe cocktail by adding 10 μl formamide, 39 μl 25% dextran sulfate, 39 μl water, 10 μl 20× SSC, and 2 μl probe.
3. Add 5–10 μl probe cocktail to a given tissue section.
4. Overlay with plastic cover slip.
5. Place slide on hot plate (95–100°C) for 5 min.
6. Place slides in humidity chamber at 37°C for 2 hr.
7. Remove cover slip; place slide in wash solution (1× SSC and 0.2% BSA) for 10 min at 50°C.
8. Wipe off excess wash solution.
9. Add 50 μl anti-digoxigenin–alkaline phosphatase conjugate per tissue section.
10. Incubate in humidity chamber for 30 min at 37°C.
11. Wash slides at room temperature for 3 min in the detection solution (0.1 M Tris, pH 9.5, 0.1 M NaCl)
12. Place slides in detection solution and nitrobluetetrazolium/bromochloroindolyl phosphate (NBT/BCIP; Oncor).
13. Incubate slides for 30 min to 2 hr, checking results periodically under microscope; counterstain with nuclear fast red.

B. Troubleshooting

1. Generalized Statements

The protocols presented here reflect the results of a large series of optimizing experiments for the different reagents (Nuovo *et al.*, 1993c). Note that the concentration of $MgCl_2$ (4.5 mM) is uniformly higher than for solution-phase PCR. This general statement applies to a wide variety of primer pairs and DNA and cDNA targets (Nuovo, 1992; Nuovo *et al.*, 1993c).

Diffusion of the amplified cDNA from the site of origin is another concern. Researchers have shown that the specific fixation chemistry has a profound effect on diffusion of the PCR product. To address this problem, some investigators hypothesized that the use of multiple primer pairs or tailed primers would result in a product greater than 1000 bp which would be too large to permeate the nuclear membrane (Haase *et al.*, 1990; Chiu *et al.*, 1992). A 450-bp product was thought to be membrane permeable. An alternative direction taken by others to solve this problem was based on the nature of cell fixation during preparation of samples, and subsequent adjustment of protease conditions. Studies examining different fixation methods for *in situ* PCR showed that fixatives such as acetone or ethanol, although excellent for solution-phase PCR, were unsuccessful for *in situ* PCR (Nuovo *et al.*, 1993c). Acetone and ethanol are denaturing fixatives that do not cross-link nucleic acids to proteins, as do fixatives such as formalin. The non-cross-linking fixatives may not allow *in situ* PCR or, alternatively, amplification may occur but the PCR product diffuses from the nucleus into the amplifying solution. Analysis of the amplifying solution showed that the PCR product was indeed located in the solution when the cells were fixed in ethanol or acetone, but not when the cells were fixed in buffered formalin (Nuovo *et al.*, 1993c). These data illustrate the need for prolonged formalin fixation and protease digestion. Formalin fixation is thought to create a migration barrier that prevents diffusion of PCR products through cell membranes. This ability is shown graphically in Fig. 1. However, this ability to prevent diffusion necessiates treatment of the samples with protease digestion to permit PCR reagent entry. The degree of inhibition of migration appears to be marked, based on studies involving cDNA from measles and a variety of human mRNAs that have showed distinct and appropriate subnuclear (i.e., nucleolar and perinucleolar) and cytoplasmic localization (Fig. 2); Nuovo *et al.*, 1992d).

2. Importance of Protease Digestion and Controls

In our experience, the most common source of difficulties for PCR *in situ* hybridization and RT *in situ* PCR is protease digestion. Different tissue samples may vary greatly in their susceptibility to protease digestion and the conditions should be determined empirically. Overdigestion results in a loss of cell morphology and the nuclei are no longer visible. Inadequate protease digestion can be recognized by an absence of signal or a weak signal in the positive control. Positive controls can be prepared in two ways. First, one can use a probe that will detect a target in every human cell. Many companies sell probes that detect the repetitive *Alu* sequence that makes up about 5% of the total human genome; we have used these probes from Digene Diagnostics (Beltsville, MD) and Oncor with good results. This positive control tests the *in situ* hybridization and detection procedures.

The other type of positive control is based on the ability to incorporate digoxigenin dUTP nonspecifically into tissue sections using PCR. If the tissue is not DNase treated and protease conditions are optimal, at least 50% of the cells will show an intense signal (Fig. 3). This figure also shows the different histological distributions of the positive control (all cell types positive) and the test section, in which only the carcinoma cells in this cervical cancer are positive for HPV 18 RNA. If few of the cells show a weak signal

1. Formalin fixation creates protein - DNA links

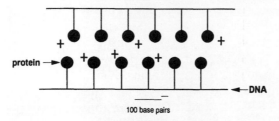

100 base pairs

2. Amplified product remains in cell - trapped on + charged amino acids

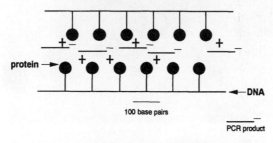

100 base pairs

PCR product

3. Non cross-linking fixative: Proteins denature - no protein - DNA links (ethanol or acetone)

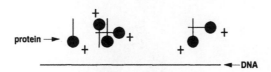

4. PCR product migrates out of cell

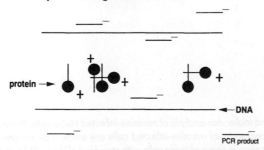

PCR product

Figure 1 Model for the effect of different fixatives on the localization of the amplified product in PCR *in situ* hybridization. In this model we hypothesize that the cross-linking induced by formalin fixation is essential for the marked limitation of diffusion of the PCR product from its site of amplification in the cell.

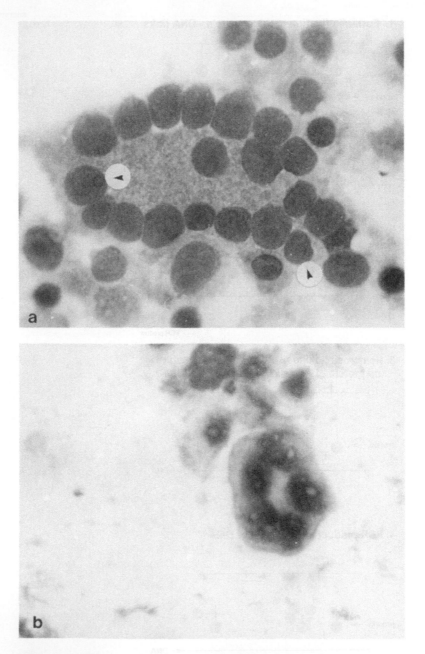

Figure 2 Cytological and molecular analysis of measles-infected HeLa cells. The multinucleation and nuclear inclusions characteristic of measles-infected cells are evident on routine H&E stain (a, arrows). A weak signal was seen in some of these cells with standard cDNA–RNA *in situ* hybridization. An intense signal was seen in each of the multinucleated cells after RT *in situ* PCR using measles-specific primers; note the sparing of the area corresponding to the inclusions (b) which were shown to be nucleoli using electron microscopic analysis (c, small arrow, nucleoli; large arrows, nucleocapsid of the virus).

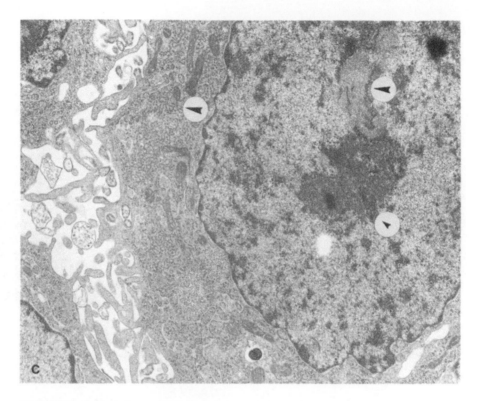

Figure 2 Continued

or there is no signal in the presence of good nuclear morphology, protease digestion is suboptimal. This problem can be overcome by performing protease digestions for varying lengths of time (e.g., 30, 60, and 90 min) to determine which time yields the strongest positive control (Fig. 3). This type of positive control tests the PCR amplification (albeit nonspecific) and detection procedures, but not the *in situ* hybridization step.

What is the nature of this positive control? Our studies have shown that it represents a combination of primer-dependent mispriming and primer-independent DNA repair. The fact that, with optimal digestion, the nonspecific signal can be eliminated with overnight DNase digestion forms the basis of the negative control for RT *in situ* PCR, and ensures that the signal observed after DNase digestion and RT-PCR with direct digoxigenin incorporation is specific. These explanations indicate that direct incorporation of the reporter molecule in tissue sections for DNA is *not* as reliable since the nonspecific signal would remain. Interestingly, freshly fixed cytospins are much less susceptible to this nonspecific signal, perhaps because less DNA damage (nicks) occurs than in paraffin-embedded tissues (Nuovo *et al.,*

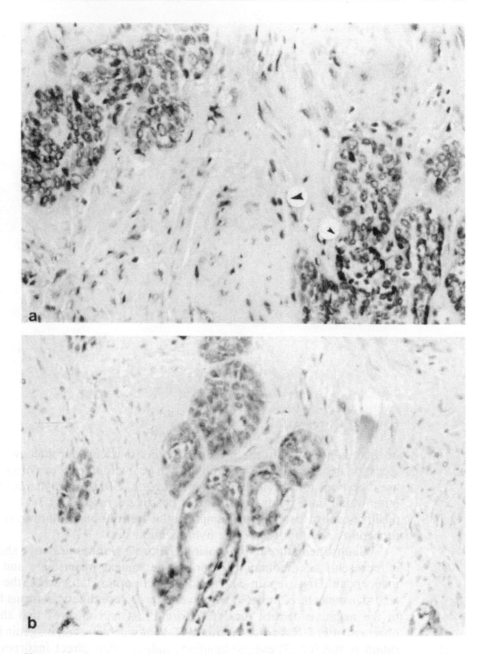

Figure 3 Comparison of the localization of the signal with the positive control and the test section for RT *in situ* PCR. The tissue is a cervical cancer that contains HPV 18 RNA. The virus is found only in the carcinoma cells. (a) Note how, in the positive control (no DNase), over 50% of the cells have a signal, including the cancer cells (small arrow) and the intervening stroma (large arrow). (b) After DNase digestion and RT-PCR with HPV 18-specific primers, the signal localizes to just the cancer cells.

1993c). However, increasing amounts of nonspecific signal can be observed in cytospins with increasing protease digestion time. This observation emphasizes the intricate association of protease digestion and fixation conditions (Nuovo *et al.,* 1993c).

We cannot overemphasize the need to perform multiple reactions on a single slide. Using a hydrophobic pen to separate different mixtures so the control reactions (e.g., nonsense primers or no *Taq* polymerase as negative control for PCR *in situ* hybridization; no DNase with direct incorporation of digoxigenin dUTP as the positive control for RT *in situ* PCR) can be done on the same slide as the test.

3. Mixed Cell Experiments

Mixing cell experiments can provide another useful control to consider in PCR *in situ* hybridization and RT *in situ* PCR. In these experiments, various concentrations of cells containing a specific target and cells that do not contain the target are mixed together. Clearly, the specificity of the *in situ* PCR test would be demonstrated by having the signal occur only in the appropriate cells. This distinction can be made on cytologic grounds, for example, when mixing multinucleated measles-infected HeLa cells and much smaller lymphocytes. When using measles-specific primers, only the larger cells should have a signal (Fig. 4). Similarly, when Epstein–Barr virus (EBV)-positive lymphocytes from a Burkitt's lymphoma cell line and squamous cells are mixed, the larger EBV-negative squamous cells should have no signal when EBV-specific primers are used (Fig. 5). Alternatively, when HPV-negative lymphocytes and the cytologically similar SiHa cells (HPV-positive epithelial cells) are mixed, they can be differentiated by colabeling for leukocyte markers that detect only the lymphocytes. Such experiments have shown only the SiHa cells have a signal with the HPV-specific primers, and the cells without signal are labeled with leukocyte common antigen (Nuovo *et al.,* 1991b).

III. APPLICATIONS

A. The Equivocal Penile or Vulvar Biopsy

About 40% of papillary lesions of the vulva or penis that are clinically suggestive of an HPV-induced condyloma (i.e., low-grade SIL) lack the diagnostic features of this lesion on histological examination (Nuovo *et al.,*

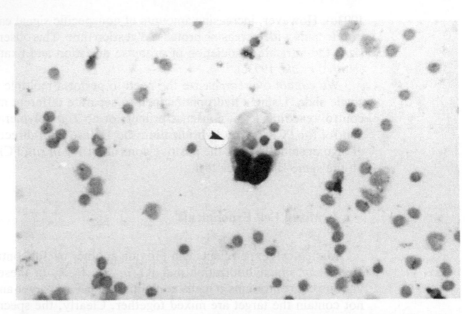

Figure 4 The specificity of RT *in situ* PCR as demonstrated by cell mixing experiments. In a cytospin, normal lymphocytes were mixed with measles-infected HeLa cells, which can be recognized by their large size, multinucleation, and inclusion. With measles-specific primers, only the HeLa cells (arrow) showed a signal, even when only 1 to a few cells were present with many lymphocytes.

1989,1990). Terms such as "equivocal for condyloma," "suggestive of condyloma," or "borderline condyloma" may be used for such lesions. Although these terms reflect the difficulty inherent in making a histological diagnosis, they may be confusing to the clinician and, more importantly, to the patient, who may be understandably confused about whether he or she has a sexually transmitted disease.

About 10–20% of these equivocal penile and vulvar lesions are HPV positive as determined by Southern blot hybridization or PCR (Nuovo *et al.*, 1989,1990). However, the histological markers of HPV detection in such tissues cannot be ascertained because DNA extraction prior to molecular analysis precludes histological correlation. This problem is eliminated with PCR *in situ* hybridization. Figure 6 shows a representative case. Note that a weak signal is evident with standard *in situ* hybridization for HPV 6/11. A much more intense signal is evident in many more cells if *in situ* hybridization is preceded by PCR. Such analyses have shown that a focally thickened granular layer in conjunction with epithelial crevices and para- or hyperkeratosis is a marker for HPV infection in such equivocal cases (Nuovo *et al.*, 1992a,d).

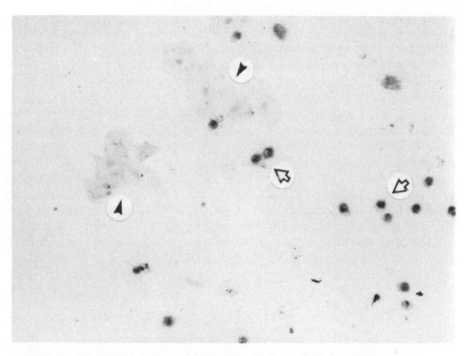

Figure 5 The specificity of *in situ* PCR as demonstrated by cell mixing experiments. An EBV-positive Burkitt's lymphoma cell line was mixed with virus-negative oral squamous cells. With the manual hot start maneuver and EBV-specific primers, all the lymphocytes displayed a signal with direct incorporation of digoxigenin-labeled nucleotide into the PCR product (open arrows). None of the squamous cells, which are easily distinguished cytologically by their ample cytoplasm (solid arrows), had nuclei with detectable signal.

B. In Situ *Detection of PCR-Amplified HIV-1 Nucleic Acids and Tumor Necrosis Factor cDNA*

The histological distribution of PCR-amplified HIV-1 DNA and RNA was investigated in cervical biopsies from women with AIDS (Nuovo *et al.*, 1993b). Interest has grown in the identification of cell types in the genital tract that may be infected by HIV-1, since they may play a role in transmitting the disease. However, HIV-1 nucleic acids are often present at levels below the threshold of standard *in situ* hybridization or immunohistochemical detection methods (Pomerantz *et al.*, 1988; Nuovo, 1992, 1993). Alternatively, infection may be active when up-regulation of viral transcripts, and perhaps production of viral particles, occurs (Pomerantz *et al.*, 1990). Use the RT *in situ* PCR methodology and PCR *in situ* hybridization is effective in de-

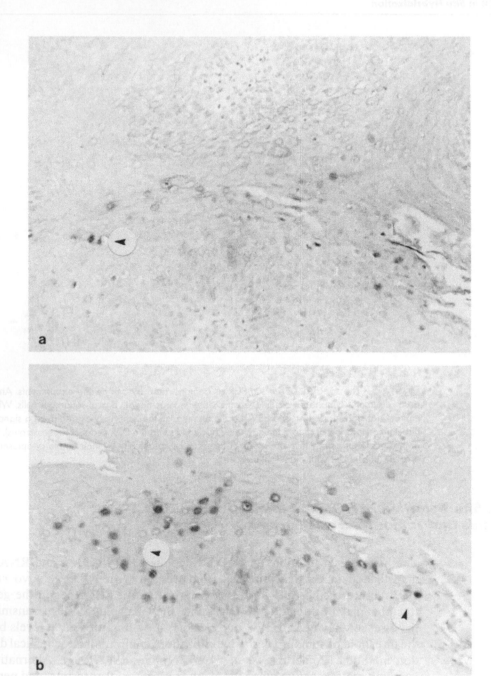

Figure 6 Histological and molecular analysis of a vulvar lesion. (a) Perinuclear halos and nuclear atypia were not evident although a thickened granular layer was noted in epithelial crevices. HPV 6 DNA was rarely detected in such cells by *in situ* analysis. (b) A hybridization signal was observed in many more cells in these areas after amplification by PCR.

termining when infection is latent (when HIV-1 DNA alone is detected) or active (when both viral RNA and DNA are present). Further colabeling experiments allow determination of the phenotype of the infected cells.

Each of 21 cervical biopsies from seropositive women had detectable HIV-1 DNA and RNA. The specificity of these results was confirmed by the absence of signal in 10 controls, the loss of signal by omission of *Taq* DNA polyermase, DNase pretreatment (but not RNase pretreatment) for HIV-1 DNA and omission of RT, predigestion in RNase (but not DNase), or use of "nonsense" primers for HIV-1 RNA (Fig. 7). The detection rate was less for standard *in situ* hybridization for HIV-1 DNA or RNA. Further, for the cases positive with standard *in situ* hybridization, the number of cells with detectable viral nucleic acid was less than after PCR amplification. This result is consistent with the observation that HIV-1 DNA and RNA tend to be present in relatively low copy numbers in infected cells, often below the detection threshold for standard *in situ* hybridization analysis.

An important finding of this study was the markedly specific histological distribution of amplified HIV-1 nucleic acids in cervical biopsies from seropositive adults (Nuovo *et al.*, 1993b). The virus was detected primarily in the endocervical aspect of the transformation zone, in stromal cells near the base of the endocervical glands, and in more scattered cells in the deep submucosa around microvessels (Fig. 7). This distribution pattern was in marked contrast to that of HPV DNA, which localizes (after PCR-amplification) only in the atypical squamous cells of the SILs (Nuovo *et al.*, 1991a). Clearly, HIV-1 and HPV infect mutually exclusive areas of the transformation zone.

The distribution pattern of the amplified HIV-1 nucleic acids was similar to that noted by immunohistochemistry for leukocytes in general and macrophages in particular. It was evident in routine histological stains that few of the HIV-1-infected cells were lymphocytes. The majority of the virus-positive cells colabeled with tumor necrosis factor cDNA, showing that these cells were macrophages. The infected macrophages may transmit the virus from the cervix to the lymphatic system and then to the regional lymph nodes, thus initiating a generalized infection.

C. Localization of PCR-Amplified Hepatitis C cDNA

Various histological changes have been associated with HCV, the etiologic agent in most cases of post-transfusion hepatitis. Although some histological changes may be suggestive of HCV, they may not appear together in such cases and they may be associated with other conditions such as auto-immune liver disease (Nuovo *et al.*, 1993a).

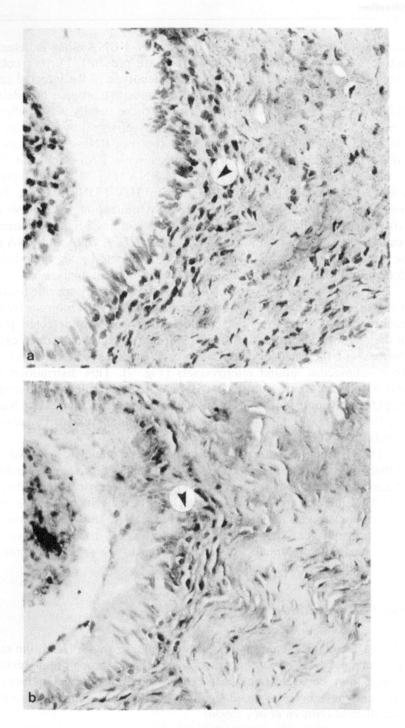

Figure 7 Molecular analyses of cervical tissue from an adult woman seropositive for HIV-1. (a) An endocervical gland (arrow) near the transformation zone is evident. HIV-1 DNA (arrow)

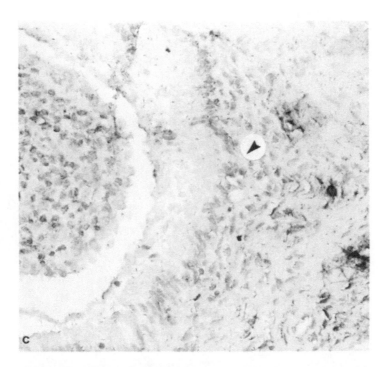

is evident after PCR *in situ* hybridization in many of the stromal cells at the base of the gland. (b) Most cells positive for HIV-1 DNA contain RT-PCR-amplified tumor necrosis factor (TNF) RNA (not shown) and HIV-1 RNA. (c) The hybridization signal for HIV-1 DNA was lost if nonsense primers were used for PCR or if *Taq* polymerase was omitted from the amplifying solution. Positive nuclei are dark because of the action of the alkaline phosphatase–anti-digoxigenin antibody conjugate on the chromagen NBT/BCIP.

Studies of HCV RNA with cDNA–RNA *in situ* hybridization have suggested that HCV RNA can be found in scattered periportal hepatocytes, but not in inflammatory cells or damaged bile duct epithelium. A study by Negron *et al.* (1992) noted that HCV RNA was detectable in the livers of HCV-infected chimpanzees by *in situ* hybridization soon after infection, but was not detectable with this methodology 3 wk after the onset of infection (Negron *et al.*, 1992). This finding suggests that HCV copy number may decrease as the disease progresses. A study comparing RNA–cDNA *in situ* hybridization and PCR-amplified cDNA suggested that RT *in situ* PCR can detect the presence of viral cDNA even when the copy number is below the detection threshold for *in situ* hybridization (Nuovo *et al.*, 1993a). PCR-amplified viral cDNA was noted in many hepatocytes in a perilobular distribution and in scattered Kupffer cells (Fig. 8). This pattern was noted after RT *in situ* PCR with direct incorporation of the labeled nucleotide, and when *in situ*

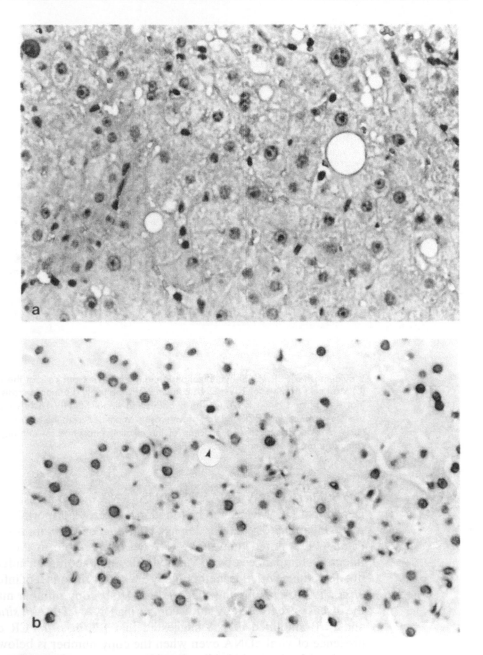

Figure 8 Comparison of the signal with the negative control and the test section for RT *in situ* PCR. (a) The tissue is a liver biopsy from a patient seropositive for infection by HCV. (b) PCR-amplified HCV cDNA was detected in hepatocytes and Kupffer cells (arrow) but not endothelial cells using RT *in situ* PCR. (c) The signal was lost if the RT step was omitted.

C

hybridization was done after the RT and PCR steps. Other experimental controls included omitting the RT step, using unrelated primer sequences called "nonsense" primers, and preceding RT *in situ* PCR with an RNase treatment. These controls resulted in an absence of signal in the tissue (Fig. 8).

Positive HCV signal was usually located in a perinuclear pattern within cells, with occasional cytoplasmic signal (Fig. 8). Kupffer cells, liver macrophages, and occasional bile duct epithelial cells located in portal tracts with an inflammatory infiltrate also displayed positive signals. No signal was noted in lymphocytes or endothelial lining cells. This concept of an internal negative control (specific cell types in a tissue that lack the target of interest) is very important in interpreting the results of PCR *in situ* hybridization, because it demonstrates the specificity of the results, as it does for immunohistochemistry.

ACKNOWLEDGMENTS

This work was supported by grants from the Lewis Foundation and Perkin-Elmer Corporation.

REFERENCES

Bagasra, O., Hauptman, S. P., Lischer, H. W., Sachs, M., and Pomerantz, R. J. (1992). Detection of human immunodeficiency virus type 1 provirus in mononuclear cells by *in situ* polymerase chain reaction. *N. Engl. J. Med.* **326,** 1385–1391.

Chiu, K. P., Cohen, S. H., Morris, D. W., and Jordan, G. W. (1992). Intracellular amplification of proviral DNA in tissue sections using the polymerase chain reaction. *J. Histochem. Cytochem.* **40,** 333–341.

Chou, Q., Russell, M., Birch, D. E., Raymond, J., and Bloch, W. (1992). Prevention of pre-PCR mis-priming and primer dimerization improves low copy number amplifications. Nucleic Acid Res. **20,** 1717–1723.

Crum, C. P., and Nuovo, G. J. (1991). "Genital Papillomaviruses and related neoplasms." Raven Press, New York.

Crum, C. P., Nuovo, G. J., Friedman, D., and Silverstein, S. J. (1988). A comparison of biotin and isotope labeled ribonucleic acid probes for in situ detection of HPV 16 ribonucleic acid in genital precancers. *Lab. Invest.* **58,** 354–359.

Embretson, J., Zupancic, M., Beneke, J., Till, M., Wolinsky, S., Ribas, J. L., Burke, A., and Haase, A. T. (1993a). Analysis of human immunodeficiency virus infected tissues by amplification and in situ hybridization revelas latent and permissive infections at single cell resolution. *Proc. Natl. Acad. Sci. USA* **90,** 357–361.

Embretson, J., Zupancic, M., Ribas, J. L., Burke, A., Racz, P., Tenner-Racz, T., and Haase, A. T. (1993b). Massive covert infection of helper T lymphocytes and macrophages by HIV during the incubation period of AIDS. *Nature* **362,** 359–362.

Erlich, H. A., Gelfand, D., and Sninsky, J. J. (1991). Recent advances in the polymerase chain reaction. *Science* **252,** 1643–1650.

Haase, A. T., Retzel, E. F., and Staskus, K. A. (1990). Amplification and detection of lentiviral DNA inside cells. *Proc. Natl. Acad. Sci. USA* **87,** 4971–4975.

Harper, M. E., Marselle, L. M., Gallo, R. C., and Wong-Staal, F. (1986). Detection of lymphocytes expressing human T-lymphotrophic virus type III in lymph nodes and peripheral blood from infected individuals by in situ hybridization. *Proc. Natl. Acad. Sci. USA* **83,** 772–776.

Lawrence, J. B., Marselle, L. M., Byron, K. S., Johnson, C. V., Sullivan, J. L., and Singer, R. H. (1990). Subcellular localization of low-abundance human immunodeficiency virus nucleic acid sequences visualized by fluorescence in situ hybridization. *Proc. Natl. Acad. Sci. USA* **87,** 5420–5424.

Lorincz, A. T. (1989). Human papillomavirus testing. *Diagn. Clin. Test.* **27,** 28–37.

Negron, F., Pacchioni, D., Shimizu, Y., Miller, R. H., Bussolati, G., Purcell, R. H., and Bonino, F. (1992). Detection of intrahepatic replication of HCV virus RNA by in situ hybridization and comparison histopathology. *Proc. Natl. Acad. Sci. USA* **89,** 2247–2251.

Nuovo, G. J. (1989). A comparison of slot blot, Southern blot and in situ hybridization analyses for human papillomavirus DNA in genital tract lesions. *Obstet. Gynecol.* **74,** 673–677.

Nuovo, G. J. (1992). "PCR in Situ hybridization: Protocols and Applications." Raven Press, New York.

Nuovo, G. J. (1993). "Cytopathology of the Cervix and Vagina: An Integrated Approach." Williams and Wilkins, Baltimore.

Nuovo, G. J., O'Connell, M., Blanco, J. B., Levine, R. U., and Silverstein, S. J. (1989). Correlation of histology and human papillomavirus DNA detection in condyloma acuminatum and condyloma-like vulvar lesions. *Am. J. Surg. Pathol.* **13,** 700–706.

Nuovo, G. J., Hochman, H., Eliezri, Y. D., Comite, S., Lastarria, D., and Silvers, D. N. (1990). Detection of human papillomavirus DNA in penile lesions histologically negative for condylomata: Analysis by in situ hybridization and the polymerase chain reaction. *Am. J. Surg. Pathol.* **14,** 829–836.

Nuovo, G. J., MacConnell, P., Forde, A., and Delvenne, P. (1991a). Detection of human papillomavirus DNA in formalin fixed tissues by in situ hybridization after amplification by PCR. *Am. J. Pathol.* **139,** 847–854.

Nuovo, G. J., Gallery, F., MacConnell, P., Becker, J., and Bloch, W. (1991b). An improved technique for the detection of DNA by *in situ* hybridization after PCR-amplification. *Am. J. Pathol.* **139,** 1239–1244.

Nuovo, G. J., Gallery, F., and MacConnell, P. (1992a). Analysis of the distribution pattern of PCR-amplified HPV 6 DNA in vulvar warts by in situ hybridization. *Modern Pathol.* **5,** 444–448.

Nuovo, M. A., Nuovo, G. J., MacConnell, P., and Steiner, G. (1992b). Analysis of Paget's disease of bone for the muscles virus using the reverse transcriptase in situ polymerase chain reaction technique. *Diagn. Mol. Pathol.* **1,** 256–265.

Nuovo, G. J., Margiotta, M., MacConnell, P., and Becker, J. (1992c). Rapid *in situ* detection of PCR-amplified HIV-1 DNA. *Diagn. Mol. Pathol.* **1,** 98–102.

Nuovo, G. J., Gorgone, G., MacConnell, P., and Goravic, P. (1992b). *In situ* localization of human and viral cDNAs after PCR-amplification. *PCR Meth. Appl.* **2,** 117–123.

Nuovo, G. J., Becker J., MacConnell, P., Margiotta, M., Comite, S., and Hochman, H. (1992e). Histological distribution of PCR-amplified HPV 6 and 11 DNA in penile lesions. *Am. J. Surg. Pathol.* **16,** 269–275.

Nuovo, G. J., Lidonocci, K., MacConnell, P., and Lane, B. (1993a). Intracellular localization of PCR-amplified HCV cDNA. *Am. J. Pathol.* **17,** 683–690.

Nuovo, G. J., Forde, A., MacConnell, P., and Fahrenwald, R. (1993b). In situ detection of PCR-amplified HIV-1 nucleic acids and tumor necrosis factor cDNA in cervical tissues. *Am. J. Pathol.* **143,** 40–48.

Nuovo, G. J., Gallery, F., Hom, R., MacConnell, P., and Bloch, W. (1993c). Importance of different variables for optimizing in situ detection of PCR-amplified DNA. *PCR Meth. Appl.* **2,** 305–312.

Patterson, B. K., Till, M., Otto, P., Goolsby, C., Furtado, M. R., McBride, L. J., and Wolinsky, S. M. (1993). Detection of HIV-1 DNA and messenger RNA in individual cells by PCR-driven in situ hybridization and flow cytometry. *Science* **260,** 976–979.

Pomerantz, R. J., de la Monte, S. M., Donegan, S. P., Rota, T. R., Vogt, M. W., Craven, D. E., and Hirsch, M. S. (1988). Human immunodeficiency virus (HIV) infection of the uterine cervix. *Ann. Intern. Med.* **108,** 321–327.

Pomerantz, R. J., Trono, D., Feinberg, M. B., and Baltimore, D. (1990). Cells nonproductively infected with HIV-1 exhibit an aberrant pattern of viral RNA expression: A molecular model for latency. *Cell* **61,**1271–1276.

Shapshak, P., Sun, N. C. J., Resnick, L., Hsu, M. Y. K., Tourtellotte, W. W., Schmid, P., Conrad, A., Fiala, M., and Imagawa, D. T. (1990). The detection of HIV by in situ hybridization. *Modern. Pathol.* **3,** 146–153.

Walboomers, J. M. M., Melchers, W. J. G., Mullink, H., Meijer, C. J. L. M., Struyk, A., Quint, W. G. J., van der Noordaa, J., and Ter Schegget, J. (1988). Sensitivity of in situ detection with biotinylated probes of human papillomavirus type 16 DNA in frozen tissue sections of squamous cell carinoma of the cervix. *Am. J. Pathol.* **131,** 587–594.

for condylomatos. Analysis by in situ hybridization and the polymerase chain reaction. Am. J. Surg. Pathol. 14, 829–836.

Nuovo, G.J., MacConnell, P., Forde, A., and Delvenne, P. (1991a). Detection of human papillomavirus DNA in formalin-fixed tissues by in situ hybridization after amplification by PCR. Am. J. Pathol. 139, 847–854.

Nuovo, G.J., Gallery, F., MacConnell, P., Becker, J., and Bloch, W. (1991b). An improved technique for the detection of DNA by in situ hybridization after PCR amplification. Am. J. Pathol. 139, 1239–1244.

Nuovo, G.J., Gallery, F., and MacConnell, P. (1992a). Analysis of the distribution pattern of PCR-amplified HPV 6 DNA in vulvar warts by in situ hybridization. Modern Pathol. 5, 444–448.

Nuovo, M. A., Nuovo, G.J., MacConnell, P., and Steiner, G. (1992b). Analysis of Paget's disease of bone for the measles virus using the reverse transcriptase in situ polymerase chain reaction technique. Diagn. Mol. Pathol. 1, 256–265.

Nuovo, G.J., Margiotta, M., MacConnell, P., and Becker, J. (1992c). Rapid in situ detection of PCR-amplified HIV-1 DNA. Diagn. Mol. Pathol. 1, 98–102.

Nuovo, G.J., Gorgone, G., MacConnell, P., and Gorevic, P. (1992d). In situ localization of human and viral cDNAs after PCR-amplification. PCR Meth. Appl. 2, 117–123.

Nuovo, G.J., Becker, J., MacConnell, P., Margiotta, M., Comite, S., and Hochman H. (1992e). Histological distribution of PCR-amplified HPV 6 and 11 DNA in penile lesions. Am. J. Surg. Pathol. 16, 269–275.

Nuovo, G.J., Lidonicci, K., MacConnell, P., and Lane, B. (1993a). Intracellular localization of PCR amplified HCV cDNA. Am. J. Pathol. 17, 683–690.

Nuovo, G.J., Forde, A., MacConnell, P., and Fahrenwald, R. (1993b). In-situ detection of PCR-amplified HIV-1 nucleic acids and tumor necrosis factor cDNA in cervical tissues. Am. J. Pathol. 143, 40–48.

Nuovo, G.J., Gallery, F., Hom, R., MacConnell, P., and Bloch, W. (1993c). Importance of different variables for optimizing in situ detection of PCR amplified DNA. PCR Meth. Appl. 2, 305–312.

Patterson, B.K., Till, M., Otto, P., Goolsby, C., Furtado, M.R., McBride, L.J., and Wolinsky, S.M. (1993). Detection of HIV-1 DNA and messenger RNA in individual cells by PCR-driven in situ hybridization and flow cytometry. Science 260, 976–979.

Pomerantz, R.J., de la Monte, S., Donegan, S.P., Rota, T.R., Vogt, M.W., Craven, D.E., and Hirsch, M.S. (1988). Human immunodeficiency virus (HIV) infection of the uterine cervix. Ann. Intern. Med. 108, 321–327.

Pomerantz, R.J., Trono, D., Feinberg, M.B., and Baltimore, D. (1990). Cells nonproductively infected with HIV-1 exhibit an aberrant pattern of viral RNA expression: A molecular model for latency. Cell 61, 1271–1276.

Shibata, D., Sun, N.C.J., Nezet, L., Hsu, M.Y.K., Tourtellote, W.W., Schmid, P., Conrad, A., Fiala, M., and Imagawa, D.T. (1990). The detection of HIV by in situ hybridization. Mod. Pathol. 3, 344–352.

Walboomers, J.M.M., Melchers, W.J.G., Mullink, H., Meijer, C.J.L.M., Struyk, A., Quint, W.G.J., van der Noordaa, J., and Ter Schegget, J. (1988). Sensitivity of in situ detection with biotinylated probes of human papillomavirus type 16 DNA in frozen tissue sections of squamous cell carcinoma of the cervix. Am. J. Pathol. 131, 587–594.

12

Nucleic Acid Sequence-Based Amplification

Roy Sooknanan
Cangene Corporation
Mississauga, Ontario
Canada L4V 1T4

Bob van Gemen
Organon Teknika
5281RM Boxtel, The Netherlands

Lawrence T. Malek
Cangene Corporation
Mississauga, Ontario
Canada L4V IT4

I. BACKGROUND

A. General Description of the Process

Nucleic acid sequence-based amplification (NASBA™)[1] is a homogeneous, isothermal *in vitro* amplification process (Davey and Malek, 1989), shown schematically in Fig. 1. The steps of the process are indicated in the boxed

[1] NASBA™ is a registered trademark of Cangene Corporation.

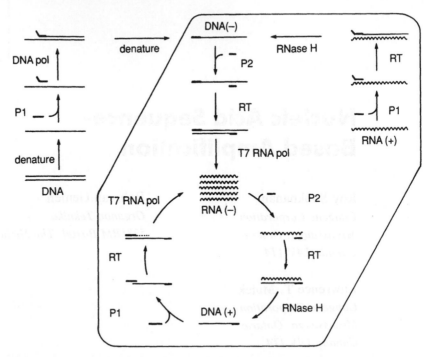

Figure 1 Schematic of the NASBA™ process. Abbreviations: RT, reverse transcriptase; P1, primer 1; P2, primer 2; pol, polymerase. Symbols: (+), plus-sense or coding strand; (−), minus-sense or noncoding strand; bold lines, primers; light lines, DNA; wavy lines, RNA; dotted line, incomplete DNA strand.

region of Fig. 1. The reaction mixture contains three enzymes—reverse transciptase (RT), RNase H, and T7 RNA polymerase—and two primers. Primer 1 (P1) contains a 3′ terminal sequence that is complementary to a sequence on the analyte nucleic acid and a 5′ terminal (+)sense sequence of a promoter that is recognized by T7 RNA polymerase. Primer 2 (P2) contains a sequence complementary to the P1-primed DNA strand. The enzymes and primers operate in concert to amplify a specific nucleic acid sequence exponentially.

The addition of an RNA begins the NASBA™ process, with annealing of P1 to the RNA analyte [RNA(+)], followed by extension of P1 with RT (Verma, 1977) to synthesize a cDNA. The original RNA analyte of the resulting RNA : cDNA hybrid is degraded by RNase H, to allow the annealing of P2 to the single-stranded cDNA [DNA(−)]. Single-stranded RNA is not a substrate for RNase H (Crouch and Dirksen, 1982). After annealing to the cDNA, P2 is extended with RT to form a DNA containing a template for transcription and a double-stranded T7 promoter. T7 RNA polymerase (Davanloo *et al.*, 1984) then recognizes the completed promoter and generates multiple copies of an RNA product [RNA(−)], which terminates at the 5′ end of P2 and is antisense to the original RNA analyte.

Each newly synthesized RNA product may then anneal to P2, from which a DNA product is synthesized with RT. The RNA strand of the resulting RNA : DNA hybrid is degraded by RNase H, to allow the annealing of P1 to the 3′ end of the DNA product [DNA(+)]. The DNA product represents the template for transcription of the RNA product. On annealing of P1, the DNA product is extended with RT using P1 as the template to form a double-stranded T7 promoter. The DNA product, now with a completed T7 promoter, directs the synthesis of many copies of the RNA product [RNA(−)]. Each newly transcribed RNA product can be used as a template for synthesis of a DNA product, which can be used as template for the further transcription of RNA product. The continuous repetition of this process results in the exponential synthesis of RNA and DNA products.

A specific nucleic acid sequence of a DNA analyte may also be amplified by NASBA™. In general, the DNA is first prepared for addition to the reaction as indicated by the steps outside the boxed region of Fig. 1. In a separate reaction, the duplex DNA is first denatured to allow one of its strand to anneal to P1. The annealed P1 is then extended with a DNA polymerase to synthesize a cDNA. The P1-primed cDNA [DNA(−)] is liberated from its DNA template by denaturation, and the entire mixture is added to a NASBA™ reaction.

Thus, the NASBA™ process, as described in Fig. 1, can be used for the amplification of internal sequences of any RNA or DNA analyte. For special cases, such as the 5′ terminal sequence of RNA or the 3′ terminal sequence of DNA, the nucleic acid template can be introduced directly into the cyclic phase. For example, a DNA sequence flanking a restriction endonuclease site may be amplified using a P1 primer that is complementary to a discrete 3′ end of one strand of a DNA restriction fragment. This DNA fragment is extended using the annealed P1 primer as a template to form a double-stranded T7 promoter and template for the transcription of the RNA product. The transcribed RNA product would then be used as template to synthesize a DNA product by hybridization and extension of a P2 primer.

The RNA products of one NASBA™ reaction may be added directly to a second reaction as an RNA analyte. The primers of this second NASBA™ reaction are typically nested within those of the first reaction. In this case, the P1 primer of the second reaction is complementary to the added RNA product of the first reaction. Thus, the RNA product of the second NASBA™ reaction would be antisense to the RNA product of the first reaction, but the same sense as the original RNA analyte of the first reaction. Nested primers in serial NASBA™ reactions are used to increase the sensitivity or specificity of detecting particular analytes.

B. Oligonucleotide Primers

A unique characteristic of the NASBA™ P1 primer that distinguishes it from the primers used in other DNA amplification processes such as the polymerase chain reaction (PCR; Mullis *et al.,* 1986) is the T7 RNA polymerase promoter sequence. This region consists of a highly conserved 23-nucleotide(nt) consensus sequence (-17 to $+6$; 5'-TAATACGACTCAC-TATAGGGAGA; Dunn and Studier, 1983; Moffat *et al.,* 1984) positioned near the 5' end of the P1 primer. The AGA sequence ($+4$ to $+6$), which is frequently present in the initiation sites of strong promoters, gives a 2- to 4-fold increase in RNA synthesis when included in the P1 primer. In addition, an extra 5 nt at the distal 5' end of the P1 primer (-18 to -22, 5'-AATTC) are absolutely required for efficient amplification. The preference for a particular sequence at this position was not studied, but an AT-rich region is common to consensus sequences of strong T7 promoters.

The sequence of the P1 primer immediately flanking the initiation site is also important in minimizing abortive cycling, a process that results in the synthesis of prematurely terminated RNA transcripts under normal (Martin *et al.,* 1988) or limiting (Ling *et al.,* 1989) nucleoside triphosphate concentrations. For T7 RNA polymerase, abortive cycling has been observed in the first 8–12 nt following the incorporation of UMP (Martin *et al.,* 1988) and before or after the incorporation of a pyrimidine nucleotide (Ikeda, 1992). In NASBA™, this event results in correspondingly fewer full-length transcripts and slower kinetics. Abortive cycling is increased by the presence of two or more consecutive pyrimidines within the first 12 nt, although a cytosine flanked by purines does not pose a problem. Whenever possible, the $+7$ to $+12$ sequence of the NASBA™ P1 primer, composing part of the hybridizing sequence, is selected to best reduce abortive cycling.

Flanking the transcription initiation site and at the 3' end of the P1 primer is a sequence that hybridizes to the analyte nucleic acid. The P2 primer has a 3' terminal sequence that hybridizes to a P1-primed DNA synthesized using the analyte nucleic acid as template. As in PCR (Watson, 1989; Williams, 1989; Kwok *et al.,* 1990), selection of these hybridizing sequences of the NASBA™ primers is dependent on nucleotide length, composition, and sequence. Although the typical length is 20 nt, primers with hybridizing sequences from 17 to 30 nt have also been used successfully. A GC content of 45–60% is recommended but, more importantly, tracts of the same nucleotide should be avoided. Primers with 3' terminal complementarity of ≥ 2 nt may result in primer dimerization. Computer programs such as OLIGO (National Biosciences, Hamel, MN) are recommended for screening primer sequences for potential interactions. As in PCR, it may be necessary to select and test more than one primer pair for each target to find the one that gives the desired performance.

C. Reaction Conditions

The NASBA™ reaction conditions and reagents that are generally optimal for most primers are shown in Table 1. The optimal concentration of ribonucleoside triphosphates (rNTPs), reported as high as 6 mM each (Fahy et al., 1991), is dependent on the concentration of KCl. By taking into account the Na$^+$ concentration from the tetrasodium nucleoside triphosphates and the K$^+$ from KCl, the optimal monovalent cation concentration is approximately 100 mM. Amplification efficiencies are equivalent for reaction conditions of 2 mM rNTPs plus 50 mM KCl, 3 mM rNTPs plus 34 mM KCl, or 4 mM rNTPs plus 18 mM KCl, but were 2.5-fold lower for 1 mM rNTPs plus 66 mM KCl. Higher KCl concentrations are inhibtory, whereas lower concentrations promote the formation of nonspecific primer-related products. Although 50 mM KCl is normally best for most primer sets, some primers may require an adjustment of the KCl concentration to maximize specificity, sensitivity, and product yield. The optimal MgCl$_2$ concentration (12 mM) is equal to the total concentration of nucleoside triphosphates. The optimal pH of the Tris-HCl buffer ranges from 8.3 to 8.5 at 25°C.

The optimal temperature is inversely related to the dimethylsulfoxide (DMSO) concentration. At 15% DMSO, the optimal reaction temperature is 40–41°C. Higher temperatures and DMSO concentrations are inhibitory, whereas lower temperatures and DMSO concentrations promote the formation of nonspecific primer-related products. The concentration of each primer in NASBA™ can vary from 0.1 to 0.2 μM but amplification is optimal when both are used in equimolar amounts. Nonspecific primer-related products

TABLE 1
NASBA™ Reaction Conditions[a]

A	B
40 mM Tris-HCl (pH 8.5)	0.2 μM P1
50 mM KCl	0.2 μM P2
12 mM MgCl$_2$	1 mM dATP
10 mM DTT	1 mM dCTP
15% DMSO	1 mM dGTP
100 μg/ml BSA	1 mM dTTP
40 units T7 RNA polymerase	2 mM ATP
8 units AMV reverse	2 mM CTP
transcriptase	
0.2 units RNase H	2 mM GTP
12.5 units RNA Guard	2 mM UTP

[a] Generally a 25-μl reaction is incubated at 40°C for 90 min.

are also formed at higher primer concentrations, whereas lower than optimal concentrations proportionally reduce the product yield and reaction kinetics.

The optimal levels of avian myeloblastosis virus (AMV) reverse transcriptase and T7 RNA polymerase in NASBA™ reactions are relatively broad. A 2- to 3-fold increase in either enzyme can produce a slight increase in the levels of amplified products, whereas a lower than optimal activity of either causes a dramatic reduction in the kinetics of amplification. A similar reduction was observed with suboptimal activities (≤0.1 U) of *Escherichia coli* RNase H. However, greater than optimal levels (≥0.2 U) of RNase H promoted the formation of primer-related nonspecific products. The optimal activities are reported for enzymes from the recommended suppliers. However, some variation has been observed in the activities of enzymes from other suppliers, particularly with *E. coli* RNase H. The optimal reaction conditions for NASBA™ are very similar to the standard assay conditions for AMV reverse transcriptase. The activities of T7 RNA polymerase and *E. coli* RNase H are approximately 3 times higher under NASBA™ conditions than under the suppliers' suggested assay conditions.

D. Kinetics

The rates of accumulation of RNA and DNA products in NASBA™ reactions were determined under optimized conditions. In separate reactions containing 10^4 molecules of added template, the amounts of RNA or DNA were quantified by monitoring the incorporation of radioactivity from [α-^{32}P]CTP or [α-^{32}P]dCTP, respectively. Analysis of the labeled products by polyacrylamide gel electrophoresis (PAGE) (Fig. 2) indicated that both RNA and DNA products were of the anticipated size. Two forms of DNA product, one the same size as the RNA product and the other approximately 20 bases longer, agreed with expected cDNA products without and with the T7 promoter, respectively.

The DNA containing the T7 promoter could represent either strand of a possible double-stranded DNA product, but was composed exclusively of the P2-primed DNA strand. The P1 primer, in contrast, remained relatively unextended throughout the course of the NASBA™ reaction. Although the priming function of the P1 primer is necessary to initiate amplification from an internal RNA or DNA sequence, it functions later only as a template for synthesis of an active double-stranded promoter (Milligan *et al.*, 1987).

Quantification of incorporated nucleotides (Fig. 3A) shows that both RNA and DNA products accumulated exponentially during the first 45 min. Thereafter, the DNA synthesis stopped but the RNA synthesis continued, albeit at a much slower rate. Nonetheless, the efficiency of transcription between 45 and 60 min was at least 10 pmol RNA per pmol transcriptionally

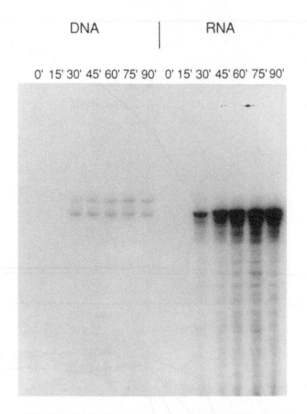

Figure 2 Characterization of labeled NASBA™ products. DNA and RNA products labeled in reactions containing [α-^{32}P]dCTP and [α-^{32}P]CTP, respectively, and sampled at the indicated times were analyzed by electrophoresis on a 7 M urea–7% polyacrylamide gel.

active template. The negative deflection from linear RNA synthesis after 45 min probably resulted from a steady loss of T7 RNA polymerase activity at 40°C. At 90 min, the amount of RNA product (approximately 4 μg or 100 pmol) exceeded the amount of DNA product approximately 20-fold.

The kinetics of P2 primer utilization in the NASBA™ process was investigated using 10^4, 10^3, 10^2, or 10 molecules of added template. Aliquots of each reaction were withdrawn at 15-min intervals. The DNA products formed by the extension of the [5'^{32}P]P2 primer in NASBA™ were separated by PAGE and quantified. As shown in Fig. 3B, DNA synthesis was nearly complete by 30 min in a reaction starting with 10^4 molecules of template, and by 45 min in one starting with 10 molecules. Within the first 30 min, the rate of DNA synthesis was roughly proportional to the logarithm of the initial template. During this exponential phase, the DNA product was amplified at least 10^{10}-fold in the first 30 min using 10 or 100 molecules of initial template. The kinetics using other primer sets may vary but, in general, a 90-min

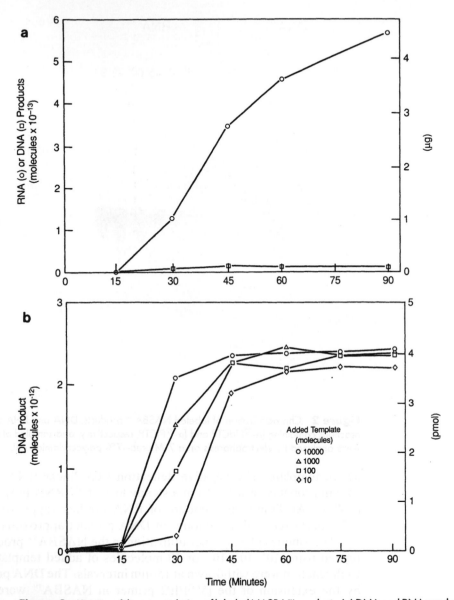

Figure 3 Kinetics of the accumulation of labeled NASBA™ products. (a) DNA and RNA products that were labeled and sampled as described in Fig. 2 were quantified for acid-insoluble cpm, as indicated. (b) DNA products formed by the extension of [5'-^{32}P]P2 primer in reactions containing the indicated amounts of initial DNA template were monitored at the indicated times. The amounts of DNA product were quantified by Cerenkov counting of excised bands from a polyacrylamide gel, as in Fig. 2.

reaction is sufficient for all primer sets without compromising sensitivity. In fact, specific product is usually detected after 60 min with most primer sets.

By 60 min, the maximum synthesis of DNA (4 pmol) is reached regardless of the amount of initial template. This level of DNA amplification represents the consumption of 80% of the P2 primer in the NASBA™ reaction. The efficient utilization of the P2 primer in the NASBA™ process was probably due to the annealing of the P2 primer being driven by an excess of RNA product. At 30 min, the RNA product was present at a 50-fold molar excess over the residual P2 primer. This condition contrasts with the relatively slow and incomplete conversion of the DNA product into the form containing the T7 promoter. Competition between the P1 primer and the same excess of RNA product to anneal with the DNA product may explain why this process never exceeds 40% completion. Hence, the excess RNA product, which augments the synthesis of its own transcription template, may ultimately limit the synthesis of its own double-stranded promoter.

E. Advantages and Disadvantages

NASBA™ offers several advantages over other amplification methods currently in use. NASBA™ is intrinsically suited to the amplification of RNA analytes because of the integration of reverse transcriptase into the amplification process. In contrast with RT-PCR, which requires a preliminary DNA synthesis step before amplification can begin, RNA analytes may be added directly to a NASBA™ reaction without special preparation. The single-stranded RNA product of NASBA™ is an ideal target for detection by various methods including solution probe hybridization. NASBA™ RNA products can be sequenced directly with a dideoxy method using reverse transcriptase and a labeled oligonucleotide primer. Similarly, the DNA product can be sequenced directly following RNase digestion of a NASBA™ reaction. The quality of the direct sequencing data is comparable to that achieved using purified plasmid DNA. In addition, the DNA product can be made double stranded, ligated into plasmid vectors, and cloned. DNA sequencing analysis of approximately 4800 bases indicated an error frequency of less than 0.3% in the cloned amplified sequences (Sooknanan, Howes, Read, and Malek, unpublished observations). Thus, the fidelity of NASBA™ is comparable to that of other amplification processes that use DNA polymerases lacking a 3'-exonuclease activity. The destruction of the original RNA analyte sequence in NASBA™ provides unique advantages for applications such as site-directed mutagenesis.

Other advantages of NASBA™ include the use of a single temperature, which obviates the need for special thermocycling equipment. The constant conditions maintained throughout the amplification reaction allow each step of the reaction to proceed as soon as an amplification intermediate becomes available. This feature not only shortens the time required for each step but also minimizes side reactions (such as the synthesis of cDNA hairpins from mRNA) while an intermediate awaits the next step. The augmented exponential kinetics of the NASBA™ process, which are caused by multiple transcription of RNA copies from a given DNA product, are intrinsically more efficient than DNA amplification methods that are limited to binary increases per cycle. Unlike amplification processes such as PCR, in which the maximum yield of product is limited by the initial primer level, the amount of RNA product that is amplified in NASBA™ exceeds the level of primers by at least one order of magnitude.

The primary disadvantage of NASBA™ relates to the use of three enzymes that must operate at a lower temperature relative to amplification methods using thermal cyclers. A practical consequence of the lower temperature may be an increase in nonspecific interactions of the primers. These interactions are minimized by the inclusion of DMSO. The lack of temperature cycles in NASBA™, although generally convenient, does remove any external control over the extent of amplification, that is, the number of cycles. In general, DNA analytes must be pretreated in a step consisting of P1-primed DNA synthesis and denaturation before being added to NASBA™ reactions. This pretreatment is roughly analogous to that used in RT-PCR for the amplification of RNA analytes.

F. Applications

Although the potential applications of NASBA™ are similar to those of other DNA amplification processes, in some instances NASBA™ offers unique advantages. Since NASBA™ uses RNA analytes directly (e.g., viral genomic RNA, mRNA, or rRNA), it can be used to detect virus production, gene expression, or cell viability. In addition, the presence of a DNA sequence, whether active or latent, can be detected by NASBA™ with an extra pretreatment of the analyte nucleic acid.

NASBA™ reactions containing an HIV-1 *gag* primer set are able to detect fewer than 10 molecules of *in vitro*-generated RNA template. NASBA™ can also detect viral RNA from as few as five HIV-1-infected cells in a background of 5×10^4 uninfected cells (Kievits *et al.*, 1991). For the detection of HIV-1 RNA in the plasma of clinical samples, the concordance of NASBA™ and RT-PCR was 91% (Bruisten *et al.*, 1993). By competitive co-amplification of an internal control RNA in NASBA™, quantification of HIV-1 RNA from plasmid was achieved with a dynamic range of 10^2 to 10^6 molecules (van Gemen *et al.*, 1993).

Serial NASBA™ reactions containing nested primers were used for the detection of *bcr-abl* mRNA in peripheral blood cells of Philadelphia chromosome-positive [Ph^1(+)] chronic myelocytic leukemia (CML) patients (Sooknanan *et al.*, 1993). This method was fully concordant with RT-PCR for detection of mRNA with both *bcr3-abl2* and *bcr2-abl2* junctions. With this strategy, the chimeric *bcr-abl* mRNA of a single Ph^1(+) leukemia cell could be detected in a background of 10^5 Ph^1(−) cells.

II. APPLICATION FOR THE DETECTION OF HIV-1 RNA IN PLASMA OR SERUM

The stepwise protocol described here consists of three parts: (1) nucleic acid isolation, (2) NASBA™, and (3) detection.

NOTE: *Because of the esquisite sensitivity of NASBA™ and other amplification methodologies, contamination of reactions with only a few molecules of specific exogenous nucleic acids will lead to false positive results. Therefore, it is crucial to follow scrupulously any measures to reduce carryover contamination (Kwok and Higuchi, 1989; Cimino et al., 1990; Sarker and Sommer, 1991; Malek et al., 1993). Whenever possible, nucleic acid isolation, NASBA™, and detection procedures should be performed in physically separated areas.*

NOTE: *Use only sterile distilled H$_2$O free of nucleases to prepare all reagents and, when possible, only sterile disposable plasticware. Avoid using diethylpyrocarbonate (DEPC)-treated water and glassware unless the DEPC is completely inactivated; otherwise NASBA will be inhibited. Wear disposable gloves at all times and change them frequently.*

A. Nucleic Acid Isolation

See Section III for the composition of buffers and reagents.

1. Materials

Lysis buffer
Silica suspension
Wash buffer
70% ethanol
Acetone

Elution buffer
1.5-ml Microfuge tubes (gamma irradiated)
15-ml Conical screw-cap tubes
Disposable plastic transfer pipette
Micropipettors
Centrifuges for 1.5-ml microfuge and 15-ml screw-cap tubes
Vortex mixer
56°C Heat block

2. Procedure

1. Add 0.5–1 ml sample (serum of plasma) to a 15-ml conical screw-cap tube containing 9.0 ml lysis buffer. Invert tubes several times to mix thoroughly. If sample volume is smaller, use a 1 : 10 ratio of sample to lysis buffer in a microfuge tube.
2. Add 70 μl silica suspension and vortex the tube for 5 sec. Leave for 10 ± 1 min at room temperature (18–25°C). **Invert every minute to mix.**
3. Centrifuge at 1500 g for 2 min. Remove supernatant using a 10-ml plastic pipette, leaving about 0.5-ml residual fluid. Remove the residual fluid with a sterile disposable transfer pipette **without** disturbing the pellet.
4. Add 1 ml wash buffer and resuspend the silica pellet by vortexing. Transfer the silica suspension to a 1.5-ml microfuge tube if a 15-ml screw-cap tube was used initially.
5. Centrifuge at 10,000 g for 15 sec. Remove supernatant with a sterile disposable transfer pipette.
6. Wash the silica pellet four times: once with wash buffer, twice with 70% ethanol, and once with acetone, as described here:

   Add 1 ml appropriate fluid to the reaction tube and vortex to resuspend pellet.
   Centrifuge at 10,000 g for 15 sec.
   Remove the supernatant with a sterile disposable tranfer pipette.

7. Completely dry the silica pellet by placing the opened tube in a heating block at 56°C for 10 ± 1 min. Cover the tube with tissue to avoid aerosol contamination.
8. Add 100 μl elution buffer (or water) and resuspend the pellet by vortexing. Incubate at 56°C for 10 ± 1 min to elute the nucleic acid.
9. Centrifuge at 10,000 g for 2 min. Using a micropipettor, transfer the supernatant to a new microfuge tube **without** disturbing the pellet.
10. Store the supernatant at −70°C (optimally) or at −20°C (acceptable).

B. NASBA™ Reaction

> NOTE: *Single-stranded RNA is added directly to NASBA™ reactions as template. DNA (single- or double-stranded) must be primed with P1 and the P1-primed strand separated by thermal denaturation before adding to NASBA™ reactions (see Section II,B,3). Double-stranded RNA only needs to be denatured.*

See Section III for the composition of buffers and reagents.

1. Materials

2.5× NASBA™ buffer
Dithiothreitol (DTT)
4× Primer mixture (see Fig. 4)
Enzyme mixture
Nuclease-free H$_2$O
1.5-ml microfuge tubes (gamma irradiated)
Micropipettors
65°C Heat block
40°C Precision water bath

2. Amplification Procedure

1. Prepare a pre-mix solution for NASBA™ reactions in a sterile micro-fuge tube containing (per reaction):

10 μl 2.5× NASBA™ buffer
1 μl 250 m*M* DTT
6.25 μl 4× primer mix
H$_2$O (nuclease-free) to 18μl

> NOTE: *Always prepare sufficient pre-mix solution for at least 1 extra reaction to account for volume loss during pipetting. These activities should*

Primer 1	5'-AATTCTAATACGACTCACTATAGGGTGCTATGTCACTTCCCCTTG GTTCTCTCA-3'
Primer 2	5'-AGTGGGGGGGACATCAAGCAGCCATGCAAA-3'
Probe	5'-GAATGGGATAGAGTGCATCCAGTGCATG-3'

Figure 4 Primers and probes used for amplification and detection of *gag* sequences of HIV-1 (van Gemen *et al.*, 1993).

be performed in a clean area to avoid contamination. Adjust the volume of water according to the nucleic acid sample volume used as template so that, the total volume is 23 μL including template. In this example, 5 μL nucleic acid sample is used. Increasing the nucleic acid sample volume further might increase inhibitory substances from the sample preparation procedure.

2. Vortex the pre-mix solution for 5 sec. Aliquot enough for a single reaction (in this example, 18 μl) into sterile 1.5-ml microfuge tubes.
3. Add the desired nucleic acid sample volume (in this example, 5 μl). Mix by tapping and centrifuge briefly.
4. Incubate at 65°C for 5 min. Transfer to a 40 ± 1°C water bath and allow to equilibrate for 5 min.
5. Add 2 μl enzyme mixture to the reaction tubes at 40 ± 1°C and gently mix by tapping the tube. Centrifuge at 10,000 g for 5 sec.
6. Continue to incubate at 40 ± 1°C for 90 min.
7. Centrifuge briefly to collect condensate. Place on ice or freeze at −20°C until ready for analysis.

3. Priming DNA Template for NASBA™

1. In a sterile microfuge tube prepare a mixture containing:

8 μl 2.5× NASBA™ buffer
1 μl Primer 1 (1–5 pmol)
2 μl DNA template (single - or double-stranded)
H₂O (nuclease-free) to 18 μl

NOTE: *2.5× NASBA™ buffer or a similar buffer containing only dNTPs can be used. The DNA template added should not exceed 1 μg.*

2. Place in a boiling water bath for 5 min. Transfer immediately to 50°C and allow to equilibrate for 5 min.
3. Add 2.0 μl of a mixture containing 10 mM DTT and 10 U AMV reverse transcriptase. Continue to incubate at 50 ± 1°C for 15 min.
4. Place in a boiling water bath for 5 min. Immediately chill on ice. This material should be kept on ice during use or stored at −20°C.
5. Use 2–5 μl P1-primed material as nucleic acid template in NASBA™.

C. Detection

The first two methods will provide rapid results and are convenient to perform in most laboratory settings. However, it may also be necessary to perform a Northern transfer of the agarose gel followed by oligonucleotide probe hybridization for product confirmation. In general, any of the numerous detection formats using either radioactive or nonradioactive oligonucleotide probes (Sambrook *et al.*, 1989) can be used with NASBA™. The oligonucleotide probe(s) must be complementary to an internal sequence that is flanked by the primers and does not cross-react with the primer sequences. Normally, sequencing is not routinely done in the clinical laboratory. However, it is required for sequence identification. The single-stranded RNA product of NASBA™ provides an excellent substrate for direct sequencing.

1. Ethidium Bromide Agarose Gel Electrophoresis

See Section III for the composition of buffers and reagents.

Materials

Low-melt agarose (NuSieve™, FMC Bioproducts, Rockland, ME)
Agarose
Ethidium bromide
Gel loading buffer
Horizontal electrophoresis apparatus

The products contained in 1- to 5-μl aliquots of completed NASBA™ reactions can be visualized by electrophoresis on agarose gels (Fig. 5). Mix samples with 2 μl gel loading buffer. Gels containing either 2% agarose or a combination of 3% low-melt agarose and 1% agarose in 1× TAE buffer containing 0.2 μg/ml ethidium bromide can be used. The latter will give

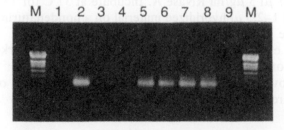

Figure 5 Detection of NASBA™ products by agarose gel electrophoresis. NASBA™ reactions containing the HIV-1 *gag* primers were performed with no template (lanes 1 and 9), 10^4 molecules of control template (lane 2), total nucleic acid from peripheral blood cells of two normal individuals (lanes 3 and 4), or total nucleic acid from four HIV-1-infected patients (lanes 5–8). The products form 5 μl-aliquots of each NASBA™ reaction were separated by electrophoresis on 3% Nu-Sieve–1% agarose gels and detected by ethidium bromide fluorescence.

better resolution of low molecular weight products. Follow standard methods for preparing and running agarose gels (Sambrook *et al.*, 1989). Do not run gels in buffer chambers that may be contaminated with ribonuclease.

2. Oligonucleotide Probe Gel Retardation Assay

See Section III for the composition of buffers and reagents.

Materials

Detection probe (see Fig. 4)
[γ-^{32}P]ATP
Polynucleotide kinase
Layer mixture
RNase A (DNase free; Boehringer Mannheim, Indianapolis, IN)
Polyacrylamide
Vertical electrophoresis apparatus
X-Ray film

1. The oligonucleotide detection probe is labeled at the 5' end with ^{32}P and polynucleotide kinase (Sambrook *et al.*, 1989).
2. Combine 2 μl labeled oligonucleotide detection probe in layer mixture and 2 μl completed NASBA™ reaction in a sterile 1.5-ml microfuge tube. Vortex to mix and centrifuge briefly.
3. Incubate at 40 ± 1°C for 15 min to hybridize. The temperature for hybridization can be increased depending on the T_m of the probe.
4. Add 1 μl 5 μg/ml RNase A. Incubate at 37°C for 15 min.
5. Load 1–2.5 μl hybridized material on a nondenaturing 7% polyacrylamide gel.
6. Start electrophoresis at 150 V. Allow the blue marker dye to reach the bottom of the gel (45–60 min).
7. Remove one glass plate. Perform autoradiography on X-ray film as required (Fig. 6).

3. Direct Sequence Analysis

See Section III for the composition of buffers and reagents.

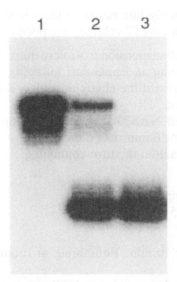

Figure 6 Detection of NASBA™ RNA product by solution hybridization and gel retardation. Polyacrylamide gel electrophoresis analysis of a [5'³²P] probe that was hybridized to RNA products of NASBA™ reactions containing nucleic acids that were isolated from plasma samples. The clinical samples were strongly positive (lane 1), weakly positive (lane 2), or negative (lane 3) to the presence of template RNA.

Materials

Sequencing primer
[γ-³²P]ATP
Polynucleotide kinase
RNase A (DNase free)
Biogel P-60 suspension (BioRad, Hercules, CA)
5× RT buffer
Deoxy:dideoxy NTPs mixture
AMV reverse transcriptase (Seikagaku, Rockville, MD)
Chase solution
Formamide dye mixture
Polyacrylamide
Microspin column (Ultrafree-MC; Millipore, Bedford, MA)
Sequencing apparatus
X-Ray film

Normally, primer 2 (P2) is labeled at the 5' end with ³²P and polynucleotide kinase (Sambrook *et al.*, 1989), and is used as the sequencing primer.

However, an internal or partially internal oligonucleotide primer may also be used when the sample contains P2-related nonspecific products.

1. Add 400 μl Biogel P-60 suspension to a microspin column. Centrifuge at 325 g for 8 min using an Eppendorf microfuge (Model 5415) or similar microfuge. Discard the eluate or replace collector tube with a fresh microfuge tube.
2. Add 10 μl completed NASBA™ reaction to top of column bed. Centrifuge at 325 g for 6 min. **Keep the eluate.**
3. Prepare a priming reaction mixture containing:

2 μl 5× RT buffer
5 μl Column-purified sample
1 μl ^{32}P-Labeled P2 (~0.5 pmol)
H_2O to 10 μl

4. Incubate at 50°C for 10 min. Equilibrate at room temperature for 5 min.
5. Add 2 μl appropriate deoxy:dideoxy NTPs mixtures to corresponding 1.5-ml microfuge tubes labeled G, A, T, or C.
6. Add 1 μl AMV reverse transcriptase (~10 U) to the priming reaction mixture. Mix by tapping gently. Centrifuge at 10,000 g for 15 sec.
7. Add 2.2 μl priming reaction mixture containing reverse transciptase to each of the four microfuge tubes (G, A, T, and C) containing the dideoxy mixtures.
8. Centrifuge at 10,000 g for 5 sec. Incubate at 48 ± 1°C for 15 min.
9. Add 2 μl chase solution. Continue to incubate at 48 ± 1°C for 10 min.
10. Add 4 μl formamide dye mixture and 1 μl 1 mg/mL RNase A. Incubate at 37°C for 15 min. The formamide dye and the RNase A solutions can be combined and added together. The RNase A digestion step is required to destroy excess RNA that causes anomalous migration of bands during electrophoresis.
11. Load 2 μl of each of the G, A, T, and C reactions onto a 7% polyacrylamide–urea sequencing gel for electrophoresis (Sambrook *et al.*, 1989). Follow with autoradiography (Fig. 7).

III. REQUIRED MATERIALS AND SOLUTIONS

Silica Suspension

1. Add sterile ddH$_2$O to 60 g silica to a final volume of 500 ml in a cylinder having a diameter of 5 cm. The height of the aqueous column is 27.5 cm.

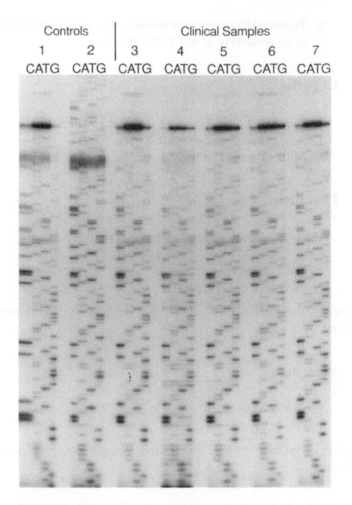

Figure 7 Direct sequence analysis of NASBA™ RNA products. Amplified products from NASBA™ reactions containing the HIV-1 *gag* primers and either 10^4 molecules of a control template (set 1) or total nucleic acid from peripheral blood cells of five HIV-1-infected patients (sets 3–7) were analyzed by dideoxy sequencing using a [5′-^{32}P]P2 primer to the RNA product. For comparison, the same sequencing procedure was used for plasmid DNA containing the HIV-1 *gag* sequence (set 2).

2. Allow to sediment at room temperature for 25 hr. Aspirate the supernatant, leaving 70 ml.
3. Add ddH$_2$O up to 500 ml and resuspend the silica particles by shaking the cylinder.
4. Allow to sediment at room temperature for 5 hr. Aspirate the supernatant, leaving 60 ml.
5. Add 600 μl 32% (w/v) HCl and resuspend silica particles by vortexing.

6. Transfer the suspension in 4-ml aliquots to 6-ml bottles, and sterilize in an autoclave at 121°C for 20 min.

L2 Buffer

1. Dissolve 12.1 g Tris-base in 800 ml water.
2. Adjust to pH 6.4 with HCl.
3. Add water to 1 liter.

Lysis Buffer

120 g guanidinium isothiocyanate (GuSCN)
2.6 g Triton X-100
100 ml L2 buffer
22 ml 0.2 M EDTA, pH 8.0

If necessary, heat to 60°C to dissolve completely. Final volume should be 222 ml.

Wash Buffer

120 g GuSCN
100 ml L2 buffer

If necessary, heat to 60°C to dissolve completely. Final volume should be 200 ml.

Elution Buffer

1. Dissolve 0.211 g Tris-base in 900 ml water.
2. Adjust to pH 8.5 with HCl.
3. Add water to 1 liter.
4. Sterilize 20 min at 120°C in an autoclave.

2.5× NASBA™ Buffer

100 mM Tris, pH 8.5
125 mM KCl
30 mM MgCl$_2$

2.5 mM each dNTP (dATP, dGTP, dCTP, dTTP)
5 mM each rNTP, (ATP, GTP, CTP, UTP)

Adjust with nuclease-free H$_2$O. Store at $-20°C$. The KCl concentration may be different for some primer sets.

The following stock solutions are required for the NASBA™ buffer:

1 M Tris, pH 8.5 (25°C)

4.42 g Tris-HCl (solid)
9.25 g Tris-base (solid)
H$_2$O to 100 ml

1 M KCl

7.45 g KCl (solid)
H$_2$O to 100 ml

1 M MgCl$_2$ · 6H$_2$O

20.33 g MgCl$_2$ · 6H$_2$O (solid)
H$_2$O to 100 ml

Alternatively, a 1 M MgCl$_2$ solution can be purchased from Sigma Chemical Co. (St. Louis, MO).

dNTPs/rNTPs

100 mM stock solutions are purchased from Pharmacia (Piscataway, NJ).

4× Primer Mixture

The oligonucleotide primers should be purified from polyacrylamide gels and desalted for consistent performance. Salts carried over in the primers may lead to inhibition of NASBA™ reactions. For a single reaction, combine:

5 pmol primer 1
5 pmol primer 2

3.75 μl 100% DMSO
H₂O (nuclease-free) to 6.25 μl

Primer mixture stocks are stable at −20°C for at least 1 yr.

Enzyme Mixture

An enzyme mixture containing bovine serum albumin (BSA), AMV reverse transcriptase, *E. coli* RNase H, and T7 RNA polymerase from the specific suppliers listed can be stored at −20°C for at least 6 mo. Enzymes from other suppliers may not function similarly in NASBA™.

For a single reaction, combine:

0.13 μL 20 mg/mL BSA (in 50% glycerol; Boehringer Mannheim)
8 U AMV reverse transcriptase (Seikagaku)
0.2 U *E. coli* RNase H (Pharmacia)
40 U T7 RNA polymerase (Pharmacia)

Store at −20°C.

A ribonuclease inhibitor such as RNAguard (Pharmacia) at 12.5 U per reaction may be included in the enzyme mixture but is not essential. When the Enzyme Mixture is to be used, remove the volume required for the number of test reactions and adjust with nuclease-free H₂O to give a final volume of 2 μl per reaction. Once H₂O is added to the enzyme mixture, it can no longer be stored at −20°C.

Dithiothreitol (DTT)

1. Prepare 1 *M* DTT stock:

   154 mg DTT powder
   H₂O (nuclease-free) to 1 ml
   Store at −20°.C.

2. Dilute stock solution to 250 m*M* in H₂O for use. Do not store dilutions for more than 1 mo at −20°C.

Agarose Gels

3% Low-melt–1% Agarose Gel

   3 g NuSieve agarose (FMC Bioproducts)
   1 g agarose
   100 ml 1× TAE

2% Agarose

2 g agarose
100 ml 1× TAE

Melt agarose, cool to 65°C, then add 4 μl of 5 mg/mL ethidium bromide.
For the formulation of 1× TAE, see Sambrook *et al.* (1989).

Gel Loading Buffer

50 mg bromophenol blue
2 ml 10 mM EDTA
30 ml 30% glycerol
68 ml H$_2$O (nuclease-free)

Layer Mixture

1 ml 100% glycerol
5 ml 20× SSC
10 mg xylene cyanol FF

Add H$_2$O to 10 ml. For the formulation of 20× SSC, see Sambrook *et al.*,
(1989).

Biogel P-60 Suspension

1. Add 10 g Biogel P-60 (BioRad) to 100 ml solution containing
   40 mM Tris (pH 7.2) and 50 mM NaCl. Mix and let stand at room
   temperature overnight.
2. Heat in autoclave at 121°C for 20 min.
3. Store at 4°C.

Always mix Biogel P-60 suspension before each use.

5× RT Buffer

400 mM Tris-HCl, pH 8.3
700 mM KCl
50 mM DTT
50 mM MgCl$_2$

Store at −20°C in 100-μl aliquots.

Deoxy:dideoxy NTPs Mixtures

200 μM:75 μM dG:ddG
200 μM:60 μM dA:ddA
200 μM:150 μM dC:ddC
200 μM:300 μM dt:ddT

Mixtures are made from commercially available stock solutions.

Chase Solution

0.5 mM dGTP
0.5 mM dATP
0.5 mM dCTP
0.5 mM dTTP

Formamide Mixture

10 mg xylene cyanol FF
10 mg bromophenol blue
200 μl 10 mM EDTA
0.5 ml 95% formamide (deionized)
300 μl H$_2$O

REFERENCES

Bruisten, S., van Gemen, B., Koppelman, M., Rasch, M., van Strijp, D., Schukkink, R., Beyer, R., Wiegel, H., Lens, P., and Huisman, H. (1993). Detection of HIV-1 distribution assays. *AIDS Res. Hum. Retroviruses* **9**, 259–265.

Cimino, G. D., Metchette, K., Isaacs, S. T., and Zhu, Y. S. (1990). More false positive problems. *Nature* **345**, 773–774.

Crouch, R. J., and Dirksen, M.-L. (1982). Ribonuclease H. *In* "Nucleases" (S. M. Linn and R. J. Roberts, eds.), pp. 211–241. Cold Spring Harbor Laboratory Press, Cold Spring Harbor, NY.

Davanloo, P., Rosenberg, A. R., Dunn, J. J., and Studier, F. W. (1984). Cloning and expression of the gene for bacteriophage T7 RNA polymerase. *Proc. Natl. Acad. Sci. USA* **81**, 2035–2039.

Davey, C., and Malek, L. T. (1989). Nucleic acid amplification process. European Patent No. EP 0329822.

Dunn, J. J., and Studier, F. W. (1983). Complete nucleotide sequence of bacteriophage T7 and the locations of T7 genetic elements. *J. Mol. Biol.* **166**, 477–535.

Fahy, E., Kwoh, D. Y., and Gingeras, T. R. (1991). Self-sustained sequence replication (3SR): An isothermal transcription-based amplification system alternative to PCR. *PCR Meth. Appl.* **1**, 25–33.

Ikeda, R. A. (1992). The efficiency of promoter clearance distinguishes T7 class II and class III promoters. *J. Biol. Chem.* **267**, 11322–11328.

Kievits, T., van Gemen, B., van Strijp, D., Schukkink, R., Dirck, M., Adriaanse, H., Malek, L. T., Sooknanan, R., and Lens, P. (1991). NASBA™ isothermal enzymatic *in vitro* nucleic acid amplification optimized for the diagnosis of HIV-1 infection. *J. Virol. Meth.* **35**, 273–286.

Kwok, S., and Higuchi, R. (1989). Avoiding false positives with PCR. *Nature* **339**, 237–238.

Kwok, S., Kellogg, D. E., Spasic, D., Goda, L., Levenson, C., and Sninsky, J. J. (1990). Effects of primer-template mismatches on the polymerase chain reaction: Human immuno-deficiency virus type 1 model studies. *Nucleic Acids Res.* **18**, 999–1005.

Ling, M.-I., Risman, S. S., Klement, J. F., McGraw, N., and McAllister, W. T. (1989). Abortive initiation by bacteriophage T3 and T7 RNA polymerases under conditions of limiting substrate. *Nucleic Acids Res.* **17**, 1605–1618.

Malek, L. T., Sooknanan, R., and Compton, J. (1993). Protocols for nucleic acid analysis by non-radioactive probes. *Meth. Mol. Biol.* **28**, 253–260.

Martin, C. T., Mueller, D. K., and Coleman, J. E. (1988). Processivity in early stages of transcription by T7 RNA polymerase. *Biochemistry* **27**, 3966–3974.

Milligan, J. F., Groebe, D. R., Witherell, G. W., and Uhlenbeck, O. C. (1987). Oligoribonucleotide synthesis under T7 RNA polymerase and synthetic DNA templates. *Nucleic Acids Res.* **15**, 8783–8798.

Moffat, B. A., Dunn, J. J., and Studier, F. W. (1984). Nucleotide sequence of the gene for bacteriophage T7 RNA polymerase. *J. Mol. Biol.* **173**, 265–269.

Mullis, K. B., Faloona, F., Scharf, S. J., Saiki, R. K., Horn, G. T., and Erlich, H. A. (1986). Specific enzymatic amplification of DNA *in vitro:* The polymerase chain reaction. *Cold Spring Harbor Symp. Quant. Biol.* **51**, 263–273.

Sambrook J., Maniatis T., and Fritsch E. (1989). "Molecular Cloning: A Laboratory Manual," 2d Ed. Cold Spring Harbor Laboratory Press, Cold Spring Harbor, NY.

Sarkar, G., and Sommer, S. S. (1991). Parameters affecting susceptibility of PCR contamination to UV inactivation. *Biotechniques* **10(5)**, 591–593.

Sooknanan, R., Malek, L. T., Wang, X-H., Siebert, T., and Keating, A. (1993). Detection and direct sequence identification of BCR-ABL mRNA in Ph+ chronic myeloid leukemia. *Exp. Hematol.* **21**, 1719–1724.

van Gemen, B., Kievitis, T., Schukkink, R., van Strijp, D., Malek, L. T., Sooknanan, R., Huisman, H. G., and Lens, P. (1993). Quantitation of HIV-1 RNA in plasma using NASBA™ during HIV-1 primary infection. *J. Virol. Meth.* **43**, 177–188.

Verma, I. M. (1977). The reverse transcriptase. *Biochim. Biophys. Acta* **473**, 1–38.

Watson, R. (1989). Formation of primer artifacts in polymerase chain reaction. *Amplifications* **1**, 5–6.

Williams, J. F. (1989). Optimization strategies for the polymerase chain reaction. *Biotechniques* **7**, 762–768.

13

The Self-Sustained Sequence Replication Reaction and Its Application in Clinical Diagnostics and Molecular Biology

Soumitra S. Ghosh and Eoin Fahy
Applied Genetics
San Diego, California 92121

Thomas R. Gingeras
Affymetrix, Inc.
Santa Clara, California 95051

I. INTRODUCTION

In the past decade, the detection and quantification of viruses such as human immunodeficiency virus type 1 (HIV-1), human T- cell lymphoma/leukemia virus types 1 and 2 (HTLV-1 and -2), and cytomegalovirus (CMV) have been challenging problems in clinical diagnostics. Traditional culture methods, complemented by nucleic acid hybridization and sequencing techniques, found early applications in the detection of such low copy number pathogens. Improvements in sensitivity over radioisotopic-hybridization detection methods were made with signal amplification strategies using enzyme reporter molecules. Although signal amplification systems are rapid compared with culture methods for detecting viruses or microbial agents, widespread adoption of these technologies in the clinical setting failed to occur because of the limited sensitivities of the detection systems. The observed threshold of detection of 10^3–10^4 target molecules for these systems, principally imposed by the nonspecific background signal amplification of the assays, is frequently higher than the viral titers of HIV-1 and CMV encountered in clinical samples. The advent of *in vitro* nucleic acid target amplification techniques (Saiki *et al.*, 1985; Lizardi *et al.*, 1988; Kwoh *et al.*, 1989; Wu and Wallace, 1989; Guatelli *et al.*, 1990; Barany, 1991) ushered in a profound change in the molecular diagnostics arena. The discovery of the polymerase chain reaction (PCR) in 1985 (Saiki *et al.*, 1985) made it possible to amplify a single copy gene over a million-fold, thereby simplifying the task of detection by hybridization detection systems.

Since the introduction of PCR, two amplification systems that rely on enzyme-mediated RNA transcription have been described. In the transcription-based amplification system (TAS) (Kwoh *et al.*, 1989), the enzymatic activities of avian myeloblastosis virus (AMV) reverse transcriptase (RT) and T7 RNA polymerase are conscripted to amplify target sequences through reiteration of a two-step cycle. The first step involves synthesis of double-stranded cDNA in which a promoter recognized by T7 RNA polymerase is inserted into the cDNA copy via one of the primers used for the RT-mediated extension reaction. In the second step, T7 RNA polymerase is used to produce 10–100 transcripts from each cDNA template. Like PCR, the TAS protocol requires temperature cycling for strand separation of RNA : DNA duplexes to allow primer extension for second-strand cDNA synthesis. The TAS reaction provides greater than 10^6-fold amplification after 4 cycles (Kwoh *et al.*, 1989) and was found to be comparable to PCR for the detection of HIV-1 in infected peripheral blood mononuclear cells (PBMCs; Davis *et al.*, 1990).

A significant improvement in the TAS methodology was achieved by introducing *Escherichia coli* RNase H as a third enzyme component in the reaction, thus eliminating the temperature cycling requirement. The resulting amplification system has been termed the self-sustained sequence replication (SSSR or 3SR) reaction (Guatelli *et al.*, 1990). The hallmarks of the 3SR reaction are its isothermal and self-cycling nature, as well as its rapid kinetics of amplification. The following sections outline this methodology in greater detail and discuss some of its unique applications.

II. CHARACTERISTICS OF THE 3SR REACTION

A. Mechanism for Exponential Amplification

The scheme for RNA replication by the 3SR reaction is shown in Fig. 1. Essentially, exponential replication of an RNA target sequence results from continuous cycles of reverse transcription and RNA transcription at a single temperature (37 to 42°C) by means of cDNA intermediates. The crucial elements of the strategy are (1) extension of an oligonucleotide primer [containing a target complementary sequence (TCS) and a T7 RNA polymerase binding site at the 5' end] to produce an RNA : DNA duplex; (2) degradation of the intermediate RNA : DNA hybrid by RNase H, a feature essential for second-strand cDNA synthesis; (3) T7 RNA polymerase-mediated transcription of the cDNA template to produce multiple copies of antisense RNA molecules; and (4) use of the RNA and cDNA products as templates for subsequent steps to sustain continuous replication. The reaction continues until the components become limiting or the enzymes become inactivated.

B. Parameters

Since the original report on the 3SR reaction (Guatelli *et al.*, 1990), the influences of the various components of the 3SR reaction have been analyzed to optimize the productivity of the amplification reaction (Fahy *et al.*, 1991). The structure and concentration of the oligonucleotide primers significantly affected the specificity and efficiency of the 3SR reaction. The T7 promoter-containing primer was designed to have four functional domains (Fig. 2); the introduction of additional bases at the 5' end of the T7 consensus sequence was found to enhance amplification. These bases are presumably necessary to ensure the formation of the full-length, double-stranded T7 promoter region in the reverse transcription step (Fahy *et al.*, 1991). Alterations in the length and composition of the transcription initiation sequence of T7 promoter-containing primers were well tolerated and provided greater than 10^9-fold amplifications in a 1-hr 3SR reaction. Although primers containing the preferred $^{+1}GGGA^{+4}$ sequence as the transcription initiation sequence (Dunn and Studier, 1983; Milligan *et al.*, 1987) do not always yield the highest levels of 3SR amplification after a 1-hr reaction, their use in 3SR reactions was shown to provide faster initial rates of synthesis. The length of the target complementary sequence (TCS) of the primers can vary between 15 and 30 bases. Use of the shorter TCS permits hybridization of the primers to be conducted under more stringent conditions. The reaction functions optimally using equimolar concentrations of primers, and when primer concentrations range is from 0.1 to 0.2 μM.

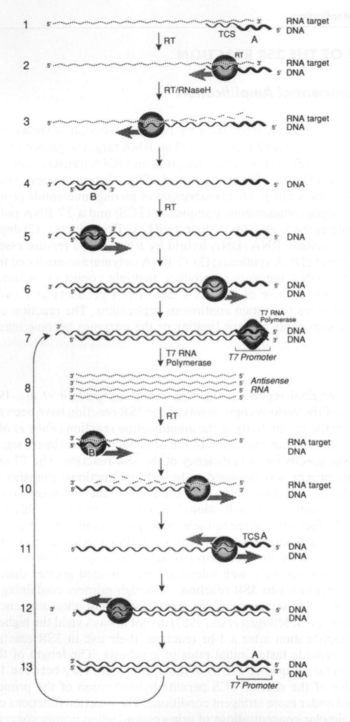

Figure 1 Strategy of 3SR amplification. The 3SR reaction consists of continuous cycles of reverse transcription and RNA transcription designed to replicate a target nucleic acid (RNA target) using a double-stranded cDNA intermediate. Oligonucleotides A and B primer DNA synthesis, producing

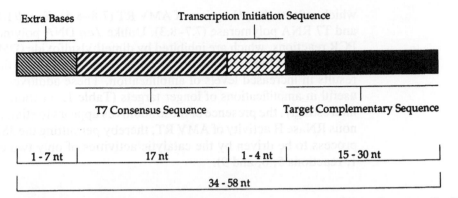

Figure 2 Diagram of a T7 promoter-containing primer, outlining the four functional domains discussed in the text. The length of each region, expressed in nucleotides, is given below the diagram. Reproduced with permission from Fahy *et al.* (1991).

The efficiency of 3SR amplification is sensitive to the total ionic strength of the reaction when chloride is present as the counterion for the monovalent and divalent salts. These 3SR reactions have a narrow optimum salt concentration; ionic strengths exceeding 550 mM result in a drastic reduction in the level of 3SR amplification (Fahy *et al.*, 1991). Studies on the substitution of chloride by acetate as the counterion for Mg^{2+} and Tris, and substitution of potassium chloride by potassium glutamate in the standard 3SR protocol (Section VI,C,1) have revealed two effects (Fahy *et al.*, 1994). First, the acetate/glutamate-containing 3SR reactions provide high levels of amplification over a wide range of potassium glutamate concentrations. This observation is consistent with a previous study on the effect of chloride-to-glutamate substitution on DNA and RNA polymerase activity (Leirmo *et al.*, 1987) in which a much broader salt optimum for these enzymes was reported. Second, the modified buffer conditions appear to enhance the thermostability of the 3SR enzymes; $\geq 10^8$-fold amplifications have been observed for acetate/glutamate-containing 3SR reactions at temperatures as high as 50°C.

The efficiency of the 3SR reaction is dependent on the concerted activities of the three enzymes. The optimal pH range of the reaction is 7.7 to 8.6,

a double-stranded cDNA containing a functional T7 promoter (Steps 1–6). Complete cDNA synthesis is dependent on the digestion of RNA in the intermediate RNA:DNA hybrid by RNase H (Step 3). Transcription-competent cDNAs are used to produce multiple (50–1000) copies of antisense RNA transcript of the original target (Steps 7–8). These antisense transcripts are immediately converted to T7 promoter-containing, double-stranded cDNA copies (Steps 9–12) and are used again as transcription templates. This process continues in a self-sustained fashion under isothermal conditions (42°C) until the components in the reaction become limiting or inactivated (enzymes). Dotted lines, RNA; thin lines, DNA; thick lines, T7 promoter sequence; circles, reverse transcriptase; diamonds, T7 RNA polymerase; TCS, target complementary sequence. Reproduced with permission from Fahy *et al.* (1991).

which reflects the pH optima of AMV RT (7.8–8.5), *E. coli* RNase (7.5–9.1), and T7 RNA polymerase (7.7–8.3). Unlike *Taq* DNA polymerase-catalyzed PCR reactions, which are inhibited by dimethylsulfoxide (DMSO), the inclusion of DMSO and polyhydric alcohols (e.g., sorbitol) in the 3SR reaction results in increased levels of amplification. These additives are particularly useful in amplifications of longer targets (Table 1, reactions 7 and 8). More interestingly, the presence of these additives appears to stimulate the endogenous RNase H activity of AMV RT, thereby permitting the 3SR amplification process to be driven by the catalytic activities of only two enzymes (Table 1, reactions 5, 6, and 9).

C. Target Specificity

Double-stranded DNA targets can be amplified as efficiently as single-stranded RNA sequences in 3SR reactions (Gingeras *et al.*, 1990). However, two thermal denaturation steps are required to separate the DNA strands. The choice of target denaturation conditions provides the 3SR reaction with the unique ability to amplify single-stranded RNA targets selectively in the presence of double-stranded DNA copies of the same sequences. This selective differentiation of targets is exemplified in Table 2. Although the single-stranded SP6 RNA was amplified after denaturation at 65°C, no amplification

TABLE 1
Effect of Additives on Two- and Three-Enzyme 3SR Reactions[a,b]

Reaction number	Enzymes (units/100 μl)			DMSO (%v/v)	Additives glycerol (%v/v)	Sorbitol (%w/v)	Amplification product size (nucleotides)	Fold amplification[c]
	AMV RT	T7 RNA Pol	RNase H					
1	30	100	4	—	—	—	343	1×10^7
2	30	100	4	10	10	—	343	5×10^7
3	30	100	4	10	—	15	343	4×10^7
4	10	20	—	—	—	—	343	$<10^4$
5	10	20	—	10	10	—	343	1×10^6
6	10	20	—	10	—	15	343	5×10^6
7	30	100	4	—	—	—	705	8×10^4
8	30	100	4	10	—	15	705	7×10^6
9	10	20	—	10	—	15	705	1×10^6

[a] Reproduced with permission from Fahy *et al.* (1991).
[b] The primer pairs 90–47/89–391 and 90–249/89–391 were used to amplify 342-base and 704-base regions, respectively, of the *pol* gene of HIV-1. Only the 89–391 (antisense) primer contained a T7 promoter sequence. Products of the 90–47/89–391 reaction were detected by BBSH using 89–534 probe and 89–535 bead sequences; those of the 90–249/89–391 reaction were detected using 89–534 probe and 89–419 beads. The *pol* sequences described in this experiment are found in Gingeras *et al.* (1991).
[c] Detection limitation of BBSH in these experiments was approximately 10^4-fold amplification.

TABLE 2
Selective Amplification of RNA over DNA Target Sequences[a]

Target sequence	Denaturation temperature (°C)	Amount of target sequence (fmol)	Volume of 3SR reaction[b] used (ml)	3SR product detected[c] (fmol)
SP6-RNA (SP65 *Env*) transcript[d]	65	6×10^{-3}	0.02	2.2
SP6-DNA (SP65 *Env*) plasmid	65	6×10^{-3}	0.02	0.02
pARV7A/2[e]	100	6×10^{-3}	0.02	3.2
pARV7A/2	65	6×10^{-3}	0.02	0.08
H_2O	65	—	0.2	0.04

[a] Reproduced with permission from Gingeras and Kwoh (1992), *in* "Jahrbuche Biotechnologie" (P. Präve, M. Schlingmann, K. Esser, R. Thauer, and F. Wagner, eds.), Band 4, pp. 403–429; Hanser Publishers, Munich.

[b] Total volume in each 3SR reaction was 100 μl.

[c] A BBSH assay using Trisacryl Oligobeads™ and ^{32}P-labeled detection probe was used to detect and quantify the 3SR products made in each reaction.

[d] The *Hind*III–*Kpn*I fragment from pARV7A/2 spanning the *env* region of HIV-1 was inserted into the multiple cloning site of the transcription vector SP65 (Promega; Madison, WI). The resulting construct (SP65ENV) was cleaved with *Hind*III and a run-off transcript of the sense strand of the HIV-1 *env* region was synthesized with SP6 RNA polymerase (New England BioLabs; Beverly, MA) under conditions recommended by the supplier.

[e] pARV7A/2 contains the complete HIV-1 genome inserted into the *Eco*RI site of pUC19 (Luciw *et al.*, 1984).

of DNA copies of the same 214-bp fragment of the HIV-1 *env* gene was observed unless the DNA targets were denatured at 95–100°C. This substrate selectivity of the 3SR reaction has immense potential for the study of biological problems relating to the onset of gene expression. Such applications include assays for the detection of mRNA to distinguish productive and latent viral infections.

D. Kinetics

The kinetics of 3SR amplification depend on reaction temperature and concentrations of the 3SR enzymes and primers, Mg^{2+}, deoxynucleotide and ribonucleotide triphosphates, and additives (Fahy *et al.*, 1991). Because there is a direct correlation between the amount of detectable 3SR product and the amount of input target, careful control of these factors and the reaction time is necessary. For example, when the kinetics rates of 3SR amplification of a 186-nucleotide region of the HIV-1 envelope (*env*) gene (10^2–10^5 molecules) were measured (Gingeras and Kwoh, 1992), the kinetic profiles showed a rapid accumulation of product during the first 10 min and

10-fold increases every 2.5 min (Fig. 3). After 30 min, there was a decrease in amplification rates; 3.75 min was needed to produce a 10-fold increase in product. A linear relationship exists between the amount of 3SR-generated product and the amount of target nucleic acid during the initial 20–30 min of the reaction. This linear correlation holds until the concentration of total 3SR product exceeds approximately 50 fmol/μl. The time course curves for the higher amounts of starting target converge as the reaction time is extended past 1 hr (data not shown).

The lack of correlation between accumulated product and input target when product concentration exceeds 50 fmol/μl most likely is a reflection of depletion of primers and enzyme instability. Other factors that can influence the kinetics of the 3SR reaction are the sequence of the nucleic acid target (Fahy *et al.*, 1991) and the extent of nonspecific product synthesis due to mispriming events. As depicted in Fig. 3, quantification of the target nucleic acid can be achieved only during the exponential phase of the 3SR reaction. For quantification of unknown samples, it is necessary to amplify a range of concentrations that provide exponential amplification during a

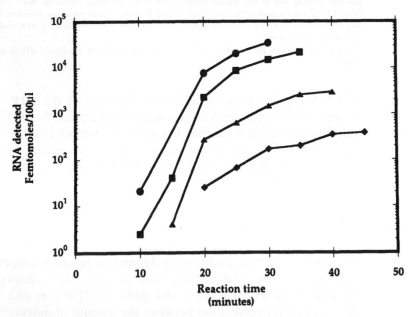

Figure 3 Kinetics of 3SR reaction during the amplification of 186 nucleotides of the *env* region of HIV-1. Initial HIV RNA concentrations in a standard 3SR reaction were 0.002 (♦) (~10^2 molecules); 0.02 (▲) (~10^3 molecules); 0.2 (■) (~10^4 molecules); and 2 attomoles (●) (~10^5 molecules). Reproduced with permission from Gingeras and Kwoh (1992), *in* "Jahrbuche Biotechnologie" (P. Präve, M. Schlingmann, K. Esser, R. Thauer, and F. Wagner, eds.), Band 4, pp. 403–429; Hanser Publishers, Munich.

defined time. Such target titrations or time considerations are not needed when semiquantitative or correlative results are desired.

III. HYBRIDIZATION AND DETECTION OF 3SR PRODUCTS

The predominant products of the 3SR reaction are single-stranded RNA and double-stranded DNA in a ratio of approximately 100 : 1 (Guatelli *et al.*, 1990; Versailles *et al.*, 1993). The synthesis of single-stranded RNA as the major species offers two important advantages: (1) the RNA products are amenable to direct sequencing (Guatelli *et al.*, 1990; Gingeras *et al.*, 1991) and (2) the single-stranded RNA is a convenient target for rapid detection, genomic characterization, and quantification by bead-based sandwich hybridization (BBSH) reactions.

Although slot-blot or Southern hybridization procedures may be employed for the detection of 3SR products (Guatelli *et al.*, 1990), the BBSH assay is attractive because it derives its sensitivity from the combined specificities of the detection oligonucleotide probe and the nucleic acid affinity support (Ranki *et al.*, 1983; Fahy *et al.*, 1993). In the BBSH assay, the 3SR product is simultaneously hybridized to an excess of a suitably labeled detection probe and the capture oligonucleotide, which is covalently linked via its 5' end to Trisacryl™ beads (Oligobeads™) (Davis *et al.*, 1990; Fay *et al.*, 1991, 1993). The fraction of the labeled oligonucleotide that is associated with the solid support permits quantification of the hybridized target. The linear range of detection of the BBSH assay is dependent on the concentration of the detection probe. The typical linear detection range of the assay using 50 fmol ^{32}P-labeled detection oligonucleotide ($\sim 2 \times 10^5$ cpm) and 25 mg Oligobeads™ is between 0.1 and 25 fmol. As expected, the linear range is smaller (0.1–10 fmol of detected product) when 20 fmol of detection probe is used in the assay. When used in conjunction with 3SR amplification, the BBSH system is capable of detecting a single HIV-1 infected cell in a background of 10^6 uninfected cells (Table 3).

Two nonisotopic BBSH methods have been described for the detection of 3SR products that provide eqiuvalent sensitivites and linear detection range as the ^{32}P/Trisacryl BBSH system just described. The first method is a particle concentration method coupled with time-resolved fluorescence detection of chelate-labeled probes (Bush *et al.*, 1991, 1992). Rare earth metal chelates have the following advantages: (1) the large Stokes shift of the chelates allows for a broad separation of the excitation and emission wavelengths and (2) gating of the long fluorescence signal minimizes the intrinsic background fluorescence of the 3SR-amplified test samples.

The second nonisotopic BBSH method uses an alkaline phosphatase-conjugated oligonucleotide probe and polystyrene oligonucleotide beads (Ishii and Ghosh, 1993). This detection system derives its sensitivity from the signal amplification achieved from the rapid turnover of a dioxetane-based substrate to a chemiluminescent (CL) product. Standard CL signals from known amounts of target RNA (Fig. 4A) are used to determine the concentrations of 3SR-amplified product. This enzyme-based BBSH assay is highly sensitive and can be used for the detection of a single HIV-1-infected cell (Fig. 4B). The coupled assay, however, is clearly inadequate for providing a linear correlation between the BBSH-detected product and the number of HIV-1-infected cells present in the samples. This lack of correlation is probably due to the variability associated with the sample preparation steps, as well as to differences in the amplification efficiencies in a 1 hour 3SR reaction of the extracted HIV-1 RNA.

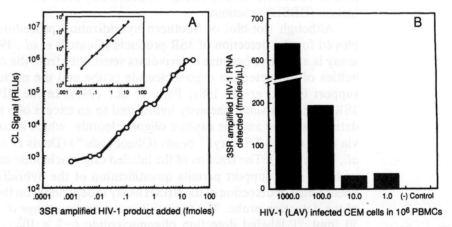

Figure 4   (A) Plot of chemiluminescent (CL) signal versus quantified amounts of 3SR-amplified HIV-1 *env* region RNA. A range of concentrations of the target RNA was assayed in bead-based sandwich hybridization (BBSH) reactions using an oligonucleotide—alkaline phosphatase conjugate probe and oilgonucleotide polystyrene beads. (*inset*) Linear portion of the curve spanning 10^{-17} to 10^{-14} moles of target. (B) Detection and quantification of HIV-1-infected-cells using a coupled 3SR/BBSH assay. Total nucleic acid from HIV-1 (LAV)-infected CEM cells in a background of 10^{-6} uninfected peripheral blood mononuclear cells (PBMCs) was isolated and subjected to 3SR amplification according to the protocol described by Ishii and Ghosh (1993). The resulting products were detected by enzyme-based BBSH. The 3SR-amplified products were quantified by correlating the observed CL signals with the linear segment of a calibration curve described earlier, and were reported as fmol/μl 3SR-amplified product. The 3SR-amplified products were also quantified independently using a BBSH assay employing the corresponding ^{32}P-labeled detection probe and Trisacryl Oligobeads™. Based on the specific activity of the detection probe, 456.0, 117.0, 29.0 and 21.0 fmolμl of 3SR product were obtained from 1000, 100, 10, and 1 HIV-1-infected cell samples, respectively. Reprinted with permission from Ishii and Ghosh (1993). Copyright © 1993 by the American Chemical Society.

Transient-state polarized fluorescence-based strategies that use oligonu-cleotide–phthalocyanine dye conjugates as detection probes have shown promise for the homogeneous detection of 3SR products (Devlin *et al.*, 1993). This detection method is based on the increase in polarized fluorescence resulting from longer rotation times of duplexes of the conjugate and its complementary targets. This system was comparable in its sensitivity to heterogeneous isotopic and nonisotopic methods.

IV. STERILIZATION OF 3SR REACTIONS

The efficacy of target amplification systems is often compromised by the occurrence of false positive reactions due to inadvertent contamination by amplification products (amplicons) derived from previous amplification reactions. This problem is of particular concern in 3SR reactions because minute volumes of 3SR amplification reactions (10^{-6} to 10^{-11} μl, containing 10^5 to ~1 molecule) can be re-amplified efficiently in subsequent 3SR reactions (Gingeras and Kwoh, 1992). Both RNA and DNA amplicons can serve as sources for "carryover" contamination. Therefore, 3SR sterilization methods must be able to modify both nucleic acid species efficiently and render them unusable as templates for subsequent 3SR amplification. A post-amplification approach using the photoactive agent 4′-aminomethyl-4,5-dimethylpsoralen (IP-10) (Cimino *et al.*, 1991; Isaacs *et al.*, 1991) was found to provide highly efficient sterilization (10^6–10^8 fold) of 3SR amplicons (Versailles *et al.*, 1993). A key feature of the strategy is the gel-based mode of delivery of IP-10 (see the procedure in Section VI). The sequestration of IP-10 in low-melting agarose in the caps of microfuge reaction tubes during the 3SR reaction prevents unwanted inhibition of the amplification process. Following amplification, a thermal melt step at 75–80°C is used to liquefy the gel to enable delivery of IP-10 to the 3SR reaction. The levels of re-amplification of 3SR products treated with various amounts of IP-10 are compared in Fig. 5. The results indicate that the sterilization effect depends on the amount of IP-10 used in the reaction. The amount of 3SR product generated from 10^{-14} moles of IP-10-modified 3SR amplicons is less than the amount of 3SR product formed from 10^{-21} moles of unmodified 3SR ampli-cons. This result corresponds to a sterilization efficiency of at least 10^7-fold. The IP-10-based sterilization method is based on the photochemical cross-linking of the isopsoralen molecule to thymine/uracil residues of 3SR ampli-cons. Although the modification impairs the nucleic acids as templates, they remain viable as targets for hybridization reactions.

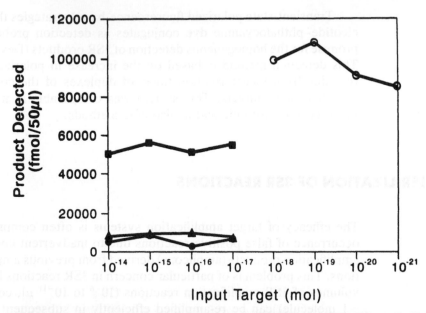

Figure 5 Relationship between input target and detected product in 3SR re-amplification of IP-10-treated HIV-1 *env* region 3SR products. Irradiated from 3SR reactions containing 0 (○), 100 (■), 200 (▲), and 300 (●) mg IP-10/ml were diluted and re-amplified in the absence of IP-10 and NuSieve™ agarose gel. Amplified products were detected by BBSH with Trisacryl Oligo-beads™ and ^{32}P-labeled detection probe.

V. APPLICATIONS OF THE 3SR REACTION

Prolonged zidovudine [formerly called azidothymidine (AZT)] therapy of HIV-1-infected patients can result in the emergence of drug-resistant viral isolates (Larder *et al.*, 1989). Point mutations in the codons for amino acid residues 67, 70, 215, and 219 of the HIV-1 *pol* gene have been primarily implicated in the reduced drug susceptibility of these viral isolates (Larder and Kemp, 1989). Gingeras *et al.* (1991) used a coupled 3SR/BBSH assay for the surveillance of these four positions in HIV-1 virion RNA samples derived from patients undergoing zidovudine treatment. The protocol consists of selective 3SR amplification of HIV-1 RNA in the presence of proviral forms followed by analysis of the 3SR-generated products by differential BBSH reactions. Specifically, 3SR reactions are used to amplify two separate regions of the HIV-1 *pol* gene that encode the 67–70 or the 215–219 amino acid loci. The 3SR-generated products are then detected by differential BBSH using probes that, under discriminating hybridization conditions, detect either wild-type or mutant sequences in the codons of interest. Interestingly, short oligonucleotide probes (15–20 nucleotides) designed to monitor the individual codons failed to hybridize to their targets, presumably because of the secondary structure of the 3SR RNA products. This result necessitated the use of longer oligonucleotide probes (25–30 nucleotides), which proved to be more successful by virtue of their greater duplex stabilities. Because of the proximity of the specified codons, these detection probes were used to monitor mutations simultaneously at two positions (codons 67 and 70 or 215 and 219). The results of a coupled 3SR/differential BBSH assay for the detection of zidovudine-related mutations present in nucleic acid targets of defined genotype (Table 3) clearly demonstrate the specificity of this hybridization correlation analysis.

The coupled 3SR/differential BBSH assay was applied to the genotype analysis of HIV-1 RNA extracted from viral stocks produced by the coculture of MT2 cells with PBMCs from infected individuals receiving zidovudine therapy. The genotype assignments for each locus were verified by nucleotide sequencing of the 3SR products, and compared with the results of genotype analysis of the same samples and of the original PBMCs by PCR/differential Southern hybridization (Gingeras *et al.*, 1991; Richman *et al.*, 1991a). Several pertinent conclusions emerged from this correlation study: (1) the sequence data and the differential BBSH assay were in agreement except for codons 67 and 70 in two samples (sequence data suggested a mixed population with the mutant as the predominant form, which was detected solely by BBSH); (2) differences at the monitored amino acid positions between PBMCs and the corresponding virions may reflect a predilection for genotypic change in culture; (3) differences in the 3SR/BBSH and PCR/Southern hybridization analyses can arise because of additional mutations at sites flanking the monitored codons and are influenced by the size and location of the hybridization probes; and (4) variability in zidovudine sensitivities of identical genotypes suggests that mutations other than the four monitored codons in the *pol* gene may influence the level of drug resistance.

The 3SR/differential BBSH assay has also been used for detection of the emergence of nevirapine-resistant HIV-1 mutants in cell culture when the virus is passaged in the presence of the drug (Richman *et al.*, 1991b). Nevirapine is a potent noncompetitive inhibitor of HIV-1 RT that interacts with the highly conserved tyrosine residues at positions 181 and 188 of the enzyme. Nevirapine resistance develops from substitution of cytosine for tyrosine at position 181. A 3SR/differential BBSH assay that monitors changes at the codon for amino acid 181 detected the presence of the mutant sequence as early as the first passage of HIV-1 in nevirapine.

The spleen fragment culture and subsequent 3SR amplification method for the production of monoclonal antibodies is a novel application of the 3SR methodology. This method has been used to produce recombinant monoclonal antibodies without hybridomas (Stillman *et al.*, 1994). The spleen fragment/3SR methodology has several advantages and avoids some of the

TABLE 3
3SR/Differential BBSH Assay Used to Determine Genotype of Nucleic Acid Targets of Defined Genotype[a]

		Femtomoles detected with probes				
Target	Stringency	90–36 (wt/wt)	90–48 (mut/mut)	90–49 (wt/mut)	90–50 (mut/wt)	89–441 (control)
Amino Acid Region 67–70[b]						
pPol 18 (wt/wt)						
	Low	26.1	2.1	27.9	11.7	11.0
	High	4.3	0.3	0.1	0.2	3.5
pHIVRTMC (mut/mut)						
	Low	1.8	42.7	45.7	21.3	33.8
	High	—[c]	5.2	0.7	—	13.7
pPol 67–18 (mut/wt)						
	Low	19.8	11.8	2.2	20.9	15.0
	High	0.2	0.4	0.4	6.8	5.3
pPol 70–18 (wt/mut)						
	Low	17.6	25.4	42.1	—	18.6
	High	—	0.6	18.8	0.1	9.0

		Femtomoles detected with probes				
Amino Acid Region 215–219[d] pARV (wt/wt)		90–303 (wt/wt)	90–304 (mut/mut)	90–305 (wt/mut)	90–302 (mut/wt)	89–535 (control)
	Low	8.0	0.2	1.0	0.5	31.2
	High	6.1	—	0.6	—	27.4
pHIVRTMC (mut/mut)						
	Low	0.3	15.8	12.3	9.6	29.7
	High	—	14.9	—	0.1	34.5
pPol 215P-18 (mut/wt)						
	Low	1.7	5.6	0.1	13.4	12.9
	High	—	0.4	0.2	2.7	18.9
pPol 215T-18 (mut/wt)						
	Low	1.7	6.3	0.9	5.8	20.6
	High	—	0.2	—	1.7	19.1
pPol 219-18 (wt/mut)						
	Low	1.0	3.4	13.3	0.1	16.3
	High	0.4	—	14.2	—	NT[e]

[a] Reproduced in part with permission from Gingeras *et al.* (1991).

[b] RNA transcripts (0.1 attomoles) were used as targets for 3SR amplification reactions. Transcription plasmids (1–10 μg) were linearized with *Eco*RI or *Hin*dIII. T7-Generated antisense transcripts were made using 1000 units of T7 enzyme. Quantification of RNA product was performed by spectroscopic measurement.

[c] Less than 0.1 femtomoles detected.

[d] *Eco*RI-cleaved DNA plasmids containing *pol* gene (Gingeras, 1991) (1.0 attomoles) were used as targets in the 3SR amplification reactions.

[e] NT, Not tested.

problems associated with the production of monoclonal antibodies using hybridoma cell line production. These advantages include the ability to prese-lect for (1) specific isotypes, (2) antibody affinities of desired levels, and (3) defined specificity using competitive analog screening. All these prese-lections can be performed during the screening of the spleen fragment cultures before carrying out the cloning, characterization, and recombinant expres-sion steps.

In addition to providing a method of producing monoclonal antibodies, the spleen fragment/3SR methodology has been useful for other types of studies. One such application involves the analysis of maturation of memory B cells (Decker *et al.* 1993). These studies presented the first *in vitro* clonal analysis of hypermutation during memory B cell development. The spleen fragment/3SR methodology confirmed that progenitors of memory B cells differ from primary B cells as originally proposed by Linton *et al.* 1989), and that progeny of B cells terminally differentiated into antibody-forming cells no longer undergo somatic mutation in response to succeeding expo-sures to antigen.

VI. EXPERIMENTAL PROCEDURES FOR DETECTION OF HIV-1

A. Design of Primers and Probes

1. Priming oligonucleotides should contain target complementary se-quences of 18–30 bases, preferably 25 bases. The GC composition should be 40–60% and the T_m values of the primers should be similar. The 6 bases at the 3′ ends should have a GC composition of less than 50% to minimize mispriming. The selected sequences should be checked for noncomplementarity (to avoid formation of primer dimers), absence of significant stem–loop structures, and absence of alternative priming sites on the target. Computer programs such as Oligo 4.0 (National Biosciences, Plymouth, MN) are useful for the selection of these primers.

2. The preferred 3SR amplification conditions comprise a T7 promoter-containing primer complementary to the target RNA and a non-promoter-containing primer homologous to the target nucleic acid. The T7 promoter-containing primer includes the 25-base sequence 5′-*AATT*AATACGACTCACTATA**GGGA**-3′ upstream from the target complementary sequence. In addition to the 17-base T7 consen-sus sequence (underlined), this primer contains the preferred tran-scription initiation sequence (bold) and an additional 4 bases at the 5′ end (italics).

3. For amplification of double-stranded DNA targets, the T7 promoter sequence may be placed on either primer. However, remember that the RNA product will have the same sense as the T7 promoter-containing primer.

4. The optimal interprimer distance for 3SR amplification is 200–400 bases.

5. Probe oligonucleotides should be approximately 30 bases in length and should be complementary to an internal region of the RNA product. The GC content should be close to 50%. Probe sequences should be checked for lack of complementarity with the priming oligonucleotides.

B. Sample Preparation of HIV-1 RNA

HIV-1 RNA may be isolated from infected CEM cells.

1. Pellet 10^5–10^6 CEM cells from 1 ml Tris-buffered saline in a microfuge tube at 5000 rpm for 10 min.

2. Resuspend pellet in 600 μl lysis buffer (20 mM Tris-HCl, pH 7.5, 150 mM NaCl, 10 mM EDTA, 0.2% SDS, 200 μg/ml proteinase K).

3. Vortex vigorously and incubate at 50°C for 45 min, vortexing for 10–15 sec every 10 min.

4. Add 600 μl phenol : chloroform : isoamyl alcohol (25 : 24 : 1). Shake and vortex to emulsify mixture. Centrifuge at 14,000 rpm for 2 min to separate phases.

5. Draw off 575 μl from the aqueous (top) phase and add 600 μl phenol : chloroform : isoamyl alcohol. Shake and vortex to emulsify mixture. Centrifuge at 14,000 rpm for 2 min to separate phases.

6. Draw off 525 μl from the aqueous (top) phase. Add 600 μl chloroform : isoamyl alcohol (24 : 1). Shake and vortex to emulsify mixture. Centrifuge at 14,000 rpm for 2 min to separate phases.

7. Draw off 400 μl from the aqueous phase. (Do not transfer any cell debris which may be at the interface.)

8. Add 40 μl of 8 M LiCl and 120 μl 100% ethanol to samples. Mix well and precipitate in a dry ice/ethanol bath for 15 min or at −20°C overnight.

9. The quantity of HIV-1 RNA in the extracted sample is determined by comparative hybridization on a slot blot with a known quantity of plasmid pARV7A/2 using a slot-blot analysis. pARV7A/2 contains the complete HIV-1 genome inserted into the *Eco*RI site of pUC19 (Luciw *et al.*, 1984).

C. 3SR Amplification of Targets

1. 3SR Amplification of RNA Targets

Stock Solutions

10× reaction buffer: 400 mM Tris-HCl, pH 8.1, 300 mM MgCl$_2$, 100 mM DTT

25 mM rNTPs: Individual rNTPs are taken up in 50 mM Tris-HCl, pH 8.1, adjusted to pH 8.00 with 10 N NaOH, and then mixed to a final concentration of 25 mM (each); a solution of 25 mM dNTPs is prepared similarly

All aqueous solutions are prepared using deionized H$_2$O treated with 0.1 M diethylpyrocarbonate (DEPC).

Procedure

1. Thaw reagents (excluding enzymes) at room temperature. Vortex and keep on ice.
2. To an autoclaved 1.5-ml microfuge tube, add

   5 μl 10× reaction buffer.
   2.5 μl each priming oligonucleotide (0.1 μM each, final)
   2 μl 25 mM dNTP mix (1 mM final)
   12 μl 25 mM rNTP mix (6 mM final)
   5 μl DMSO (10% final)
   11 μl 68.2% sorbitol (15% final)

It is recommended that a master mix composed of the 10× buffer, NTPs, primers, DMSO, and sorbitol be made and a 40-μl aliquot of this mix be pipetted into each reaction tube prior to addition of target.

3. Add 5 μl RNA target. (Use H$_2$O for negative control reactions.) Vortex and centrifuge reaction tubes briefly.
4. Prepare 3SR enzyme mix. Each reaction requires:

   15 U AMV reverse transcriptase
   1 U $E.$ $coli$ RNase H
   50 U T7 RNA polymerase

Add 40 mM Tris-HCl, pH 8.1, 10 mM DTT for a final volume of 5 μl for the enzyme mix. Prepare the enzyme master mix in microfuge tube.

Overestimate the volume of the enzyme mix by at least 10% to compensate for the higher viscosity of these solutions. Mix the enzymes with the automatic pipettor tip by carefully depressing and releasing the plunger. *Do not vortex!* The mix can be stored on ice or in a cooling block for up to 30 min.

5. Place reaction tubes at 65°C for 1 *min* to denature the target. Transfer tubes to a heating block at 42°C and incubate for at least 1 min.
6. Add 5 μl 3SR enzyme mix to each tube and gently flick 5–10 times. Incubate at 42°C for 1–2 hr. Store reactions at −20°C.

2. 3SR Amplification of DNA Targets

An additional DNA synthesis step is required to allow DNA targets to be amplified by 3SR.

1. Follow Steps 1 and 2 in the previous procedure, replacing the RNA target with DNA.
2. Place reaction tubes in a water bath at 100°C for 2 min to denature the target.
3. Transfer tubes to a heating block at 42°C and incubate for 2 min. Add 10 U AMV reverse transcriptase and incubate at 42°C for 10 min.
4. Heat at 100°C for 1 min.
5. Transfer tubes to 42°C and incubate for 1–2 min.
6. Add 5 μl 3SR enzyme mix to each tube and flick 5–10 times. Incubate at 42°C for 1–2 hr. Store reactions at −20°C.

3. Positive Control for 3SR Amplification

Prepare the control target transcript. The *Hind*III–*Kpn*I fragment from pARV7A/2 spanning the *env* region of HIV-1 is inserted in the polylinker of the transcription vector SP65 (Promega, Madison, WI). The resulting construct (SP65ENV) is cleaved with *Hind*III and a run-off transcript of the sense strand of the HIV01 *env* region is synthesized with SP6 RNA polymerase (New England BioLabs; Beverly, MA) under conditions recommended by the supplier. The amount of transcript is quantified by comparative hybridization on a slot-blot with a known amount of pARV7A/2.

The 3SR reaction target is the HIV-1 *env* transcript (0.1 amol). The HIV-1 *env*-specific sequences 88–211 (upstream primer):

5′-AATTTAATACGACTCACTATA**GGGA**TCTATTG
TGCCCCGCTGGTTTTGCGATTCTA-

3′ and 88–347 (downstream primer):

5′-AATTTAATACGACTCACTATAGGGATGTACTA
TTATGGTTTTAGCATTGTCTGTGA-

3′ are used (Ishii and Ghosh, 1993).

Use of this RNA target and primer pair yields an amplification efficiency of at least 10^8-fold in 1-hr 3SR reaction at 42°C.

D. Purification of 3SR Amplification Products

1. Synthesis of Sulfhydryl-Trisacryl Support

1. Filter 10 g Trisacryl GF-2000 (Sepracor, Marlborough, MA) in a sintered-glass funnel; wash resin with 2 liters H_2O and dry by suction.
2. Add 20 ml ethylenediamine in a round-bottom flask and equilibrate at 90°C in an oil bath.
3. Add resin to flask and react at 90°C for 1 hr with constant stirring.
4. Cool flask to room temperature by placing on ice. Transfer to sintered-glass funnel and wash with 200 ml 0.2 M NaCl, 1mM HCl.
5. Wash with 400 ml 0.1 M NaCl. An aliquot of the final filtrate may be tested for the absence of residual ethylenediamine by a standard trinitrobenzene-sulfonic acid colorimetric assay (Inman, 1994).
6. Dry resin by suction and transfer to round-bottomed flask. Add 40 ml 0.5 M NaHCO$_3$, pH 9.7, to the resin.
7. Add 2.5 g D,L-N-acetylhomocysteine thiolactone to the resin and react at room temperature for 1 hr with constant stirring. Add an additional 1 g D,L-N-acetylhomocysteine thiolactone and allow reaction to proceed overnight with constant stirring.
8. Filter resin and wash with 400 ml 0.1 M NaCl. Transfer to round-bottomed flask and add 20 ml 0.1 M NaOAc, pH 6.0. Add 200 mg succinic anhydride and react for 30 min with constant stirring.
9. Filter resin and wash with 400 ml 0.1 M Tris-HCl, pH 8.5. Transfer resin to a beaker and stir in 400 ml 0.1 M Tris-HCl, pH 8.5, for 2 hr to hydrolyze thioesters.
10. Filter resin and wash with 400 ml 10 mM Tris-HCl, pH 8.0, 1 mM EDTA, 0.02% NaN$_3$. Store in this buffer at 4°C.

2. Covalent Attachment of Oligonucleotides to Trisacryl Support

1. Prepare oligonucleotide 86–273 (5′-AGTCTACGAGAAGAAGAG-GTAGTAATTAGA-3′) containing a 5′-hexylamine group using the

AminoLink-2 reagent (Applied Biosystems, Foster City, CA) in the final coupling step of the automated synthesis.

2. Precipitate the oligonucleotide (1 A_{260} unit) in a silanized microfuge tube and resuspend in 300 μl 0.2 M HEPES, pH 7.7. Add 0.4 mg N-succinimidyl bromoacetate (Bernatowicz and Matsueda, 1986) in 30 μl N,N-dimethylformamide (DMF) and react for 2 hr at room temperature.

3. Precipitate the derivatized oligonucleotide twice with 1/10 volume of 8 M LiCl and 3 volumes 100% ethanol.

4. Resuspend in 7 ml degassed 0.1 M triethylammonium phosphate (TEAP), 1 mM EDTA, pH 9.0.

5. Add 25 ml 50 mM K_2HPO_4, 1 mM EDTA, 20 mM DTT, pH 8.0, to 10 g derivatized Trisacryl and stir at room temperature for 1 hr to reduce any disulfide bonds.

6. Wash with 4 × 25 ml degassed 50 mM K_2HPO_4, 1 mM EDTA, pH 8.0, and finally with 25 ml degassed 0.1 M TEAP, 1 mM EDTA, pH 9.0.

7. Add freshly prepared bromoacetyl oligonucleotide solution from step 4 to resin in a 50-ml conical tube. Layer argon over reactants and seal cap with parafilm. React at room temperature overnight on a rotary shaker.

8. Add 1 mmol iodoacetic acid (186 mg) to the tube and react for 1 hr on a rotary shaker (to alkylate unreacted sulfhydryl groups on the resin).

9. Wash resin with 2 × 25 ml 0.1 M Tris-HCl, pH 8.0, 0.1 M NaCl, 1 mM EDTA, 0.1% SDS; 4 × 25 ml 0.1 M sodium pyrophosphate, pH 7.5; and finally 2 × 25 ml 10 mM Tris-HCl, pH 8.0, 1 mM EDTA.

10. Store resin at 4°C in 10 mM Tris-HCl, pH 8.0, 1 mM EDTA, 0.02% NaN_3 (50 ml per 10 g Trisacryl).

Buffers are degassed by bubbling argon through the solutions for 30 min.

3. Covalent Attachment of Oligonucleotides to Polystyrene Support

1. Aliquot 200 μl 10% suspension of carboxyl-derivatized polystyrene beads (CML particles, 0.8 μM; Seradyn, Indianapolis, IN) into a 1.5-ml microfuge tube.

2. Wash with 3 × 1 ml H_2O, centrifuging at 14,000 rpm for 2 min to pellet the beads. After final wash, resuspend pellet in 200 μl 0.1 M imidazole, pH 6.0.

3. Dissolve 0.6 A_{260} units of hexylamine-derivatized oligonucleotide in

800 μl 0.1 M imidazole, pH 6.0, containing 20 mg 1-(3-dimethylamino-propyl)-3-ethylcarbodiimide.

4. Add this solution to the beads and react for 18 hr at room temperature with mechanical agitation.
5. After the reaction is complete, add 500 μl 1× SSC, 0.5% SDS to the reaction mixture and vortex. Centrifuge and remove supernatant.
6. Wash with 3 × 1 ml 1× SSC, 0.5% SDS and resuspend beads in a final volume of 200 μl of the same buffer.

E. Detection of 3SR Amplification Products

1. Detection of 3SR Products by Trisacryl Oligobead™-Based BBSH

Reagents

20× SSC: 3 M NaCl, 0.3 M Na citrate; adjust to pH 7.0 with HCl; autoclave and store at room temperature

20× SSPE: 3.6 M NaCl, 200 mM NaH$_2$PO$_4$, 20 mM Na$_2$EDTA; adjust to pH 7.4 with NaOH; autoclave and store at room temperature

2× Bead hybridization solution: 20% dextran sulfate, 10× SSPE, 0.2% SDS; in a 50-ml conical tube, add dextran sulfate and SSPE and bring to 35 ml with sterile H$_2$O; vortex and place at 42°C for 15 min; vortex until dextran sulfate is in solution, add SDS, invert to mix, and bring to 40 ml with H$_2$O

^{32}P-Labeled Probe: 90–422 (5'-AATTAGGCCAGTAGTATCAACT-CAACTGCT-3')

Procedure

1. Thaw reaction tubes at room temperature and centrifuge briefly to remove condensation.
2. Perform 10-fold serial dilutions of the 3SR product in 10 mM Tris-HCl, pH 8.1, 1 mM EDTA (TE).
3. Prewarm 2× hybridization solution at 42°C.
4. Remove bottom cap from a 2-ml microcolumn (QS-GS; Isolab, Akron, OH). Add 125 μl well-mixed 200 mg/ml Trisacryl Oligobead™ suspension in TE. Use air pressure from top to force TE buffer through frit. Place bottom cap back on column.
5. Add 30 μl pre-warmed 2× bead hybridization solution to the column.
6. Add each target dilution in a volume of 20 μl in TE. Target dilutions

should not contain more than 10 μl original 3SR reaction or hybridization will be adversely affected.

7. Add 10 μl 5 fmol/μl solution of the ^{32}P-labeled detection oligonucleotide to produce a total hybridization volume of 60 μl.

8. Place top cap on column and place at 42°C. Vortex gently, bringing beads into solution. Incubate for 90 min at 42°C in a constantly shaking bath set at approximately 250 rpm.

9. Place columns in a rack above labeled 7-ml scintillation vials. Add 1 ml 2× SSC at 42°C to each column. Use air pressure to force the wash solution through frits. Repeat wash step for a total of 6 washes.

10. Place columns into empty 7-ml scintillation vials. By Cerenkov counting, measure radioactivity associated with beads and wash for 0.1 min. Use a blank tube for counter background (BKG).

11. To determine the amount of detection probe captured, calculate: (cpm of beads − counter BKG) ÷ [(cpm of beads − counter BKG) + (cpm of wash − counter BKG)] = fraction of probe captured. Fraction of probe captured × total fmol detection oligonucleotide added = fmol 3SR product detected.

2. Detection of 3SR Products by BBSH with Alkaline Phosphatase-Conjugated Oligonucleotide Probes

1. Prepare alkaline phosphatase-conjugated oligonucleotides by the protocol of Ghosh *et al.* (1990) or by procedures in the references cited in that reference. Dilute conjugates in 50 mM NaCl, 10 mM MgCl$_2$, 0.1% gelatin, 0.1 M Tris-HCl, pH 7.5, prior to the assay.

2. Suspend the oligonucleotide–polystyrene beads (50 μg) in prehybridization buffer (5× SSC, 0.5% SDS, 0.02 μg/ml calf thymus DNA) and pellet by centrifugation at 14,000 rpm for 5 min.

3. Remove the supernatant and add 25 μl 2× hybridization buffer (10× SSC, 1% SDS, 0.04 μg/ml calf thymus DNA, 5% glycerol), pre-warmed to 50°C.

4. Perform 10-fold serial dilutions of the 3SR product in TE (10 mM Tris-HCl, pH 8.0, 1 mM EDTA).

5. Add 10μl diluted 3SR product to the beads.

6. Add 15 μl conjugate (10 fmol), vortex the mixture, and incubate at 50°C for 1 hr.

7. Add 50 μl wash solution (0.1× SSC, 0.1% SDS) and let stand for 3 min at room temperature. Centrifuge and remove supernatant.

8. Wash with 2 × 50 μl wash solution.

9. Add 200 μl Lumi-Phos 530 (Lumigen, Inc., Detroit, MI) to the bead–target–probe complex from the BBSH reaction and vortex gently.

10. Allow to react in the dark at room temperature for 1 hr.
11. Transfer the mixture to a cuvette and measure the chemiluminescent signal in a luminometer.

Use a standard curve plotting CL signal against known amounts of target RNA for quantification of 3SR reaction products. Subtract the background CL signal of dioxetane substrate from test and calibration samples before performing calculations.

3. Slot-Blot Detection of 3SR Products with ^{32}P-Labeled Probes

Reagents

4× BP: 2% (w/v) bovine serum albumin (BSA), 2% (w/v) polyvinyl-pyrrolidone (PVP, MW 40,000); dissolve in sterile H_2O and filter through 0.22-μm cellulose acetate membranes (Corning Glass Works, Corning, NY); store at −20°C in 50-ml conical tubes

DM5 solution: 10× SSC, 7.4% formaldehyde; mix 15 ml TE, 25 ml 20× SSC, and 10 ml of 37% (w/v) formaldehyde; store at room temperature

Procedure

1. Perform 10-fold serial dilutions of the 3SR RNA product in TE. Mix 5 μl each target dilution with 100 μl DM5 solution in 1.5-ml sterile microfuge tubes. Target dilutions should not contain more than 5 μl original 3SR reaction or hybridization will be adversely affected.
2. Incubate the DM5 solutions at 65°C for 10 min; then put the samples on ice.
3. To denature DNA, add TE to the sample for a final volume of 90 μl. Add 10 μl 2 N NaOH and vortex. Incubate at 65°C for 30 min, and then put the samples on ice. Neutralize the sample with 100 μl 2 M ammonium acetate.
4. Wet a piece of nitrocellulose that has been cut to fit the slot-blot apparatus in water and then in 10× SSC. Assemble the slot-blot apparatus according to the manufacturer's directions, and load the denatured samples.
5. Fix the nucleic acids to the nitrocellulose by baking at 80°C under vacuum for 1 hr or exposing the blot to 120,000 μJ/cm^2 UV light (254 nm).
6. Prehybridize the nitrocellulose filter for 10–30 min in ~5 ml 1× BP, 5× SSPE, 1% SDS at the temperature to be used for the hybridization

incubation. For 30-base probes, 55°C is usually optimal. For shorter probes or probes with low GC content, a lower temperature must be used.

7. Add at least 2×10^6 cpm of detection oligonucleotide per ml hybridization solution (typically 500 μl). Double-seal the filter in Scotchpak™ heat-sealable pouches (Kapak Corporation, Minneapolis, MN) and incubate for 90 min at 55°C.

8. Wash the filter 3 times at room temperature with 5-min washes of $1\times$ SSPE, 1% SDS on a platform shaker. For higher stringency, wash the filter once at the hybridization temperature in $1\times$ SSPE, 1% SDS for 1 min.

9. Visualize by autoradiography on Kodak XAR film at −70°C with an intensifying screen.

10. Estimate the amount of 3SR product detected by visual comparison with hybridization standards of known concentration.

4. Detection of 3SR Products by Slot-Blot Analysis with Alkaline Phosphatase-Conjugated Oligonucleotide Probes

1. Immobilize 3SR product (0.01–50 fmol) on nitrocellulose filters, as just described.

2. Prehybridize the filter in $5\times$ SSC, 0.5% BSA, 0.5% PVP, 0.1% SDS for 10 min at 50°C.

3. Hybridize in the same buffer with 10 pmol alkaline phosphatase–oligonucleotide conjugate at 50°C for 1 hr.

4. Wash filter with 3×10 ml wash buffer ($1\times$ SSC, 0.1% SDS) at room temperature and 1×10 ml at 50°C.

5. Wash filter with 3×10 ml developing buffer (0.1 M Tris-HCl, pH 9.5, 0.1 M NaCl, 0.01 M MgCl$_2$).

6. Add a solution of 0.33 mg/ml nitroblue tetrazolium (NBT) and 0.16 mg/ml 5-bromo-4-chloro-3-indolyl phosphate (BCIP) in developing buffer containing 0.33% v/v DMF to the filter and develop for 4 hr.

F. Sterilization of 3SR Reactions with IP-10

Stock Reagents

4.6 μg/μl IP-10 (4′-aminomethyl-4,5-dimethylisopsoralen; HRI Associates, Inc., Concord, CA); store in the dark at room temperature

1% NuSieve agarose (FMC, Rockland, ME) in DEPC-treated H_2O; store at room temperature

Procedure

1. Prepare a 0.5% gel containing 1200 μg/ml IP-10. Melt a 1-ml stock of 1% agarose in a microfuge tube by placing in a heating block for 1 min at 80°C.
2. Add 50 μl melted gel to 26.1 μl IP-10 stock and 23.9 μl DEPC-treated H_2O in an autoclaved microfuge tube to give 100 μl sterilizing gel. Volumes may be scaled up to 200 μl or more for use in larger (>10 reactions) amplification experiments. Store the sterilizing gel samples in the dark at room temperature.
3. Prior to 3SR amplification, melt the sterilizing gel for 1 min at 80°C. Assuming a 3SR reaction volume of 50 μl, transfer 10 μl melted gel to the inside of the reaction-tube caps (the gel solidifies immediately).
4. Cover tubes with aluminum foil. Proceed as usual for RNA amplification by adding 3SR reagents and target, heating at 65°C for 1 min, and finally adding the enzymes. Gently vortex the reaction tubes. It is important not to centrifuge the reaction tubes at this stage to avoid dislodging the gel.
5. After the 3SR reaction has been completed, centrifuge for 5 sec (to dislodge agarose from cap) and heat at 80°C for 1 min to melt the agarose and inactivate the enzymes.
6. Vortex tubes and centrifuge for 5 sec to ensure that the entire solution is at the bottom of the tube prior to irradiation. This procedure gives a final IP-10 concentration of 200 μg/ml.
7. Keep reaction tubes on ice and UV-irradiate at 4°C for 20 min (HRI-100 UV illuminator; HRI Associates, Inc., Concord, CA).
8. Store 3SR reactions at −20°C or perform BBSH analysis.

VII. CONCLUSIONS

The unique ability of the 3SR reaction to amplify RNA selectively in the presence of double-stranded DNA of the same sequence has potential for studies involving temporal gene expression and quantification of the transiently expressed mRNA species. Detection and quantification of specific RNA transcription is essential to understanding the mechanisms underlying viral and genetic diseases, embryological development, and cancer. Another exciting aspect of the 3SR reaction is the coupling of this isothermal RNA amplification methodology with an *in vitro* translation system for the expression of functional gene products. An example of such a coupled transcription–translation system using a 3SR-like protocol with an *in vitro* translation system has been described for the simultaneous amplification and expression of brome mosaic virus genomic RNA (Joyce, 1993). Finally, the 3SR/differential BBSH assay has utility for the detection of genetic mutations. However, design of the assay requires prior knowledge of mutation loci. The combination of 3SR amplification and the RNase A mismatch scanning technique (Myers *et al.*, 1985; Winter, *et al.*, 1985; Lopéz-Galindéz, *et al.*, 1991) provides a potential approach for the identification of the sites of genetic mutations.

REFERENCES

Barany, F. (1991). Genetic disease detection and DNA amplification using cloned thermostable ligase. *Proc. Natl. Acad. Sci. USA* **88,** 189–193.

Bernatowicz, M. S., and Matsueda, G. R. (1986). Preparation of peptide-protein immunogens using *N*-succinimidyl bromoacetate as a heterobifunctional crosslinking reagent. *Anal. Biochem.* **155,** 95–102.

Bush, C. E., Vanden Brink, K. M., Sherman, D. G., Peterson, W. R., Beninsig, L. A., and Godsey, J. H. (1991). Detection of *Escherichia coli* rRNA using target amplification and time-resolved fluorescence detection. *Mol. Cell. Probes* **5,** 467–472.

Bush, C. E., Donovan, R. M., Peterson, W. R., Jennings, M. B., Bolton, V., Sherman, D. G., Vanden Brink, K. M., Beninsig, L. A., and Godsey, J. H. (1992). Detection of human immunodeficiency virus type 1 RNA in plasma samples from high-risk pediatric patients by using the self-sustained sequence replication reaction. *J. Clin. Microbiol.* **30,** 281–286.

Cimino, G. D., Metchette, K. C., Tessman, J. W., Hearst, J. E., and Isaacs, S. T. (1991). Post-PCR sterilization: A method to control carryover contamination for the polymerase chain reaction. *Nucleic Acids Res.* **19,** 99–107.

Davis, G. R., Blumeyer, K., DiMichele, L. J., Whitfield, K. M., Chappelle, H., Riggs, N., Ghosh, S. S., Kao, P. M., Fahy, E., Kwoh, D. Y., Guatelli, J. C., Spector, S. A., Richman, D. D., and Gingeras, T. R. (1990). Detection of human immunodeficiency virus type 1 in AIDS patients using amplification-mediated hybridization analyses: Reproducibility and quantitative limitations. *J. Infect. Dis.* **162,** 13–20.

Decker, D. J., Linton, P.-J., Jacobs, S., Biery, M., Gingeras, T. R., and Klinman, N. R. (1994) Defining subsets of naive and memory B cells based on their ability to somatically mutate *in vitro. Science* (submitted).

Devlin, R., Studholme, R. M., Dandliker, W. B., Fahy, E., Blumeyer, K., and Ghosh, S. S. (1993). Homogeneous detection of nucleic acids by transient state polarized fluorescence. *Clin. Chem.* **39**(9), 1939–1943.

Dunn, J. J., and Studier, F. W. (1983). The complete nucleotide sequencing of bacteriophage T7 DNA, and the locations of T7 genetic elements. *J. Mol. Biol.* **166,** 477–535.

Fahy, E., Kwoh, D. Y., and Gingeras, T. R. (1991). Self-sustained sequence replication (3SR): An isothermal transcription-based amplification system alternative to PCR. *PCR Meth. Appl.* **1,** 25–33.

Fahy, E., Biery, M., Goulden, M., Ghosh S. S., and Gingeras T. R. (1994). Issues of variability, carry-over contamination and detection in 3SR-based assays. *PCR Meth. Appl.* **3,** 583–594.

Fahy, E., Davis, G. R., DiMichele, L. J., and Ghosh, S. S. (1993). Design and synthesis of polyacrylamide-based oligonucleotide supports for use in nucleic acid diagnostics. *Nucleic Acids Res.* **21,** 1819–1826.

Ghosh, S. S., Kao, P. M., McCue, A. W., and Chappelle, H. L. (1990). Use of maleimide-thiol coupling chemistry for efficient synthesis of oligonucleotide-enzyme conjugate hybridization probes. *Bioconjugate Chem.* **1,** 71–76.

Gingeras, T. R., and Kwoh, D. Y (1992). In vitro nucleic acid target amplification techniques: Issues and benefits. *In* "Jahrbuche Biotechnologie" (P. Präve, M. Schlingmann, K. Esser, R. Thauer, and F. Wagner, eds.), Vol. 4, pp. 403–429. Hanser Publishers, Munich.

Gingeras, T. R., Whitfield, K. M., and Kwoh, D. Y. (1990). Unique features of the self-sustained sequence replication (3SR) reaction in the in vitro amplification of nucleic acids. *Ann. Biol. Clin.* **48,** 498–501.

Gingeras, T. R., Prodanovich, P., Latimer, T., Guatelli, J. C., Richman, D. D., and Barringer, K. J. (1991). Use of self-sustained sequence replication amplification reaction to analyze and detect mutations in zidovudine-resistant human immunodeficiency virus. *J. Infect. Dis.* **164,** 1066–1074.

Guatelli, J. C., Whitfield, K. M., Kwoh, D. Y., Barringer, K. J., Richman, D. D., and Gingeras, T. R. (1990). Isothermal, *in vitro* amplification of nucleic acids by a multienzyme reaction modeled after retroviral replication. *Proc. Natl. Acad. Sci. USA* **87,** 1874–1878.

Inman, J. K. (1974). Covalent linkage of functional groups, ligands and proteins to polyacrylamide beads. *Meth. Enzymol.* **34,** 30–58.

Isaacs, S. T., Tessman, J. W., Metchette, K. C., Hearst, J. E., and Cimino, G. D. (1991). Post-PCR sterilization: Development and application to an HIV-1 diagnostic assay. *Nucleic Acids Res.* **19,** 109–116.

Ishii, J., and Ghosh, S. S. (1993). Bead-based sandwich hybridization characteristics of oligonucleotide-alkaline phosphatase conjugates and their potential for quantitating target RNA sequences. *Bioconjugate Chem.* **4,** 34–41.

Joyce, G. F. (1993). Evolution of catalytic function. *Pure Appl. Chem.* **65,** 1205–1212.

Klinman, N. R., Press, J. L., Pickard, A. R., Woodland, R. T., and Dewey, A. F. (1974). Biography of the B cell. *In* "The Immune System" (E. Sercarz, A. Williamson, and C. F. Fox, eds.), pp. 357–365. Academic Press, New York.

Kwoh, D. Y., Davis, G. R., Whitfield, K. M., Chapelle, H. L., DiMichele, L. J., and Gingeras, T. R. (1989). Transcription-based amplification system and detection of amplified human immunodeficiency virus type 1 with a bead-based sandwich hybridization format. *Proc. Natl. Acad. Sci. USA* **86,** 1173–1177.

Larder, B. A., and Kemp, S. D. (1989). Multiple mutations in HIV-1 reverse transcriptase confer high-level resistance to zidovudine (AZT). *Science* **246,** 1155–1158.

Larder, B. A., Darby, G., and Richman, D. D. (1989). HIV with reduced sensitivity to zidovudine (AZT) isolated during prolonged therapy. *Science* **243,** 1731–1734.

Leirmo, S., Harrison, C., Cayley, D. S., Burgess, R. R., and Record, M. T. (1987). Replacement of potassium chloride by potassium glutamate dramatically enhances protein-DNA interactions in vitro. *Biochemistry* **26,** 2095–2101.

Linton, P.-J., Decker, D., and Klinman, N. R. (1989). Primary antibody forming cells and secondary B cells are generated from separate precursor cell subpopulations. *Cell* **59,** 1049–1059.

Lizardi, P. M., Guera, C. E., Lomeli, H., Tussie-Luna, I., and Kramer, F. R. (1988). Exponential amplification of recombinant-RNA hybridization probes. *Biotechnology* **6,** 1197–1202.

Lopéz-Galindéz, C., Rojas, J. M., Najra, R., Richman, D. D., and Perucho, M. (1991). Characterization of genetic variation and 3'-azido-3'-deoxythymidine resistance mutations of human immunodeficiency virus by the RNase A mismatch cleavage method. *Proc. Natl. Acad. Sci. USA* **88**, 4280–4284.

Luciw, P. A., Potter, S. J., Steimer, K., Dina, S. D., and Levy, J. A. (1984). Molecular cloning of AIDS-associated retrovirus. *Nature* **312**, 760–763.

Milligan, J. F., Groebe, D. R., Witherell, G. W., and Uhlenbeck, O. C. (1987). Oligoribonucleotide synthesis using T7 RNA polymerase and synthetic DNA templates. Nucleic Acids Res. **15**, 8783–8798.

Myers, R. M., Larin, Z., and Maniatis, T. (1985). Detection of single base substitutions by ribonuclease cleavage at mismatches in RNA : DNA duplexes. *Science* **230**, 1242–1246.

Orlandi, R., Gussow, D. H., Jones, P. T., and Winter, G. (1989). Cloning immunoglobulin variable regions for expression by the polymerase chain reaction. *Proc. Natl. Acad. Sci. USA* **86**, 3833–3877.

Ranki, M., Palva, A., Virtanen, M., Laaksonen, M., and Söderland, H. (1983). Sandwich hybridization as a convenient method for detection of nucleic acids in crude samples. *Gene* **21**, 77–85.

Richman, D. D., Guatelli, J. C., Grimes, J., Tsiatis, A., and Gingeras, T. (1991a). Detections of mutations associated with zidovudine resistance in human immunodeficiency virus by use of the polymerase chain reaction. *J. Infect. Dis.* **164**, 1075–1081.

Richman, D. D., Shih, C.-K., Lowy, I., Rose, J., Prodanovich, P., Goff, S., and Griffin, J. (1991b). Human immunodeficiency virus type 1 mutants resistant to nonnucleoside inhibitors of reverse transcriptase arise in tissue culture. *Proc. Natl. Acad. Sci. USA* **88**, 11241–11245.

Saiki, R. K., Scharf, S., Mullis, K. B., Horn, G. T., Erlich, H. A., and Arnheim, N. (1985). Enzymatic amplification of b-globin sequences and restriction site analysis for diagnosis of sickle cell anemia. *Science* **230**, 1350–1354.

Sastry, L., Alting-Mees, M., Huse, W. D., Short, J. M., Sorge, J. A., Hay, B. N., Janda, K. D., Benkovic, S. J., and Lerner, R. A. (1989). Cloning of the immunological repertoire in *Excherichia coli* for generation of monoclonal catalytic antibodies: Construction of a heavy chain variable region-specific cDNA library. *Proc. Natl. Acad. Sci. USA* **86**, 5728–5732.

Stillman, C. A., Linton, P.-J., Koutz, P., Decker, D. J., and Klinman, N. R. (1994). Specific immunoglobin cDNA clones produced from hybridoma cell lines and murine spleen fragment cultures by 3SR amplification. *PCR Meth. Appl.* **3**, 320–331.

Versailles, J., Berckhan, K., Ghosh, S. S., and Fahy, E. (1993). Photochemical sterilization of 3SR reactions. *PCR Meth. Appl.* **3**, 151–158.

Winter, E., Yamamoto, F., Almoguera, C., and Perucho, M. (1985). A method to detect and characterize point mutations in transcribed genes: Amplification and overexpression of the mutant c-K$_i$-ras allele in human tumor cells. *Proc. Natl. Acad. Sci. USA* **82**, 7575–7579.

Wu, D. Y., and Wallace, R. B. (1989). The ligation amplification reaction (LAR): Amplification of specific DNA sequences using sequential rounds of template-dependent ligation. *Genomics* **4**, 560–569.

14

Ligase Chain Reaction for the Detection of Infectious Agents

John D. Burczak[1], Shanfun Ching, Hsiang-Yun Hu, and Helen H. Lee
Probes Diagnostic Business Unit
Abbott Laboratories
Abbott Park, Illinois 60064

I. INTRODUCTION

The routine use of DNA probe diagnostics in clinical laboratories has developed more slowly than originally predicted. Techniques such as Southern blots, dot blots, and *in situ* hybridization have proved useful, but are labor intensive and have limited sensitivities. These shortcomings are particularly apparent in the diagnosis of infectious diseases (Lew, 1991). Standard noniso-

[1] *Current address:* Department of Molecular Diagnostics, SmithKline Beecham Pharmaceuticals, King of Prussia, Pennsylvania 19406

topic nucleic acid hybridization assays can only detect target nucleic acid sequences that are present in high copy number. Detection of low copy number sequences requires the use of long assay times, highly radioactive probes, and large amounts of clinical specimen. Under these conditions, the lower limit of detection is no better than 10^4 molecules (Landegren, 1988). Consequently, the use of nonamplified probe technology for infectious diseases has been confined largely to culture confirmation. The use of probe technologies that incorporate *in vitro* amplification of target nucleic acid sequences can overcome the limitation of low sensitivity and eliminate the need for highly radioactive probes and/or biological amplification (culture) prior to detection.

The ligase chain reaction (LCR) is highly efficient and can be used to amplify specific DNA sequences *in vitro* (Barany, 1991; Birkenmeyer and Mushahwar, 1991). Similar to the polymerase chain reaction (PCR), LCR exploits two properties of nucleic acids, namely, the highly specific nature of nucleotide base pairing and the strand separation and reannealing of double-stranded DNA by heating and cooling. PCR requires two oligonucleotide primers, a thermostable DNA polymerase, and deoxynucleoside triphosphates. In contrast, LCR uses four oligonucleotide probes, a thermostable DNA ligase, and a high energy dinucleotide, NAD^+, to catalyze target sequence amplification.

II. PRINCIPLE OF THE LIGASE CHAIN REACTION

The principle of LCR is shown in Fig. 1. In the first step, a clinical sample is added to an LCR reaction mixture containing four oligonucleotide probes (two complementary pairs), the thermostable enzyme DNA ligase, and NAD^+. The oligonucleotides are chosen so that same-sense probes lie adjacent to one another when hybridized to a target DNA sequence. Oligonucleotides in this configuration represent a suitable substrate for joining by DNA ligase. To initiate the process, reactions are heated to 85–95°C to ensure strand separation of double-stranded DNA (both the target and the complementary oligonucleotide pairs). The reaction mixture is then cooled to allow hybridization of the oligonucleotides to their respective target DNA strands. The adjacent probes are then covalently linked by a thermostable DNA ligase. The ligated products, or amplicons, are complementary to the target nucleic acid sequence and function as targets in the next cycle of amplification. Thus, exponential amplification of the specific target DNA sequences is achieved through repeated cycles of denaturation, hybridization, and ligation in the presence of excess oligonucleotide probes. Exponential amplifica-

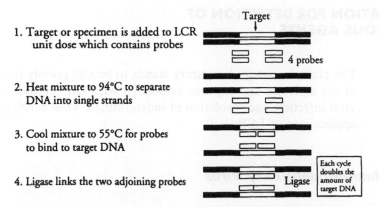

1. Target or specimen is added to LCR unit dose which contains probes

4 probes

2. Heat mixture to 94°C to separate DNA into single strands

3. Cool mixture to 55°C for probes to bind to target DNA

4. Ligase links the two adjoining probes

Ligase

Each cycle doubles the amount of target DNA

Figure 1 General outline for ligase chain reaction (LCR) amplification of DNA.

tion allows detection of minute levels of target using LCR. By varying the number of temperature cycles, it is possible to modulate the lower limit of detection by LCR, so in theory a single target molecule is detectable. The statistical distribution of low numbers of molecules in solution makes a somewhat higher limit of detection practical for reproducible sensitivity.

A problem that has been noted in LCR is the production of ligation products that are generated in a target-independent manner. This event is presumably due to blunt end ligation of the oligonucleotide duplexes. Irrespective of the mechanism of generation, the effect of this problem is the production of an amplifiable background, which limits the sensitivity of LCR assays. Several methods have been employed to reduce the generation of these target-independent ligation products. One such method is a modification of standard LCR known as gap LCR (G-LCR). In G-LCR, oligonucleotide probes are chosen that are incapable of being ligated together in a target-independent manner because the duplexes are not blunt ended (Burczak *et al.*, 1989; Dille *et al.*, 1993). As a consequence, when same-sense probes are hybridized to their respective target strands, a gap of one to several bases exists between them. This gap region is filled using a thermostable DNA polymerase and a subset of deoxyribonucleotide triphosphates (dNTPs). The dNTPs used in the reaction are determined by the nucleotide sequence of the gap region. Exponential amplification of the specific target DNA sequences is achieved through repeated cycles of denaturation, hybridization, gap filling, and ligation in the presence of excess oligonucleotide probes. The remarkable specificity of G-LCR is due to the low probability of the hybridization–extension–ligation sequence happening randomly. Thus, generation of LCR product in the absence of target, or in the presence of similar organisms, is highly unlikely.

III. APPLICATION FOR DETECTION OF INFECTIOUS AGENTS

The clinical virology laboratory stands to benefit greatly from LCR because of the difficulty and expense associated with viral culture. Confirmation of viral infection and resolution of indeterminate viral serology are important applications of LCR.

A. Human Immunodeficiency Virus

All blood donations are currently screened for human immunodeficiency viruses (HIV), the human T-lymphotropic virus type I (HTLV-1), and the hepatitis C virus (HCV) using enzyme immunoassay (EIA) methods (Burczak *et al.*, 1989). Seropositive results are confirmed by Western blots or radioimmunoprecipitation assays (RIPA) (Lee *et al.*, 1992). Depending on the viral marker, indeterminate confirmatory results can occur in 20–35% of the EIA positive specimens. Ideally, a confirmatory assay should provide direct rather than indirect evidence of viral infection. Because of its high sensitivity and specificity, LCR can be a valuable tool for the direct detection of viral infections. In addition, LCR can be used to clarify the serological status of individuals with indeterminate confirmatory test results (Jackson *et al.*, 1990).

Quantitative LCR for HIV, hepatitis B virus (HBV), or HCV may also be useful in monitoring therapeutic efficacy or disease progression. The quantitative aspect of LCR amplification is illustrated in Fig. 2. Dilutions of a plasmid containing HIV-1 DNA were detected using LCR oligonucleotides specific for a portion of the *pol* gene (J. Carrino, unpublished data). LCR probes can be designed for other proviral genes, for genomic RNA, or for both to help establish the clinical significance of changes in copy numbers during therapy.

B. Human Papillomavirus

Because most microorganisms contain both highly conserved, genus-specific DNA sequences and strain- or species-specific sequences, the appropriate target sequence can allow the user to detect a group of organisms or to differentiate species or subtypes. The importance of viral typing is illustrated by the fact that more than 60 human papillomavirus (HPV) types have been

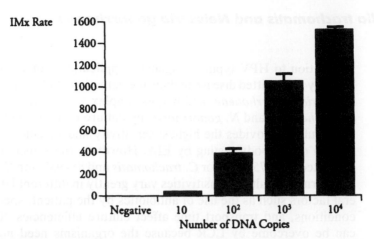

Figure 2 Histogram showing LCR results for the detection of different copy numbers of target HIV proviral DNA.

detected and at least 20 HPV types have been detected in the human genital tract (Hampl *et al.*, 1991). In addition, several HPV types are associated with the development of cervical cancer. HPV typing of women with cytologically abnormal Pap smears can contribute to the clinical management and follow-up of these individuals. Due to the specificity of LCR, it is possible to develop assays that can distinguish between different HPV types. In one study, an LCR assay specific for HPV type 16 (HPV-16) was used to correctly identify 16 of 16 tissue samples determined to be positive for HPV-16 by Southern blot. Further, a second LCR assay specific for HPV type 18 (HPV-18) was used to correctly identify 5 of 5 tissue samples determined to be positive for HPV-18 by Southern blot (data not shown).

C. Human T-Lymphotrophic Virus Type I and Type II

Another application of LCR for viral typing is the differentiation of HTLV-I from HTLV-II, the former being the cause of adult T-cell leukemia/lymphoma and a progressive neurological disorder similar to multiple sclerosis, whereas the latter virus is not yet associated with any disease (Burczak *et al.* 1989; Rosenblatt *et al.*, 1990; Lee *et al.*, 1992). Because of a high degree of amino acid homology, differentiation between the two viruses is difficult by serology. Identification and follow-up of individuals with HTLV-II infection may help define its role in pathogenesis (Lee *et al.*, 1989).

D. *Chlamydia trachomatis* and *Neisseria gonorrhoeae*

In addition to HPV typing, diagnostic applications for LCR in the area of sexually transmitted disease include the detection of *Chlamydia trachomatis, Neisseria gonorrhoaeae,* and herpes simplex virus (HSV). The isolation of *C. trachomatis* and *N. gonorrhoeae* by culture remains the "gold standard" since culture provides the highest sensitivity and specificity compared with antigen or antibody testing by EIA. However, the sensitivity of culture is estimated to be 75–85% for *C. trachomatis* and 85–95% for *N. gonorrhoaeae.* Furthermore, culture sensitivities vary greatly in different laboratories. Several factors such as the use of antibiotics by the patient, specimen transport conditions, and transport time affect culture efficiencies. These obstacles can be overcome by LCR because the organisms need not be viable for detection. The detection of *C. trachomatis* by LCR requires that the selected target sequences be present in all variants but absent from other microorganisms, particularly those commonly found in genital specimens. LCR detection of all 15 recognized *C. trachomatis* serological variants (serovars) has been demonstrated but the *C. trachomatis* target sequence was not detected in 88 other potentially cross-reacting microorganisms. The cutoff values are validated by rigorous analysis of clinical specimens, concentrating on those giving discordant results between culture and LCR, so that optimal sensitivity and specificity are achieved. *Chlamydia trachomatis* studies by LCR on 2003 female endocervical specimens demonstrated 94.7% sensitivity and 99.8% specificity compared with 70.8% sensitivity and 100% specificity (by definition) for culture (Burczak *et al.,* 1993). LCR detection of *N. gonorrhoeae* in 2235 endocervical specimens resulted in 98.9% sensitivity and 99.6% specificity compared with 88.9% sensitivity and 100% specificity (by definition) for culture (Ching *et al.,* 1993).

Asymptomatic male carriers represent a reservoir population largely untested for *C. trachomatis* infection because of the difficulty of obtaining urethral swab specimens in males. Culture of urine specimens for *C. trachomatis* is usually not practical because of low sensitivity and the variable results caused by the cytopathic effects of urine. In addition, the number of organisms is often lower in urine than in urogenital swab specimens. As a result, an easy and sensitive LCR procedure for the detection of *C. trachomatis* in urine was developed. The performance of *C. trachomatis* detection by LCR on male urine specimens compared with culture on urethral swab specimens from the same individuals is shown in Table 1. The results of 496 paired specimens demonstrated 95.7% (45/47) sensitivity for LCR compared with 44.7% (21/47) for culture on urethral swabs, with LCR detecting 51.0% more positives than culture. Of the 26 discordant specimen pairs that were positive by LCR but negative by culture, 24 (79.2%) were confirmed as

TABLE 1

**Comparison of *Chlamydia trachomatis*
Detection by LCR on Male Urine
Specimens with Culture on Matched
Urethral Specimens**[a]

	Urethral culture	
Urine LCR	Positive	Negative
Positive	19	26[b]
Negative	2	449

[a] *Chlamydia trachomatis* LCR sensitivity: 95.7% (45/47);
Chlamydia trachomatis LCR specificity: 100% (449/449);
Chlamydia trachomatis culture sensitivity: 44.7% (21/47).
[b] All 26 confirmed as positive.

true positives by direct fluorescent antibody (DFA) analysis of a cytospin sediment from the culture specimen, and the remaining 2 (20.8%) discordant specimen pairs were confirmed by LCR using probes directed against the gene encoding the major outer membrane protein (MOMP) of *C. trachomatis*. The screening of urine by LCR should have a positive impact on the control of *C. trachomatis* infections, particularly with respect to the identification of asymptomatic male carriers.

IV. METHODOLOGY

The application of LCR in the clinical laboratory combines three essential steps: specimen preparation, amplification, and detection of amplified products. Many modifications specific to a particular organism or specimen type have been developed.

A. Specimen Preparation

A variety of clinical specimen types may be encountered in a diagnostic laboratory including blood, cerebrospinal fluid, sputum, fecal material, urine, and endocervical or urethral swabs. Optimized specimen preparation is critical for the efficient amplification of specific target DNA sequences and is a step that is often overlooked. The primary requirements of sample preparation are that the target DNA sequence be accessible and intact. This means that LCR specimen preparation is more robust than that for culture, for which microorganism viability is required. DNA may be released by simply lysing the cells through detergent treatment, repeatedly freezing and thawing, or boiling in a hypotonic solution (Lew, 1991). LCR uses relatively short target sequences, approximately 45–50 nucleotides. Therefore the risk of disrupting the DNA within the target sequence by physical or enzymatic cleavage is minimized. These features contribute to the robustness of the LCR sample preparation procedure. LCR amplification reactions do not require fresh specimen or highly purified DNA. However, like PCR, some clinical specimens such as urine contain inhibitors of enzymes required for the amplification step. In these cases, additional sample preparation steps are required to remove endogenous inhibitors that might be present in the specimen. These inhibitors may be removed during specimen preparation by a variety of steps including centrifuging the sample through 50% sucrose and resuspending in an appropriate buffer.

An example of sample preparation is illustrated by the detection of *C. trachomatis* or *N. gonorrhoeae*. Urethral or endocervical swabs are collected and placed in specimen tubes containing 5 mM N-(2-hydroxyethyl)-piperazine-N'-(3-propane-sulfonic acid) (EPPS) buffer at pH 7.8 The DNA is released by heating at 97°C for 15 min. For detection of these microorganisms in urine, the sample must first be centrifuged prior to extraction, so 1 ml urine is centrifuged at approximately 13,000 g for 10 min. Then the pellet resuspended in 1 ml buffer consisting of 5 mM EPPS at pH 7.8, and heated at 97°C for 15 min. The prepared specimens may be stored at 2–8°C or frozen at −70°C until tested. For the detection of bloodborne viruses such as HBV, the serum or plasma must first be treated by proteinase K, followed by heating (Williams and Kwok, 1992). For whole blood or sputum, a simple and reproducible sample preparation procedure remains elusive.

B. DNA Amplification

For the amplification process in LCR to be both sensitive and specific, the selection of appropriate target sequences, the probe design, the choice of DNA ligase, the thermal cycling conditions, and the concentration of probes are all critical elements.

1. Target Sequence Selection

The ideal nucleic acid target sequence should be unique to the organism of interest and should be conserved among different isolates. Mismatches between hybridized probes and target sequences, particularly near the point of ligation, can affect the efficiency of probe ligation. Computer searches of nucleic acid sequence libraries can facilitate the choice of the appropriate target sequence, but the specificity of the target sequence must ultimately be proven empirically. Target sequences with significant internal and flanking secondary structure should be avoided because these structures may restrict hybridization of the probes. Although GC-rich target sequences can often form stable secondary structures that are more difficult to amplify, the use of glycerol and dimethylsulfoxide (DMSO) in the hybridization solution may alleviate some of these effects (Williams and Kwok, 1992). To achieve high sensitivity, it is often desirable to choose a target sequence that is present in multiple copies. Highly conserved nucleic acid sequences tend to produce a broader spectrum of detection, whereas the selection of a sequence unique to a given subtype or species leads to more restricted detection.

2. Probe Design

There are no hard and fast rules in the selection of LCR probes. Typically, the probes are 18–30 bases in length. Probes shorter than 18 bases may require lower hybridization temperatures, which can compromise specificity. Probes with a sequence capable of forming secondary structures or dimers (e.g., those with high GC content) should be avoided. Figure 3 shows the probes that are used to amplify a 47-bp sequence in the *C. trachomatis* plasmid using the G-LCR format (Birkenmeyer and Mushahwar, 1991; Dille *et al.*, 1993).

3. Selection of Thermal Cycling Conditions

In addition to specific target selection, the specificity of the LCR reaction for a given organism can be achieved by the selection of an appropriate hybridization temperature. The use of thermostable enzymes allows the annealing and enzymatic steps to be carried out at temperatures approaching the melting temperatures of the oligonucleotides, thereby increasing the

Probe 1:	5'-GATACTTCGCATCATGTGTTCC-3'
Probe 2:	3'-CTATGAAGCGTAGTACACAAp-5'
Probe 3:	5'-pAGTTTCTTTGTCCTCCTATAACG-3'
Probe 4:	3'-CCTCAAAGAAACAGGAGGATATTGC-5'
Gap Fill:	dGTP

Figure 3 G-LCR probes and gap fill for the detection of *Chlamydia trachomatis*.

specificity of the hybridization reaction. The level of sensitivity of LCR can be controlled by varing the number of amplification cycles. This ability allows considerable freedom in modulating the desired sensitivity when developing assays for different pathogens. Thermostable ligases survive the rigors of cycling at high temperatures and function optimally at the temperatures that are used to control the stringency of the hybridization of the LCR oligonucleotides. Although all ligases seem to demonstrate blunt-end ligation to a greater or lesser degree, differences in blunt-end ligation activity have been noted for various ligases. A ligase with reduced blunt-end ligation activity is preferred because blunt-end ligation can compromise the sensitivity of the system.

A typical G-LCR reaction mixture consists of 100 mM EPPS, pH 7.8, 30 mM MgCl$_2$, 1.0 mM EDTA, 20 μM NAD$^+$, 1 μl [^{32}P]GTP (3000 Ci/ mmol; Amersham, Arlington Heights, IL), 10 μg/ml acetylated bovine serum albumin (BSA), 180,000 U/ml thermostable DNA ligase, 20 U/ml thermostable DNA polymerase, and equimolar amounts (1.0–2.0 × 10^{13} molecules/ ml) of the four probes described in Fig. 3. Microcentrifuge tubes containing 100 μl reaction mix and 100 μl treated sample are placed into a thermal cycler for 38–40 cyles at 85°C for 30 sec and 50°C for 20 sec (Dille *et al.*, 1993).

C. Detection of Amplified Products (Amplicons)

A variety of methods can be used to detect the amplified products of LCR including gel electrophoresis, autoradiography, and immunological methods. The detection method should be designed with ease of use, speed, and automation in mind to best fit the needs of a clinical laboratory. Ligated probes may be separated from unligated probes prior to detection by electrophoresis and can be detected by direct visualization, Southern blotting, or autoradiography. In the procedure already described in which the ligated probes are labeled with ^{32}P, 25 μl amplification mixture is added to an equal volume of loading buffer (80% formamide, 50 mM Tris-borate, pH 8.3, 1 mM EDTA, 0.1% xylene cyanol, 0.1% bromophenol blue) and electrophoresed through a 10% polyacrylamide/8 M urea gel. Following electrophoresis, gels are exposed at −70°C to X-OMAT AR film (Kodak, Rochester, NY) with intensifying screens to detect radiolabeled bands. The incorporation of a radioactive label into LCR amplification products is identified by comparison against radiolabeled molecular weight markers. Alternatively, one or several of the oligonucleotides used for amplification can be end labeled (i.e., with polynucleotide kinase and [γ-^{32}P]ATP), and reaction products can be identified by a mobility shift due to ligation.

Although the methods described here work well, most clinical laboratories prefer to use nonisotopic detection methods. LCR ligation products can be readily separated from nonligated probes and detected in an immunocapture format using LCR oligonucleotides labeled with specific haptens (Fig. 4). Hapten labeling can be achieved by attaching a primary amine (3'-Amine-ON CPG, Aminomodifier II; Clonetech, Palo Alto, CA) to the 3' or 5' end of an oligonucleotide, followed by reaction with any of a number of different reactive haptens using conventional activation and linking chemistries. In this case, probes 1 and 2 are labeled with hapten A at the 5' and 3' end, respectively, while probes 3 and 4 are labeled with a different hapten (hapten B) at the 3' and 5' end, respectively. In the presence of specific target, probes 1 and 3 are ligated, as are probes 2 and 4. Only the ligated product contains both haptens. A solid phase coated with antibody against hapten A is used to capture the ligated product as well as the unligated probes labeled with hapten A. After washing, enzyme-conjugated (e.g., horseradish peroxidase or alkaine phosphatase) antibodies against hapten B are added. On addition of appropriate substrate, a positive signal is generated in the form of a color reaction. The solid phase can be a microtiter plate or microparticles, the latter having the advantage of a larger surface area for better sensitivity and easier quality control during the manufacturing process. Irrespective of the solid phase format, the use of haptenated probes transforms the detection of LCR products into a standard immunocapture assay format, which can be easily automated using current EIA detection instruments. Figure 5 demonstrates the correlation between the detection of a 48-bp LCR amplicon by autoradiography and a microparticle immunoassay (MEIA). MEIA detection eliminates the need for radioisotopes and is well suited to routine use in the clinical laboratory.

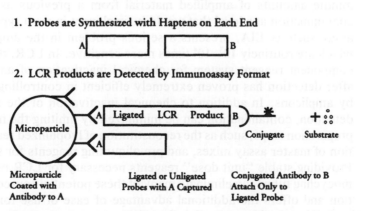

Figure 4 Detection of LCR amplicon by microparticle enzyme immunoassay.

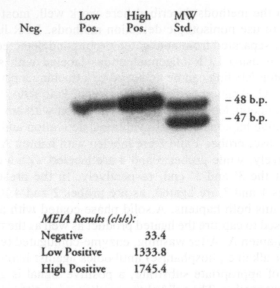

MEIA Results (c/s/s):

Negative	33.4
Low Positive	1333.8
High Positive	1745.4

Figure 5 Correlation of *Chlamydia trachomatis* LCR amplicon (48 bp) detection by autoradiography and microparticle enzyme immunoassay (MEIA). Neg. is a negative control of human placental DNA, Low Pos. is a low *Chlamydia trachomatis*-positive specimen. High Pos. is high *Chlamydia trachomatis*-positive specimen, and MW Std. are molecular weight markers of 48 and 47 bp.

D. Contamination Control

A potential problem with any *in vitro* target amplification system is the accidental contamination of clinical specimens or reaction reagents with minute amounts of amplified material from a previous assay. Carryover contamination at a level that would be irrelevant in most types of diagnostic assay such as EIA, presents a serious problem in the amplified systems, which are routinely 10^3–10^4 times more sensitive. In LCR, the use of a two-component reagent system for chemical inactivation of amplified product after detection has proven extremely efficient in controlling contamination by amplicons. In addition to chemical inactivation of the amplicons after detection, contamination can be minimized by limiting the number of assay preparation steps, such as the reconstitution of lyophilized reagents, preparation of master assay mixes, and pre-aliquoting reagents for subsequent use. Providing stable "unit dose" reagents necessary for LCR reactions in thermocycling tubes can eliminate many of these potential sources of contamination and offers the additional advantage of ease of use for the laboratory technicians in a routine clinical setting.

V. CONCLUSIONS

Nonamplified DNA probe diagnostic assays are limited in sensitivity and are often tedious to perform. *In vitro* target amplification methods such as PCR and LCR provide a means for greatly increasing sensitivity while maintaining the high degree of specificity inherent to nucleic acid hybridization. LCR has been applied successfully to the detection of a number of infectious disease organisms using simple sample preparation methods, easy detection of amplicons by automated EIA instruments, and contamination control by chemical inactivation of amplicons and a unit-dose reagent format. LCR-based diagnostics will facilitate the transfer of amplification-based DNA probe technology from the research setting to the clinical laboratory, and holds great promise for the detection of infectious microorganisms, particularly those that are difficult to culture.

REFERENCES

Barany, F. (1991). The ligase chain reaction in a PCR world. *PCR Meth. Appl.* **1,** 5–16.

Birkenmeyer, L. G., and Mushahwar, I. K. (1991). DNA probe amplification methods. *J. Virol. Meth.* **35,** 117–126.

Burczak, J. D., Canavaggio, M. C., Bates, P. F., Steaffens, J. W., and Lee, H. H. (1989). Retroviruses in animals and humans. *Clin. Lab. Sci.* **2(5),** 294–298.

Burczak, J. D., Quinn, T. C., Schachter, J., Stamm, W. E., and Lee, H. H. (1993). Ligase chain reaction for the detection of *Chlamydia trachomatis* in endocervical specimens. Abstracts of the 6th European Congress of Clinical Microbiology and Infectious Diseases (Abstract 837).

Ching, S. F., Ohhashi, Y., Birkenmeyer, L., Armstrong, A., and Lee, H. (1993). A ligase chain reaction method for the detection of *Neisseria gonorrohoeae*. Abstracts of the 6th European Congress of Clinical Microbiology and Infectious Diseases (Abstract 374).

Dille, B. J., Butzen, C. C., and Birkenmeyer, L. G. (1993). Amplification of *Chlamydia trachomatis* DNA by ligase chain reaction. *J. Clin. Microbiol.* **31(3),** 729–731.

Hampl, H., Marshall, R. L., Perko, T., and Solomon, N. (1991). Alternative methods for DNA probing in diagnosis: ligase chain reaction (LCR). *In* "PCR Topics: Usage of Polymerase Chain Reaction in Genetics and Infectious Diseases" (A. Rolfs, H. Schumacher, and P. Marx, eds.), pp. 15–22. Springer Verlag, Berlin.

Jackson, J. B., MacDonald, K. L., Cadwell, J., Sullivan, C., Kline, W. E., Hanson, M., Sannerud, K. J., Stramer, S. L., Fildes, N. J., Kwok, S. Y., Sninsky, J. J., Bowman, R. J., Polesky, H. F., Balfour, H. H., and Osterholm, M. T. (1990). Absence of HIV infection in blood donors with indeterminate western blot tests for antibody to HIV-1. *N. Engl. J. Med.* **322,** 217–221.

Landegren, U., Kaiser, R., Caskey, C. T., and Hood, L. (1988). DNA diagnostic molecular techniques and automation. *Science* **242,** 229–237.

Lee, H., Swanson, P., Shorty, V. S., Zack, J. A., Rosenblatt, J. D., and Chen, I. S. Y. (1989).

High rate of HTLV-II infection in seropositive IV drug abusers in New Orleans. *Science* **244,** 471–475.

Lee, H. H., Canavaggio, M., and Burczak, J. D. (1992a). Immunoblotting. *In* "Laboratory Diagnosis of Viral Infections" (E. H. Lennette, ed.), 2d Ed., pp. 195–210. Marcel Dekker, New York.

Lee, H. H., Shih, J., Burczak, J. D., and Steaffens, J. (1992b). Human T-lymphotropic viruses: types I and II. *In* "Laboratory Diagnosis of Viral Infections" (E. H. Lennette, ed.), 2d Ed., pp. 495–513, Marcel Dekker, New York.

Lew, A. M. (1991). The polymerase chain reaction and related techniques. *Curr. Opin. Immunol.* **3,** 242–246.

Rosenblatt, J. D., Plaeger-Marshall, S., Giorgi, J. V., Swanson, P., Chen, I. S. Y., Chin, E., Wang, H. J., Canavaggio, M., Black, A. C., and Lee, H. (1990). A clinical, hematologic and immunologic analysis of 21 HTLV-II infected intravenous drug users. *Blood* **76,** 409–417.

Williams, S. D., and Kwok, S. (1992). Polymerase chain reaction: Application for viral detection. *In* "Laboratory Diagnosis of Viral Infection" (E. H. Lennette, ed.), 2d Ed., pp. 147–173. Marcel Dekker, New York.

15

A Chemiluminescent DNA Probe Test Based on Strand Displacement Amplification

G. T. Walker, C. A. Spargo, C. M.
Nycz, J. A. Down, M. S. Dey, A. H.
Walters, D. R. Howard, W. E. Keating,
M. C. Little, J. G. Nadeau, S. R.
Jurgensen, and V. R. Neece
Becton Dickinson Research Center
Research Triangle Park, North Carolina
27709

P. Zwadyk, Jr.
Veterans Administration Hospital
Durham, North Carolina 27705

329

I. INTRODUCTION

This chapter introduces a DNA probe test that is based on strand displacement amplification (SDA) and chemiluminescence detection (Spargo *et al.,* 1993). A step-by-step protocol is presented for amplification and detection of a DNA sequence specific to *Myobacterium tuberculosis,* the causative agent of tuberculosis. The chapter also covers application of the protocol to clinical sputum specimens. The system can be performed in about 6 hr and provides sensitivity to a few genomes of *M. tuberculosis.* The *M. tuberculosis* system provides a model for development of SDA and chemiluminescence detection systems for viruses and other infectious agents.

II. DESCRIPTION OF STRAND- DISPLACEMENT AMPLIFICATION

SDA is an isothermal method of amplifying target that can produce a 10^8-fold amplification during a single 2-hr incubation at 41°C. Single temperature incubation lowers the instrumentation expense of SDA compared with methods that use temperature cycling, such as the polymerase chain reaction (PCR; Erlich *et al.,* 1991) and the ligase chain reaction (LCR; Barany, 1991). Isothermal incubation also minimizes some of the variability associated with handling multiple clinical specimens. Reagent costs are somewhat higher for SDA than for PCR and LCR because SDA requires two enzymes, a DNA polymerase and a restriction enzyme, as opposed to a single enzyme for the other methods. However, SDA requires fewer enzymes than other isothermal techniques such as self-sustained sequence replication (3SR; Fahy *et al.,* 1991) or nucleic acid sequence-based amplification (NASBA™; Malek *et al.,* 1992) that generally use three enzymes (see Chapters 12 and 13). Another undesirable characteristic of SDA is its inability to efficiently amplify long (>200 nucleotides) target sequences (Walker, 1993). Although this limitation is not significant in clinical diagnostic laboratories where the length of the amplified product is generally not important, it may restrict its use in the research laboratory (e.g., in gene cloning).

Detailed mechanistic descriptions and empirical discussions of SDA have been provided elsewhere (Walker *et al.,* 1992b; Walker, 1993). The following discussion is intended to provide a practical understanding of SDA and its underlying mechanisms. The main reaction events are depicted in Fig. 1. In the first step, a primer bearing a *Hinc*II restriction site (5'-GTTGAC) binds to a complementary target nucleic acid. The primer and target are extended by the exo⁻ Klenow fragment of DNA polymerase in the presence of dGTP, dCTP, dUTP, and dATP containing an alpha-thiol group (dATPαS). This

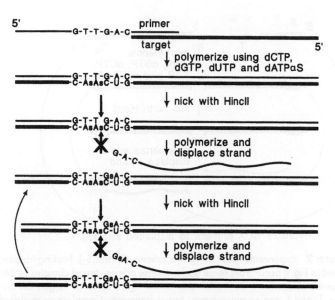

Figure 1 Linear SDA using a single primer. The primer, which contains a *Hinc*II recognition sequence (5'-dGTTGAC), binds a target fragment at its 3' end. The primer–target complex is extended by DNA polymerase in the presence of dGTP, dCTP, dUTP, and dATPαS, forming a hemi-thiolated *Hinc*II site. *Hinc*II nicks the recognition site after which DNA polymerase initiates replication at the nick and displaces the downstream strand. The nicking/displacement steps cycle because a nickable *Hinc*II site is regenerated.

reaction forms a double-stranded *Hinc*II recognition site, one strand of which contains phosphorothioate linkages located 5' to each dA (5'-GUC$_S$A$_S$AC). *Hinc*II nicks the hemiphosphorothioate recognition site between the T and the G in the sequence 5'-GTT↓GAC without cutting the complementary thiolated strand (5'-GUC$_S$A$_S$AC). After *Hinc*II dissociates from the nicked site, exo⁻ Klenow initiates DNA synthesis at the nick, and a new downstream DNA strand is synthesized. This extension–displacement step regenerates an unnicked *Hinc*II site that can be nicked again. The appearance of the phosphorothioate linkage in the sequence 5'-GTTG$_S$AC does not inhibit subsequent nicking because the phosphorothioate linkage is not at the cleavage position. Consequently, the nicking and strand displacement steps cycle continuously, producing a nickable *Hinc*II site.

The steps in Fig. 1 produce a target strand approximately every 3 min in the linear amplification mode. However, the amplification system becomes exponential with a *doubling* time of about 3 min when sense and antisense reactions are coupled using a large concentration of two primers (S$_1$ and S$_2$), each of which contains a *Hinc*II site (Fig. 2). The left side of Fig. 2 represents the series of events in Fig. 1 using S$_1$. Each strand displaced here serves as a target for S$_2$ on the right side due to the complementary nature of DNA. Likewise each displaced product from the right side is a target for S$_1$.

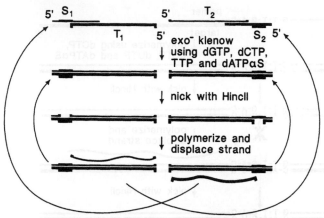

hybridize SDA primers to displaced strands

Figure 2   Exponential SDA using two primers (S_1 and S_2). Nicking/displacement steps (as described in Fig. 1) occur from sense and antisense primer–target complexes. Strands displaced from the series of reactions on the left side of the figure serve as targets on the right side and vice versa. T_1 and T_2 are complementary target fragments. Complete or unnicked HincII recognition sequences are represented by ▬▬▬. Partial or nicked HincII recognition sequences are represented by ▪ and ▪.

This description of SDA pertains to target DNA fragments with defined 5′ and 3′ ends, a situation not encountered with a target sample of genomic DNA. We therefore developed a target generation step that uses two additional primers and again exploits the strand-displacing activity of exo⁻ Klenow to generate target fragments with defined 5′- and 3′ ends (Fig. 3; Walker et al., 1992b). During target generation, the sample is heat denatured in the presence of four primers (B_1, B_2, S_1, and S_2). S_1 and S_2 bind to the target at positions flanking the fragment to be amplified. B_1 and B_2 bind to the target at positions upstream of S_1 and S_2. Exo⁻ Klenow simultaneously extends all four primers and initiates a cascade of extension and displacement steps. For example, S_1 is extended and its extension product (S_1-ext) is displaced through extension of the upstream primer B_1. S_1-ext then serves as target for B_2 and S_2, as did the original target DNA. Subsequently, S_2 is extended and displaced from the template S_1-ext, forming a target for S_1. A series of completely analogous events originates from the other target strand. The end result is a number of target fragments with hemiphosphorothioate HincII sites at one or both ends (Fig. 3). These fragments directly enter the amplification cycle in Fig. 2 on nicking by HincII.

SDA operates by a very simple protocol. All the SDA reagents except exo⁻ Klenow and HincII are added to the clinical sample. The sample is heated for 2 min in a boiling water bath to lyse organisms and denature the

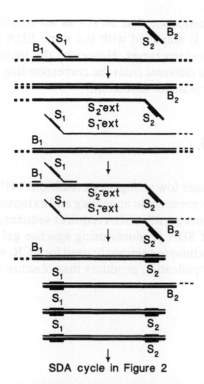

Figure 3 The target generation step of SDA prior to entering the cyclic phase (Fig. 2). Genomic target DNA is heat denatured in the presence of four primers (B_1 B_2, S_1, and S_2). Addition of DNA polymerase initiates a cascade of primer extension and strand displacement steps, resulting in fragments containing hemi-thiolated *Hinc*II recognition sites at one or both ends. On nicking by *Hinc*II, these fragments enter the cyclic phase of SDA shown in Fig. 2. *Hinc*II recognition sequences are represented by ▄▄▄.

target DNA. After returning the sample to 41°C, exo⁻ Klenow and *Hinc*II are added; SDA proceeds for 2 hr at this temperature.

A. Use of an Amplification Control Sequence to Monitor SDA Inhibition with Clinical Specimens

Clinical specimens may contain a variety of amplification inhibitors. In addition, specimens such as sputum contain human and bacterial DNA, which reduces SDA sensitivity by promoting background amplification (Walker *et al.*, 1992a,b). Respiratory specimens typically contain 0.1–1.0 μg nonmycobacterial DNA (J. A. Down, unpublished results). Since inhibitors can produce false negative results, we routinely co-amplify and separately detect an oligodeoxynucleotide sequence that serves as an amplification control. The amplification control is amplified with the same SDA primers (S₁ and S₂) used for the *M. tuberculosis* target. However, the internal sequence of the amplification control is different from the corresponding *M. tuberculosis* sequence to allow its independent detection.

III. DETECTION OF SDA REACTIONS

Because SDA operates under low stringency conditions (41°C), background amplification can occur by nonspecific annealing and extension of SDA primers on nontarget DNA. Consequently, SDA requires sequence-specific detection methods. Analysis of SDA products using agarose gel electrophoresis and ethidium staining, a technique commonly used for PCR, routinely reveals a number of nontarget amplification products that obscure identification of target amplification.

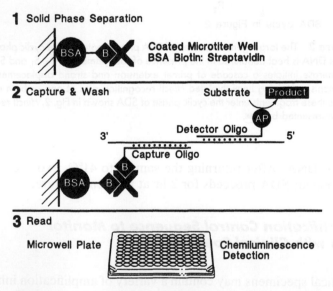

Figure 4 Schematic of the microtiter assay for detection of SDA products. Step 1: Solid phase is coated with biotinylated bovine serum albumin (BSA) and a secondary layer of streptavidin. Step 2: Free biotin binding sites on the streptavidin bind the complex formed by capture and detector probe hybridization to the SDA product. Unbound AP–detector probe is washed away. A chemiluminescent substrate for alkaline phosphate (AP) is added. Step 3: Luminescence indicates the presence of amplified DNA.

A. Detection Using a Chemiluminescence-Based Sandwich Assay

Amplified *M. tuberculosis* target DNA and the control are detected by a chemiluminescence-based sandwich assay using a biotinylated capture probe and a detector probe that is conjugated to alkaline phosphatase (AP) for use with a chemiluminescent substrate (Fig. 4; Spargo *et al.*, 1993). This assay offers significant advantages with respect to gel electrophoresis, autoradiography, and Southern blots. It requires only 1.5 hr, accommodates 96 detections per microwell plate, and does not have the regulatory, safety, or environmental concerns associated with the use of radioactivity. Results are quantitative, and the method has sensitivity equivalent to that of ^{32}P detection (Walker *et al.*, 1992a,b; Walker, 1993).

IV. PERFORMANCE OF SDA WITH CLINICAL *MYCOBACTERIUM TUBERCULOSIS* SPECIMENS

This section describes amplification and detection of DNA from *M. tuberculosis*. This process includes co-amplification of a control to test the efficiency of SDA. Although the details of the sample processing, SDA, and detection protocols are specific to *M. tuberculosis*, the method provides a framework for general application of SDA in the clinical laboratory.

A. Target DNA Sequence for M. tuberculosis

The IS6110 insertion element is a DNA sequence that is specific to species representing the *M. tuberculosis* complex (*M. tuberculosis, M. bovis, M. bovis-BCG, M. africanum*, and *M. microti;* Thierry *et al.*, 1990). We amplify a subsequence of the IS6110 element, as described elsewhere (Walker *et al.*, 1992b; Spargo *et al.*, 1993).

Each SDA reaction contains four primers—S_1 (5′-dTTGAATAGTCGG-TTACTT<u>GTTGAC</u>GGCGTACTCGACC), S_2 (5′-dTTGAAGTAACCGAC-TATT<u>GTTGAC</u>ACTGAGATCCCCT), B_1 (5′-dTGGACCCGCCAAC), and B_2 (5′-dCGCTGAACCGGAT)—that are purified by denaturing gel electrophoresis and electroelution (Ausubel *et al.*, 1989). The *Hinc*II recognition sites in S_1 and S_2 are underlined. SDA reactions (50 μl) contain final concentrations of 500 nM S_1 and S_2, 50 nM B_1 and B_2, 50 mM K_2HPO_4, pH 7.6, 0.1 mg/ml bovine serum albumin (BSA), 0.5 mM dUTP, 0.2 mM each dGTP, dCTP, and dATPαS, 7 mM $MgCl_g$, 14% glycerol, *M. tuberculosis* DNA (or clinical specimen), 150 U *Hinc*II, 3 U exo⁻ Klenow, and 25,000 molecules of the amplification control (5′-dACTGAGATCCCCTAGCGACGATGTCT-GAGGCAACTAGCAAAGCTGGTCGAGTACGCC). The 14% glycerol concentration includes the glycerol provided by the stock solutions of *Hinc*II and exo⁻ Klenow.

1. Stock Reagent Solutions

0.5 *M* K$_2$ HPO$_4$, pH 7.6

dNTP solution: 10 m*M* each dATPαS, dGTP, dCTP; 25 m*M* dUTP (Pharmacia, Piscattaway, NJ)

BSA solution: 5 mg/ml acetylated BSA (New England Biolabs, Beverly, MA)

55% (v/v) glycerol

Primer solution: 25 µ*M* each S$_1$ and S$_2$, 2.5 µ*M* each B$_1$ and B$_2$

0.35 *M* MgCl$_2$

Amplification control: amplification control oligodeoxynucleotide is present in 25,000 copies/µl in 25 m*M* K$_2$HPO$_4$, pH 7.6, 250 ng/µl ultra pure human placental DNA (Sigma)

75 U/µl *Hinc*II (contains 50% glycerol) (New England Biolabs)

15 U/µl exo⁻ Klenow (United States Biochemical, Cleveland, OH)

2. SDA Protocol

1. Prepare an SDA "master mix" appropriate for the number of SDA reactions to be performed. Master mix for a single SDA reaction consists of:

   1.37 µl H$_2$O
   1 µl amplification control
   1 µl primer solution
   3.63 µl 0.5 *M* K$_2$HPO$_4$, pH 7.6
   1 µl dNTP solution
   10 µl 55% (v/v) glycerol
   1 µl BSA solution
   1 µl 0.35 *M* MgCl$_2$

   Mix by vortexing. When preparing the concentrated master mix, MgCl$_2$ is added last so that MgCL$_2$ and K$_2$HPO$_4$ do not mix at higher concentrations because magnesium phosphate readily precipitates.

2. Add 20 µl master mix to 25 µl processed clinical sample and mix by vortexing.

3. Heat sample for 2.5 min in a boiling water bath. Cool for 2 min in a 41°C water bath.

4. Dilute exo⁻ Klenow 1:5 in 50 mM K$_2$HPO$_4$, pH 7.0, 1 mM DTT, 50% (v/v) glycerol to a concentration of 3 U/μl.

5. Prepare an SDA "enzyme mix" appropriate for the number of SDA reactions. The enzyme mix for a single SDA reaction consists of:

   2 μl H$_2$O
   2 μl 75 U/μl $Hinc$II
   1 μl 3 U/μl exo⁻ Klenow

6. Add 5 μl enzyme mix to each SDA reaction and mix.

7. Incubate for 2 hr at 41°C, followed by 1 min at 95°C to inactivate the enzymes.

Each SDA sample is split four ways for (duplicate) detection of the amplified *M. tuberculosis* target and the control sequence using the chemiluminescence detection assay described later. For the *M. tuberculosis* target, the sequence of the biotinylated capture probe is 5'-BBB-dCCTGAAAGAC-GTTAT (BBB represents 3 biotins) whereas the detector probe sequence is 5'-dCCACCATACGGATAG-AP (AP represents alkaline phosphatase). For the amplification control, the capture and detector probes are 5'-BBB-dGCTTTGCTAGTTGCC and 5'-dTCAGACATCGTCGTC-AP, respectively.

SDA and detection results from culture-positive *M. tuberculosis* specimens are shown in Fig. 5A. All *M. tuberculosis* signals are above those obtained from control samples containing three *M. tuberculosis* genomes. Thus, these culture-positive specimens are easily detected by SDA. At very high *M. tuberculosis* concentrations (sample P8, Fig. 5A), the amplification control signal was decreased because of competitive inhibition by *M. tuberculosis* (Walker *et al.,* 1992a,b).

For culture-negative clinical specimens, *M. tuberculosis* signals were not significantly above (2 standard deviations) the level of the negative control (0 genomes; Fig. 5B). However, before deciding whether they are truly negative, SDA inhibition in these specimens must be ruled out. Comparison of the amplification control signal from these specimens with that of the negative control sample revealed an absence of inhibition. Thus, the negative clinical specimens were determined to be negative by SDA. The amplification control signal also provided an indication of the consistency of SDA and detection among samples (Fig. 5B).

The *M. tuberculosis* and amplification control signals can be combined in a single analysis to differentiate positive and negative clinical specimens. For each clinical specimen, the *M. tuberuculosis* signal is normalized with the corresponding amplification control signal (Fig. 5C). The normalized signals from all the negative clinical specimens are below that of the positive

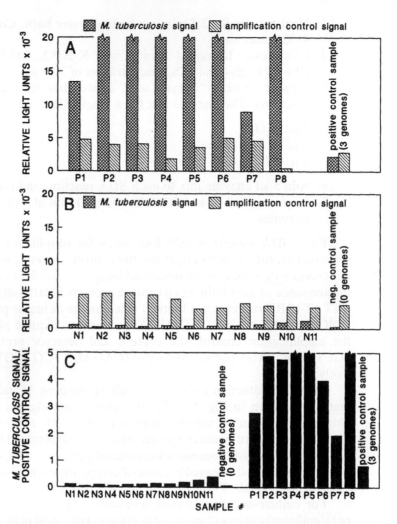

Figure 5 SDA and detection of clinical *Mycobacterium tuberculosis* samples. (A) Luminescent signals (relative light units) after SDA and detection for an array of clinical samples that were previously identified as positive for *M. tuberculosis* by culture-based detection. For each sample, signals from amplified *M. tuberculosis* DNA and from the amplified control sequence are indicated. The dynamic range of the luminometer limits the signals of samples P2–P6 and P8 to 20,000 relative light units. The positive control sample (3 genomes) is a prepared sample containing three genomes of purified *M. tuberculosis* DNA. (B) Luminescent signals (relative light units) after SDA and detection for an array of clinical samples that were previously identified as negative for *M. tuberculosis* by culture-based detection. The negative control sample (0 genomes) is a prepared sample that does not contain *M. tuberculosis* DNA. (C) Normalized signals for the positive and negative clinical samples. For each sample, the *M. tuberculosis* signal was divided by the corresponding amplification control signal. For samples P2–P6 and P8, their normalized signals are artificially limited because their *M. tuberculosis* signals (A) are limited by the dynamic range of the luminometer. Therefore, differences in the normalized values for these samples are not relevant.

control sample (3 genomes) whereas the normalized signals from the positive clinical specimens are above it.

V. PROCESSING CLINICAL SPECIMENS PRIOR TO SDA

This section discusses general considerations and methods for preparing clinical specimens for SDA, as well as a specific protocol for respiratory mycobacterial specimens. The sample processing protocol must accomplish several goals. The first is removal or inactivation of components that inhibit the SDA reaction. Inhibitors include substances contained in the clinical specimen and reagents commonly used in sample processing protocols, for example, detergents, metal chelators, and salts (Dey *et al.*, 1993). Polysaccharides (Demeke and Adams, 1992), heparin (Beutler *et al.*, 1990), and hemoglobin (Ruano *et al.*, 1992) are known inhibitors of PCR, but their effects on SDA have not been determined. The current method for mycobacterial specimens uses multiple washes to remove inhibitors and leaves the processed sample in a solution that is compatible with SDA (25 mM K$_2$HPO$_4$, pH 7.6).

The processing procedures must also concentrate and lyse the organisms to release the target DNA. Release of DNA from mycobacteria is difficult because of the resilience of the peptidoglycan layer of their cell walls. Other investigators have used combinations of biochemical and mechanical methods such as proteolysis (Plikaytis *et al.*, 1991; Cousins *et al.*, 1992; Kolk *et al.*, 1992; Savic *et al.*, 1992); glycosidase treatment (Del Portillo *et al.*, 1991; Eisenach *et al.*, 1991; Plikaytis *et al.*, 1991; Cousins *et al.*, 1992); detergents (DeWit *et al.*, 1990), Del Portillo *et al.*, 1991; Pierre *et al.*, 1991; Plikaytis *et al.*, 1991; Shankar *et al.*, 1991; Sritharin and Barker, 1991; Brisson-Noel *et al.*, 1989; Cousins *et al.*, 1992; Kolk *et al.*, 1992; Thierry *et al.*, 1992; Shawar *et al.*, 1993), alkaline treatment (Pierre *et al.*, 1991; Plikaytis *et al.*, 1991; Sritharin and Barker, 1991; Brisson-Noel *et al.*, 1989; Thierry *et al.*, 1992), sonication (Sjobring *et al.*, 1990; Buck *et al.*, 1992; Savic *et al.*, 1992), freeze-thaw techniques (Buck *et al.*, 1992), and heating (Sritharin and Barker, 1991) to lyse mycobacteria. Many of these procedures also utilize other purification steps such as phenol : chloroform extraction (Brisson-Noel *et al.*, 1989; De Wit *et al.*, 1990; Del Portillo *et al.*, 1991; Pierre *et al.*, 1991; Shankar *et al.*, 1991; Cousins *et al.*, 1992; Savic *et al.*, 1992; Thierry *et al.*, 1992) or solid-phase adsorption (Eisenbach *et al.*, 1991; Plikaytis *et al.*, 1991) prior to nucleic acid amplification and detection.

The third goal of a sample processing procedure is to sterilize the sample because *M. tuberculosis* is a highly infectious pathogen. After exploring several disinfection methods including disinfectants, proteases, surfactants, autoclaving, and heat, we determined that heating was the most efficient and convenient method for killing mycobacteria. Heat-induced killing of mycobacteria also lyses the organisms and denatures the DNA prior to SDA. The fourth criterion for *Mycobacterium* sample processing is that the procedure must be compatible with culture-based diagnosis because antibiotic susceptibility testing must still be performed using culture methods.

A. Processing Respiratory M. tuberculosis Specimens

This sample processing protocol is compatible with a standard protocol for culture-based diagnosis that is recommended by the Centers for Disease Control and Prevention and by Becton Dickinson for use with the BACTEC™ liquid culture system. The procedure should be performed by workers wearing protective clothing that includes mist respirator masks (Moldex 2300, Culver City, CA), disposable gloves, safety glasses, and disposable gowns. Work should be done in a Class 2B (externally vented) laminar flow biohazard containment hood that is situated in a laboratory with negative pressure air flow.

1. Reagents

- Liquefication solution: 2% (w/v) NaOH, 1.45% (w/v) sodium citrate, 0.5% (w/v) *N*-acetyl cysteine (NALC); solution should be prepared fresh daily by mixing equal volumes of autoclaved 4% sodium hydroxide and 2.9% sodium citrate and then adding solid NALC to 0.5% (w/v)
- Neutralization buffer: 68 m*M* sodium/potassium phosphate, pH 6.8; prepared by combining 68 m*M* KH_2PO_4 and 68 m*M* Na_2HPO_4 to pH 6.8; prepared reagents can be obtained from Remel (Lenexa, KS).

2. Sample Processing Protocol

1. Add an equal volume of the liquefication solution to 5–10 ml sputum in a 50-ml polypropylene tube and vortex. Larger volumes of patient sputum are processed in multiple 50-ml tubes and the final processed samples are pooled. Treatment with NALC and sodium hydroxide also kills the other bacteria in the specimen. Mycobacteria survive because of their resilient cell wall structure.
2. Incubate the sample 20 min at room temperature.
3. Dilute the sample to 50 ml with 68 m*M* sodium/potassium phosphate, pH 6.8 to lower the pH to 10–11. Vortex.

4. Centrifuge 20 min at 3000 g.
5. Decant the supernatant into biohazardous waste and retain the pellet (white or beige sample debris containing mycobacteria).
6. Transfer ~0.25 ml NALC pellet to a 2-ml screw-capped microfuge tube. Reserve the remainder of the NALC pellet for culture-based testing.
7. Add 1 ml 25 mM K$_2$HPO$_4$, pH 7.6, vortex, and centrifuge at 13,000 g for 1 min.
8. Decant supernatant into biohazardous waste and retain the pellet.
9. Repeat Steps 7 and 8.
10. Resuspend the pellet in 1 ml 25 mM K$_2$HPO$_4$, pH 7.6, by vortexing.
11. Heat the sample for 30 min in a boiling water bath. Note that dry heating blocks usually do not consistently kill mycobacteria (■ Zwadyk, unpublished results). Therefore we recommend using only a boiling water bath.
12. Cool the sample to room temperature; use 25 μl processed sample for SDA.

VI. PROTOCOL FOR CHEMILUMINESCENT DETECTION

The microwell plate procedure uses two oligodeoxynucleotide probes for the detection of amplified target DNA. In a single-step protocol, a biotinylated capture probe and an alkaline phosphatase-conjugated detector probe hybridize to the amplification product and anchor it to a streptavidin-coated microwell. Single-step hybridization and capture is more convenient than assays involving separate DNA hybridization and microwell capture steps. After a series of wash steps to remove nonhybridized detector probe, a chemiluminescent substrate is added and light output is quantified using a luminometer.

A. Preparation of Coated Microwell Plates

1. Reagents

5 μg/ml biotinylated bovine serum albumin (biotin *BSA; Pierce, Rockford, IL) in carbonate buffer, pH 9.6
50 mM carbonate buffer: Na$_2$CO$_3$/NaHCO$_3$, pH 9.6
LabSystems microwell strips (Cat. #SBD9502-440; LabProducts™, Research Triangle Park, NC
FTA hemagglutination buffer (Becton-Dickson Microbiology Systems, Cockeysville, MD), pH 7.2

100 μg/ml streptavidin (Bethesda Research Laboratories, Gaithersburg, MD) in FTA hemagglutination buffer, pH 7.2

Blocking buffer hemagglutination buffer, pH 7.2, 0.05% w/v BSA (fraction V; Sigma Chemical Company, St. Louis, MO)

2% (w/v) trehalose (pharmaceutical grade, #PM TRE2-08; Quadrant, Trumpington, Cambridge, England) in FTA hemagglutination buffer, pH 7.2

Aluminized Mylar pouches (Duralamb, Appleton, WI)

Silica desiccants (DriCap™, Cat. #02000-33AG01; MultiForm Desiccants, Buffalo, NY)

2. Procedure for Plate Coating

1. Add 100 μl/well 5 μg/ml biotin*BSA to the microwells.
2. Cover the plates with parafilm and incubate overnight (16–24 hr) at room temperature (22–25°C).
3. Invert the plates and tap dry on a paper towel. Wash twice, 375 μl/wash, using FTA hemagglutination buffer. Invert and tap plate dry. (Residual solution on top of the microwell plate is removed by blotting on a paper towel.
4. Add 100 μl streptavidin to each microwell.
5. Cover the plates as before and incubate overnight at room temperature.
6. Invert the plates and tap dry. Add 300 μl/well blocking buffer.
7. Cover the plates and incubate overnight as described.
8. Invert the plates and tap dry. Wash twice using 375 μl/well FTA hemagglutination buffer for each wash. Wash once with 375 μl/well 2% trehalose/FTA hemagglutination buffer.
9. Invert the plates and tap dry.
10. Dry the plates for 1 hr 37°C in cirulating air incubator.
11. Seal the dried plates in Mylar pouches using 8 desiccants/plate. Store overnight at room temperature prior to use.
12. Once the Mylar bag has been opened, store the plates at 4°C. The plates will remain stable for at least 2 months at 4°C.

B. *Oligodeoxynucleotide Probes*

After SDA, a central region of amplified target is available for hybridization to the capture and detector probes. These probes must be desinged to minimize competition with SDA primers and formation of undesirable secondary structures. Since the microwell assay is performed at 37°C, probes must be designed so they hybridize with melting temperatures (T_m) of 40–50°C. This T_m range ensures specificity and sensitivity at 37°C. A number of methods are available for assessment of probe design. Especially useful are software packages such as Oligo™ (National Biosciences, Plymouth, MN) or PC-Gene™ (Intelligenetics, Inc., Geneva, Switzerland).

The oligodeoxynucleotide capture probe, which contains three biotin groups at its 5' end, is synthesized using Biotin-On Phosphoramidite (Clontech, Palo Alto, CA). Purification is achieved by reverse-phase high pressure liquid chromatography (HPLC) (Aquapore RP 300 Column—220 × 4.6 mm, C8 column 7 particle, 300 Å pore size; Brownlee Lab, Santa Clara, CA) with a UV monitor at 254 nm, a gradient of 7–22% buffer B over 1 hr (buffer B: 100% acetonitrile; buffer A: 0.1 M triethylamine-acetate, pH 7), and a flow rate of 1 ml/min. The elution point should occur at approximately 18% buffer B; however, this point may vary depending on the HPLC system.

The oligodeoxynucleotide detector probe is synthesized using a 3'-amino-modifier C3 column (Glenn Research, Steriling, VA). This produces an oligodeoxynucleotide with a 3' amino group for subsequent conjugation with alkaline phosphatase. Before AP conjugation, the 3' amino-oligodeoxynucleotide is purified by denaturing acrylamide gel electrophoresis and electroelution (Ausubel *et al.*, 1987).

A specific example of the AP conjugation procedure follows.

1. Calf intestine alkaline phosphatase (AP) (EIA grade; Boehringer Mannheim, Indianapolis, IN) was dialyzed overnight at 4°C against 50 mM potassium phosphate (pH 7.5) and subsequently centrifuged to remove aggregates.
2. 4 ml 10 mg/ml AP was combined with 40 μl 50 mM succinimidyl-4-(p-maleimidophenyl)butyrate (SMPB; Pierce) dissolved in N,N'-dimethylformamide (DMF, Aldrich, Milwaukee, WI) and allowed to react in the dark at room temperature for 30 min.
3. Derivatized AP and excess SMPB were separated using a NAP-25 column (Pharmacia, Piscattaway, NJ) and 50 mM potassium phosphate (pH 7.5, degassed and purged with N_2).
4. The absorbance of the NAP-25 column fractions was read at 260 and 280 nm, and the void volume was collected. The concentration of derivatized alkaline phosphatase was determined by absorbance at 280 nm using an extinction coefficient of 750 M^{-1} cm^{-1}.
5. The derivatized AP (175.5 mmol) was obtained and stored on ice (less than 2 hr) until conjugation with the derivatized oligodeoxynucleotide.
6. The 3' amino-oligodeoxynucleotide (98.4 μl 508.2 μM solution; 50 nmol) was added to 13.4 μl 1 M potassium phosphate, pH 7.2, and mixed with 26.8 μl 50 mM N-succinimidyl-3-(2-pyridyldithio)-

propionate (SPDP; Pierce) in DMF. This mixture was incubated in the dark for 1 hr at room temperature.

7. Dithiothreitol (DTT; 1 M in 50 mM potassium phosphate, pH 7.5) was added to the oligodeoxynucleotide/DMF mixture to a final concentration of 0.1 M and allowed to incubate for 15 min at room temperature.

8. Excess DTT and SPDP were separated from the derivatized oligodeoxynucleotide by elution over a NAP-25 column with 50 mM potassium phosphate (pH 7.5, degassed and purged with N_2).

9. The derivatized oligodeoxynucleotide eluted in the void volume, as judged by absorbance at 260 and 280 nm. The concentration of the derivatized oligodeoxynucleotide was determined by absorbance at 260 nm.

10. The reduced oligodeoxynucleotide was reacted with the derivatized AP within 10 min to avoid oxidation. Solutions containing the derivatized oligodeoxynucleotide and the derivatized AP (in 50 mM potassium phosphate, pH 7.5) were mixed and incubated for 2–4 hr at room temperature and then overnight at 4°C. The AP concentration in this mixture should be 2–2.5 times that of the oligodeoxynucleotide. A higher ratio may lead to poor HPLC resolution of free AP from the AP–oligodeoxynucleotide conjugate.

11. The reaction was quenched using 1/100 volume 50 mM beta-mercaptoethanol in 50 mM potassium phosphate, pH 7.5.

12. The crude conjugate was concentrated to a volume of approximately 2 ml using a Centriprep 30 (Amicon, Beverly, MA) concentrator with 20 mM Tris, pH 7.5.

13. The crude conjugate was purified by HPLC using a DEAE-5PW column (7.5 mm × 7.5 cm) and a gradient of 0–66% buffer B (buffer B: 20 mM Tris, 1 M NaCl, pH 7.5; buffer A: 20 mM Tris, pH 7.5) and a flow rate of 1 ml/min while monitoring the absorbance at 254 nm. Elution occurs at approximately 60% buffer B.

14. Absorbances at 260 and 280 nm were recorded for the HPLC fractions using a spectrophotometer. We pool fractions with $A_{260}/A_{280} = 1$ for probes 13–15 nucleotides in length.

15. The AP concentration of the conjugated oligodeoxynucleotide was determined (BCA Protein Assay Kit; Pierce).

16. The AP detector probe was diluted to 2 μM in 20 mM Tris (pH 7.5), 1 M NaCl, 50 μg/ml sonicated salmon sperm DNA, 0.05% sodium azide, and stored thereafter at 4°C. The 20 mM Tris, pH 7.5, solution containing 1 M NaCl was autoclaved before the addition of the other components.

It is useful to develop and test capture/detection assays with a target oligodeoxynucleotide that serves as a model for the SDA target. Synthetic target is used to optimize the working concentrations of the probes and to serve as a general assay control. In this case, two synthetic targets were prepared to represent amplified *M. tuberculosis* target:

5′-dGACACUGAGAUCCCCUAUCCGUAUGGUGGAUAA-
CGUCUUUCAGGUCGAGUACGCCGUCUUUUU

and the amplified control:

5′-dGACACUGAGAUCCCCUAGCGACGAUGUCUGAGGCAA-
CUAGCAAAGCUGGUCGAGUCAGCCGUCUUUUU

Because of the sensitivity of SDA and the chemiluminenscence detection system, accidental contamination of amplified reactions with these synthetic targets is a major concern. Consequently, they should be used with great care to avoid contamination of samples about to undergo SDA. We keep them segregated from laboratory areas in which SDA is performed. These synthetic targets contain five uracils at the 3′ end that are not representative of the amplified *M. tuberculosis* target or the amplification control. This modification reduces, but does not eliminate, their ability to serve as targets should they accidentally contaminate an SDA reaction.

C. Preparation of SDA Samples for Detection

Completed SDA reactions, containing amplified target DNA and amplified control, are processed as described here. First, 200 μl autoclaved deionized water is added to each 50-μl SDA reaction in siliconized microfuge tubes. This 250-μl volume allows duplicate detection of both amplified *M. tuberculosis* target and the amplified control in separate assay wells using the appropriate capture and detector oligodeoxynucleotides. Duplicates are recommended to monitor detection reproducibility. Dilution of SDA samples prior to detection also produces higher signal by decreasing assay inhibitory factors, namely magnesium and SDA primers. In this system, the binding regions of SDA primer S_1 and the *M. tuberculosis* capture probe overlap by 2 nucleotides. Therefore, reduction in the concentration of S_1 prior to detection facilitates capture.

Assay control curves are prepared using synthetic target DNA representing the *M. tuberculosis* target and the amplification control. We prepare an assay control by diluting the target oligodeoxynucleotides (*M. tuberculosis* sequence and the amplified control sequence) to 32000 amol/50 μl of 50 mM K$_2$HPO$_4$, pH 7.5, 0.1 mg/ml BSA, 7 mM MgCl$_2$, 14% glycerol, 500 nM S_1 and S_2 primers, and 50 nM B_1 and B_2 primers. This sample is then diluted in the same solution to target oligodeoxynucleotide concentrations of 8000, 2000, 500 and 0 attomoles/50 μL. Each of these 50-μl assay controls (0–32000 amol/50 μl) is then diluted with 200 μl sterile deionized water, as are the SDA samples.

D. Detection Assay Procedure

Each lot of capture and detector probes is optimized to produce the maximum specific signal by using synthetic target DNA. Generally, the working concentrations are 40 nM for capture probes and 5–20 nM for detector probes.

1. Reagents

Stringency wash: 100 mM Tris, pH 7.5, 250 mM NaCl, 0.1% BSA (Fraction V), 0.1% sodium azide, 0.01% v/v Nonidet 40
Lumiphos™ 530 (Lumigen, Inc., Detroit, MI)
Synthetic target diluent: SDA buffer
Hybridization buffer: 100 mM Tris, pH 7, 1.8 M NaCl, 0.2% w/v acetylated BSA, 0.1 mM ZnCl$_2$, 0.1% w/v sodium azide
Capture probe, 2μM (see text for sequence)
Detector probe, 2μM (see text for sequence)
SDA or assay control sample

2. Procedure for Microwell Assay

1. Prepare hybridization mixture (50 μl/microwell) at the predetermined optimal concentrations of capture and detector probes. For example, 500 μl hybridization mix containing 40 nM capture probe and 10 nM detector probe would be prepared by mixing 10 μl 2 μM capture probe, 2.5 μl 2 μM detector probe, and 487.5 μl hybridization buffer. This solution is stable at room temperature for 24–48 hr.
2. Denature target (diluted SDA or assay control sample) at 95°C for 3 min. Remove from heat and cool at room temperature for 5 min.
3. Add 50 μl/well denatured target to the coated microwell and then add 50 μl/well hybridization mixture. Four aliquots from each SDA sample are used: two for the detection of *M. tuberculosis* and two for the amplification control. For each synthetic target sample, three aliquots are used for detection in triplicate.
4. Cover and incubate 37°C for 45 min.
5. Invert the plates and tap dry. Wash the wells three times for 1 min each with 300 μl stringency wash. Invert and tap dry.

6. Add $100 \mu l$ Lumiphos™ 530 to each well. Cover the plates and incubate for 30 min at 37°C.
7. Read luminescence (relative light units) on a microwell luminometer (LabProducts™).

Settings for the luminometer are: 37°C, 2.0 sec/well read, no blanking, no calculation, no pre-incubation. We read luminescence 30–45 min after addition of the Lumiphos™ 530 to maximize sensitivity and accuracy. With some extremely high signals, the signal strength may drift or decrease after 45 min due to substrate consumption.

E. General Comments on Chemiluminescent Detection

Commercial plates coated with streptavidin are available from companies such as Xenopore (Cat. # XPPOO100; Saddlebrook, NJ). Precoated plates are convenient, but some plates are not compatible with luminescence because they are transparent. A selection of uncoated microwell plates is commercially available for chemiluminescent detection (Dynatech, Chantilly,VA; LabProducts™). When selecting microwell plates, one should consider their availability, ease of use, luminometer compatibility, optimal coating parameters, and opaqueness. White plates provide greater sensitivity whereas gray or black plates decrease light transmission and, hence, sensitivity. Plate color dictates "cross-talk" levels (the bleeding of light from one well to an adjacent well). A balance must be achieved between end point sensitivity and cross-talk.

X-Ray film (TMG; Kodak, Rochester, NY) provides an inexpensive alternative to luminometers. However, X-ray film procedures are not very sensitive or quantitative. Exposure times vary from 5 to 60 min, depending on the system and luminescent substrate used.

VII. CONCLUSION

We have presented a practical description of a DNA probe test for *M. tuberculosis* that is based on SDA and chemiluminescence detection. Application of this system to other targets would naturally require modifications of the sample processing protocol, the DNA sequences, the reaction conditions of SDA, and the detection assay. Nonetheless, this chapter should provide a general framework for development of other viral or bacterial systems in the clinical laboratory.

ACKNOWLEDGMENTS

We thank P. Scott and C. Dean for preparation of the oligodeoxynucleotide conjugates, M. O'Connell and P. Haaland for statistical design and analysis, and Salmon Siddiqi for supplying the *Mycobacterium* cultures.

REFERENCES

Ausubel, F. M., Brent, R., Kingston, R. E., Moore, D. D., Smith, J. A., Seidman, J. G., and Struhl, K. (eds). (1987). "Current Protocols in Molecular Biology." Wiley and Sons, New York.

Barany F. (1991). The ligase chain reaction in a PCR world. *PCR Meth. Appl.* **1**, 5–16.

Beutler, E., Gelbart, T., and Kuhl, W. (1990). Interference of heparin with the polymerase chain reaction. *Biotechniques* **9(2)**, 166.

Brisson-Noel, A., Lecossier, D., Nassif, X., Gicquel, B., Levy-Frebault, V. and Hance, A. J. (1989). Rapid diagnosis of tuberculosis by amplification of mycobacterial DNA in clinical samples. *Lancet ii*, 1069–1071.

Buck, G. E., O'Hara, L. C., and Summersgill, J. T. (1992). Rapid, simple method for treating clinical specimens containing *Mycobacterium tuberculosis* to remove DNA for polyermase chain reaction. *J. Clin. Microbiol.* **30**, 1331–1334.

Cousins, D. V., Wilton, S. D., Francis, B. R., and Gore, B. L. (1992). Use of polymerase chain reaction for rapid diagnosis of tuberculosis. *J. Clin. Microbiol.* **30**, 255–258.

Del Portillo, P., Murillo, L. A., and Patarroyo, M. E. (1991). Amplification of a species-specific DNA fragment of *Mycobacterium tuberculosis* and its possible use in diagnosis. *J. Clin. Microbiol.* **29**, 2163–2168.

Demeke, T., and Adams, R. (1992). The effects of plant polysaccharides and buffer additives on PCR. *Biotechniques* **12**, 333–334.

DeWit, D., Shoemaker, S., and Sogin, M. (1990). Direct detection of *Mycobacterium tuberculosis* in clinical specimens by DNA amplification. *J. Clin. Microbiol.* **28**, 2437–2441.

Dey, M., Howard, A., Keating, W., Howard, D., Down, J., and Little, M. (1993). Inhibition of DNA amplification by sample collection devices. *Am. Soc. Microbiol. Abstr.* C224.

Eisenach, K. D., Sifford, M. D., Cave, M. D., Bates, J. H., and Crawford, J. T. (1991). Detection of *Mycobacterium tuberculosis* in sputum samples using a polymerase chain reaction. *Am. J. Rev. Resp. Dis.* **144**, 1160–1163.

Erlich, H. A., Gelfand, D., and Sninsky, J. J. (1991). Recent advances in the polymerase chain reaction. *Science* **252**, 1643–1651.

Fahy, E., Kwoh, D. Y., and Gingeras, T. R. (1991). Self-sustained sequence replication (3SR): An isothermal transcription-based amplifiction system alternative to PCR. *PCR Meth. Appl.* **1**, 25–33

Kolk, A. J. H., Schuitema, A. R. J., Kuijuper, S., Leeuwen, J., Hermans, P. W. M., Embden, J. D. A., and Hartskeerl, R. A. (1992). Detection of *Mycobacterium tuberculosis* in clinical samples by using polymerase chain reaction and a nonradioactive detection system. *J. Clin Microbiol.* **30**, 2567–2575.

Malek, L. T., Davey, C., Henderson, G., and Sooknanan, R. (1992). Enhanced nucleic acid amplification process. United States Patent Number 5, 130, 238.

Pierre, C., Lecossier, D., Boussougnat, T., Bocart, D., Joly, V., Yeni, P., and Hance, A. J. (1991). Use of a reamplification protocol improves sensitivity of detection of *Mycobacte-*

rium tuberculosis in clinical samples by amplification of DNA. *J. Clin Microbiol.* **29,** 712–717.

Plikaytis, B. B., Eisenach, K. D., Crawford, J. T., and Shinnick, T. M. (1991). Differentiation of *Mycobacterium tuberculosis* and *Mycobacterium bovis* BCG by a polymerase chain reaction assay. *Mol. Cell. Probes* **5,** 215–219.

Ruano, G., Pagliaro, E. M., Schwartz, T. R., Lamy, K., Messina, D., Gaensslen, R. E., and Lee, H. C. (1992). Heat-soaked PCR: An efficient method for DNA amplification with applications to forensic analysis. *Biotechniques* **13,** 266–274.

Savic, B., Sjobring, U., Alugupalli, S., Laison, L., and Miorner, H. (1992). Evaluation of polymerase chain reaction, tuberculostearic acid analysis, and direct microscopy for the detection of *Mycobacterium tuberculosis* in sputum. *J. Infect. Dis.* **166,** 1177–1180.

Shankar, P., Manjunath, N., Mohan, K. K., Prasad, K., Behari, M., Shriniwas, and Ahuja, G. K. (1991). Rapid diagnosis of tuberculosis meningitis by polymerase chain reaction. *Lancet* **33,** 5–7.

Shawar, R. M., El-Zaatari, F. A. K., Nataraj, A., and Clarridge, J. E. (1993). Detection of *Mycobacterium tuberculosis* in clinical samples by two-step polymerase chain reaction and non-isotopic hybridization methods. *J. Clin. Microbiol.* **31,** 61–65.

Sjobring, U., Mecklenberg, M., Anderson, A. B., and Miorner, H. H. (1990). Polymerase chain reaction for detection of *Mycobacterium tuberculosis. J. Clin. Microbiol.* **28(10),** 2200–2204.

Spargo, C. A., Haaland, P. D., Jurgensen, S. R., Shank, D. D., and Walker G. T. (1993). Chemiluminescent detection of strand displacement amplified DNA from species comprising the *Mycobacterium tuberculosis* complex. *Mol. Cell. Probes* **7,** 395–404.

Sritharin, V., and Barker, R. (1991). A simple method for diagnosing *M. tuberculosis* infection in clinical samples using PCR. *Mol. Cell. Probes* **5,** 385–395.

Thierry, D., Cave, M. D., Eisenhach, K. D., Crawford, J. T., Bates, J. H., Gicquel, B., and Guesdon, J. L. (1990). IS*6110,* an IS-like element of *Mycobacterium tuberculosis* complex. *Nucleic Acids Res.* **18,** 188.

Thierry D., Chureau, C., Aznar, C., and Guesdon, J. L. (1992). The detection of *Mycobacterium tuberculosis* in uncultured clinical specimens using the polymerase chain reaction and a non-radioactive probe. *Mol. Cell. Probes* **6,** 181–191.

Walker, G. T. (1993). Empirical aspects of strand displacement amplification (SDA). *PCR Meth. Appl.* **3,** 1–6.

Walker, G. T., Little, M. C., Nadeau, J. G., and Shank, D. D. (1992a). Isothermal *in vitro* amplification of DNA by a restriction enzyme/DNA polymerase system. *Proc. Natl. Acad. Sci. USA* **89,** 392–396.

Walker, G. T., Fraiser, M. S., Schram, J. L., Little, M. C., Nadeau, J. G., and Malinowski, D. P. (1992b). Strand displacement amplification—An isothermal, *in vitro* DNA amplification technique. *Nucleic Acids Res.* **20,** 1691–1696.

16

Ligation-Activated Transcription Amplification: Amplification and Detection of Human Papillomaviruses

David M. Schuster, Mark S. Berninger, and Ayoub Rashtchian

Life Technologies, Inc.

Gaithersburg, Maryland 20844-9980

I. INTRODUCTION

The specificity and quantitative nature of nucleic acid hybridization has been exploited in the past three decades for a variety of purposes. With the advent of nonradioactive detection methods, nucleic acid hybridization has become a major tool in detection of infectious agents or genetic abnormalities. The sensitivity required for diagnosis of many infectious diseases has not been achieved by nucleic acid hybridization assays that do not have an amplification step. Numerous approaches have been developed to overcome this shortcoming, some of which are described in this volume. Enzymatic signal amplification systems have been described that use alkaline phosphatase as

351

the detection system (Schaap *et al.*, 1989; Schaap and Akhaven-Tafti, 1990). Other investigators have used nucleic acid probes that carry large numbers of detectable moieties such as alkaline phosphatase, thereby increasing the detectable signal per hybridizable target sequence (Urdea *et al.*, 1987,1990). Despite some success in the use of these approaches, none has been universally applicable.

An efficient solution to this problem has been the development of *in vitro* methods for amplifying the nucleic acid sequences of interest. The polymerase chain reaction (PCR; Saiki *et al.*, 1985; Mullis an Faloona, 1987) was the first method for sequence-specific amplification of target molecules in a sample. This method has rapidly become one of the most important tools in molecular biology. The widespread utility of amplification in molecular biology and molecular diagnostic procedures has stimulated the development of other nucleic acid amplification methods (Kwoh *et al.*, 1989; Guatelli *et al.*, 1990). Although some of these methods are dependent on the presence of a DNA target and amplify a target sequence; other methods are based on replicatable probes that are amplified after hybridization to reveal the presence of the probe (Lizardi *et al.*, 1988). Target amplification systems such as PCR, the ligase-based amplification system (LAR; Barringer *et al.*, 1990), the ligase chain reaction (LCR; Barany, 1991a,b), and the transcription-based amplification system (TAS; Kwoh *et al.*, 1989) use repeated temperature cycling to achieve amplification. In contrast, the probe amplification systems (Lizardi *et al.*, 1988) and the self-sustained sequence replication reactions (3SR; Guatelli *et al.*, 1990) do not require thermal cycling. In this chapter we describe an isothermal method for amplification of nucleic acids based on ligation of a promoter sequence to a target nucleic acid, resulting in target amplification through a transcribed RNA intermediate. This target amplification methodology has been combined with solution hybridization-based enzymatic detection of amplified target nucleic acid to produce a highly sensitive method for the detection of nucleic acids. Application of these technologies to detection of human papillomaviruses (HPV) is described.

II. PRINCIPLE OF LIGATION-ACTIVATED TRANSCRIPTION AMPLIFICATION

The ligation-activated transcription (LAT) reaction requires the simultaneous action of four enzymes in a single isothermal reaction (Fig. 1). The enzymes required for this system include DNA ligase, RNA-dependent DNA polymerase (reverse transcriptase), RNA polymerase, and ribonuclease H (RNase H). The LAT procedure is based on the ligation of a double-stranded T7 RNA polymerase promoter sequence to the 3' end of the DNA to be

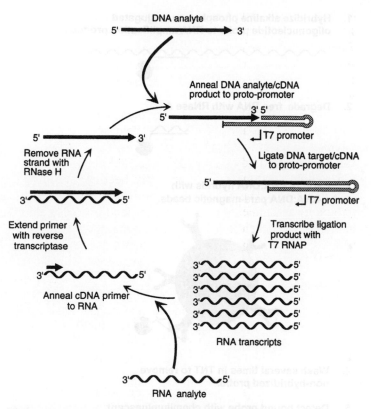

Figure 1 Schematic representation of LAT amplification.

amplified. This novel oligonucleotide, or proto-promoter, contains a blocked 3' single-stranded region that is complementary to the 3' end of the target sequence and a double-stranded T7 RNA polymerase promoter. After ligation of the proto-promoter to the target, the RNA polymerase produces multiple copies of RNA complementary to the DNA. These RNAs are, in turn, converted to cDNA by reverse transcriptase after binding a synthetic cDNA primer. The RNA strand of the RNA : DNA hybrid is digested by RNase H, resulting in single-stranded cDNA. The cDNA can then ligate to the proto-promoter to initiate another cycle of amplification. An RNA target can be amplified beginning at the reverse transcription step.

III. ANTIBODY-CAPTURE SOLUTION HYBRIDIZATION

Antibody-capture solution hybridization is a rapid nonradioactive method for detecting target RNA. The principle of this methodology is shown sche-

1. **Hybridize alkaline phosphatase conjugated oligonucleotide probe to RNA amplification product.**

2. **Degrade free RNA with RNase A.**

3. **Capture RNA:DNA hybrids with α-RNA:DNA para-magnetic beads.**

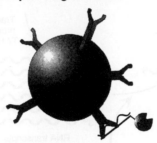

4. **Wash several times in TNT to remove non-hybridized probe.**

5. **Detect bound probe with chemiluminescent substrate.**

Figure 2 Antibody capture solution hybridization.

matically in Fig. 2. The method uses hybridization of a DNA probe to an RNA target in solution, followed by separation of RNA : DNA hybrids from unhybridized probe onto a suitable solid support (Rashtchian *et al.*, 1987,1990; Stollar and Rashtchian, 1987). RNA : DNA hybrids are captured using magnetic beads carrying immobilized antibody against RNA : DNA hybrids. The enzyme-labeled probe can be detected using an appropriate colorimetric, fluorescent, or chemiluminescent substrate. The probes were designed to hybridize to the RNA product of the LAT amplification reaction because RNA is the predominant product of the amplification.

IV. PROTOCOLS

For routine laboratory procedures, a reagent mix should be prepared that contains the required buffers, ribo- and deoxyribonucleoside triphosphate substrates, cDNA primer, and proto-promoter for amplification of specific RNA or DNA sequences. The required enzymes—T7 RNA polymerase, T4 DNA ligase, Superscript™ II reverse transcriptase (GIBCO/BRL, Life Technologies, Inc., Gaithersburg, MD), and *Escherichia coli* RNase H (GILBO/BRL)—should be added as a mixture from a concentrated preparation. Enzyme and reaction mixes are stable for at least 1 month when stored at −20°C. Amplification reactions are assembled by adding the reagent mix, a ribonuclease inhibitor, and the enzyme mix to produce a master mix that is aliquoted for individual reactions. LAT amplification reactions are typically performed in 25-μl volumes.

A. Stock Solutions

All stock solutions must be prepared as sterile RNase-free solutions. Where appropriate, solutions may be treated with diethylpyrocarbonate (DEPC) to inactive RNases. It is sufficient to prepare solutions using virgin chemicals and plasticware with DEPC-treated sterile distilled water, followed by 0.2-μm filtration.

LAT amplification is extremely sensitive to the activity of each enzyme in the reaction. Unit values given for commercially available enzyme preparations are often inaccurate and cannot be used to determine the amounts of each enzyme to add to the mixture. Enzymes were obtained as concentrated preparations from GIBCO/BRL.

Enzymes

T7 RNA polymerase, 12,000 U/μl
Superscript™ II reverse transcriptase, 1050 U/μl
T4 DNA ligase, 16.2 U/μl
E. coli RNase H, 1 U/μl

Reagents

2 *M* potassium glutamate
2 *M* Tris-HCl, pH 8.3
1 *M* MgCl$_2$
40% (w/v) polyethylene glycol (PEG) 8000 (Sigma Chemical Co., St. Louis, MO)
10% Triton® X-100
1 *M* dithiothreitol (DTT)

Nucleotide mix: 3.125 mM each dNTP, 12.5 mM each NTP; prepared from 100 mM stock solutions (Pharmacia, Piscataway, NJ) using DEPC-treated sterile water

5× Reaction buffer: 0.25 M Tris-HCl, pH 8.3, containing 0.875 M potassium glutamate, 30 mM MgCl$_2$, 0.05% Triton® ×-100, and 50 mM DTT

Enzyme storage buffer: 20 mM Tris-HCl, pH 7.7, 100 mM potassium glutamate, 0.1 mM EDTA, 1 mM DTT, 0.01% Triton® ×-100, 50% glycerol

Reagent mix (sufficient for 100 reactions)

5× reaction buffer, 500 μl
nucleotide mix, 200 μl
proto-promoter oligo, 400 pmol
cDNA primer, 1250 pmol
40% PEG-8000, 375 μl
DEPC-treated water, 1550 μl, final volume

Enzyme mix (final volume, 400.0 μl; sufficient for 100 reactions)

T7 RNA polymerase (12,000 U/μl), 3.3 μl (400 U/25-μl reaction)
Superscript™ II RT (1050 U/μl), 9.5 μl (100 U/25-μl reaction)
RNase H (1 U/μl), 5.0 μl (0.05 U/25-μl reaction)
T4 DNA ligase (16.2 U/μl), 6.2 μl (1.0 U/25-μl reaction)
enzyme storage buffer, 376.0 μl

NOTE: Use 4 μL enzyme mix per 25-μl reaction.

B. OLIGONUCLEOTIDES

Oligonulceotides were synthesized by phosphoramidite chemistry using an Applied Biosystems (Foster City, CA) Model 380A synthesizer. Primers and oligonucleotides were purified by anion exchange HPLC or denaturing acrylamide gel electrophoresis.

1. Preparation of Proto-promoter

The proto-promoter is a unique T7 promoter-containing oligonucleotide that has two distinct regions. The 5′ region is composed of a double-stranded

T7 RNA polymerase promoter sequence bridged by a hairpin loop. The 3′ region is single-stranded and is complementary to the 3′ end of the cDNA synthesized during the amplification process. The 3′ residue of the proto-promoter is cordycepin (3′-deoxyadenosine 5′-monophosphate) and therefore cannot function as a primer.

We have synthesized proto-promoter using a single 82-bp oligonucleotide or using a ligation method to assemble the proto-promoter from two smaller oligonucleotides. These strategies are diagrammed in Fig. 3 for the HPV16 proto-promoter. Although both methods have produced functional proto-promoter, our best LAT amplification results were obtained using the ligation protocol. The ligation method offers an added benefit because a single generic T7 promoter oligonucleotide can be used to construct a variety of sequence-specific proto-promoters, thus saving time and reducing the cost of oligonucleotide synthesis.

Oligonucleotides 1123 and 742 were synthesized from cordycepin CPG (Cat No. N2024SC; Peninsula Laboratories, Inc., Belmont, CA). The 5′

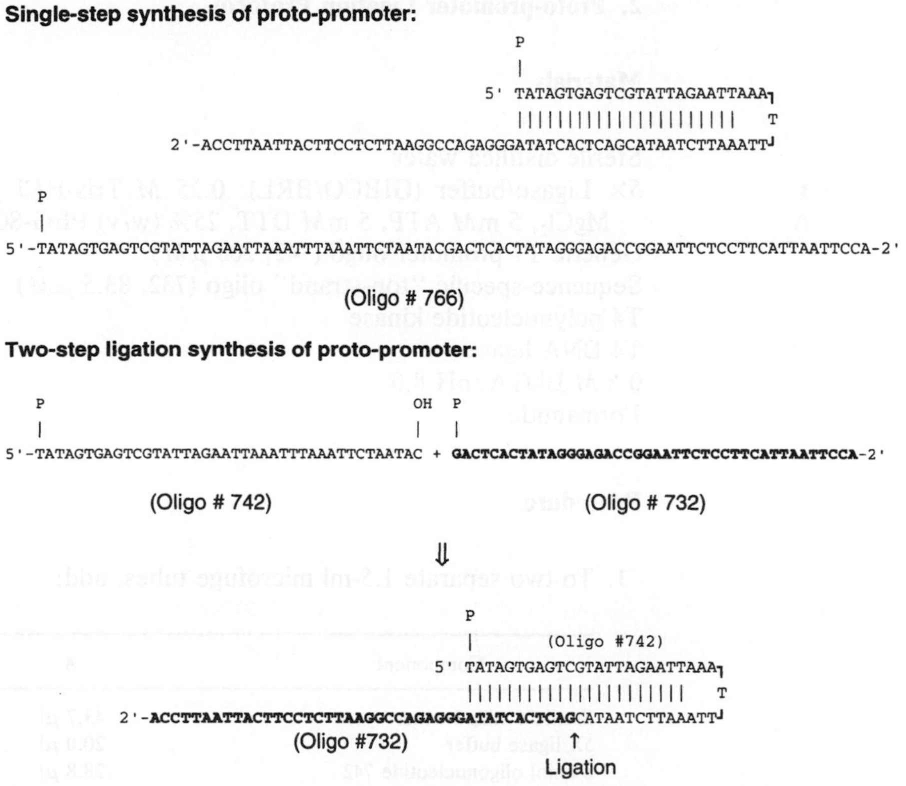

Single-step synthesis of proto-promoter:

```
                                              P
                                              |
                        5' TATAGTGAGTCGTATTAGAATTAAA┐
                           |||||||||||||||||||||||||  T
         2'-ACCTTAATTACTTCCTCTTAAGGCCAGAGGGATATCACTCAGCATAATCTTAAATT┘
```

```
    P
    |
5'-TATAGTGAGTCGTATTAGAATTAAATTTAAATTCTAATACGACTCACTATAGGGAGACCGGAATTCTCCTTCATTAATTCCA-2'

                            (Oligo # 766)
```

Two-step ligation synthesis of proto-promoter:

```
    P                                          OH  P
    |                                          |   |
5'-TATAGTGAGTCGTATTAGAATTAAATTTAAATTCTAATAC + GACTCACTATAGGGAGACCGGAATTCTCCTTCATTAATTCCA-2'

            (Oligo # 742)                                 (Oligo # 732)

                                    ⇓

                                              P
                                              |        (Oligo #742)
                        5' TATAGTGAGTCGTATTAGAATTAAA┐
                           |||||||||||||||||||||||||  T
         2'-ACCTTAATTACTTCCTCTTAAGGCCAGAGGGATATCACTCAGCATAATCTTAAATT┘
                        (Oligo #732)          ↑
                                             Ligation
```

Figure 3 Nucleotide sequence of and strategies for synthesis of HPV16 proto-promotor oligonucleotide.

termini of oligonucleotides used for proto-promoter construction were chemically phosphorylated using 5′ Phosphate-ON CE-Phosphoramidite (Clontech Laboratories, Palo Alto, CA) or treated with T4 polynucleotide kinase (GIBCO/BRL) in an excess of ATP). For ligation we used 6 nmol each of oligo 742 and 1124. Ligated product was electrophoresed on an 8 M urea, 8% polyacrylamide gel. Proto-promoter product was excised from the gel and eluted with 0.5 M NH$_4$OAc, 0.1% SDS, 10 mM Tris-HCl, pH 8.0, 1 mM EDTA. Approximately 80–90% (as judged by UV shadowing of the gel) of the two oligos ligated. Gel-purified proto-promoter was desalted by Sephadex® G-50 (Pharmacia) chromatography, eluted with DEPC-treated sterile water, and then dried. Proto-promoter was resuspended in 200 μl DEPC-treated sterile water and quantified by A$_{260}$ determination.

The single-step synthesis proto-promoter, oligo 1123, used cordycepin CPG and was chemically phosphorylated with 5′ Phosphate-ON CE-Phosphoramidite. Material from a crude synthesis was gel purified as described.

2. Proto-promoter Ligation Protocol

Materials

Sterile distilled water
5× Ligase/buffer (GIBCO/BRL): 0.25 M Tris-HCl pH 7.6, 50 mM MgCl$_2$, 5 mM ATP, 5 mM DTT, 25% (w/v) PEG-8000
Generic T7 promoter oligo (742, 208 μM)
Sequence-specific "top-strand" oligo (732, 83.5 μM)
T4 polynucleotide kinase
T4 DNA ligase
0.5 M EDTA, pH 8.0
Formamide

Procedure

1. To two separate 1.5-ml microfuge tubes, add:

Component	A	B
Sterile distilled water	43.7 μl	—
5X ligase buffer	20.0 μl	20.0 μl
6 nmol oligonucleotide 742	28.8 μl	—
6 nmol oligonucleotide 732	—	72.5 μl
T4 polynucleotide kinase (10 U/μl)	2.5 μl	2.5 μl
Final volume	95.0 μl	95.0 μl

2. Incubate both reactions 5 min at 37°C.
3. Incubate both reactions 5 min at 94°C to heat inactivate the kinase.
4. Transfer the contents of both reactions to a single 1.5-ml tube.
5. Add 10 μl T4 DNA ligase (1 U/μl).
6. Incubate 1 hr at 37°C.
7. Add 10 μl 0.5 M EDTA to stop the reaction.
8. Lyophilize in a Speed-Vac® (Savant Instruments, Farmingdale, NY).
9. Resuspend concentrated ligation product with 50 μl deionized formamide.
10. Electrophorese the ligation products on a denaturing 8% polyacrylamide gel according to standard protocols (Sambrook *et al.*, 1989).
11. Visualize proto-promoter product by UV shadowing.
12. Excise proto-promoter band from the gel. Crush the gel matrix and elute with 2–3 volumes 0.5 M NH₄OAc, 0.1% SDS, 10 mM Tris-HCl, pH 8.0, 1 mM EDTA at 37°C for 16 hr.
13. Desalt by gel-exclusion chromatography (PD-10 column; Pharmacia), eluting with DEPC-treated sterile water.
14. Lyophilize as in Step 8.
15. Resuspend proto-promoter in 200 μl DEPC-treated sterile water and quantify by A_{260} determination.

C. LAT Amplification

1. Prepare a master mix by combining 15.5 μl reagent mix, 0.5 μl placental ribonuclease inhibitor (RNasin®; Promega Corporation, Madison, WI), and 4 μl enzyme mix for each reaction. Always make sufficient mix for one additional reaction, for example, if 20 reactions are to be performed, prepared a master mix sufficient for 21 reactions. To a sterile RNase-free 1.5-ml microfuge tube on ice, add in order:

325.5 μl	reagent mix
10.5 μl	RNasin® (10 U/μl)
84.0 μl	enzyme mix

for a final volume of 420.0 μl.

NOTE: *The requirement for placental ribonuclease inhibitor is dependent on the quality of the analyte sample. In "clean" model systems, RNase inhibitor is not needed; however, LAT amplification of material isolated from clinical specimens requires RNasin®. Exercise caution when using ribonuclease inhibitors. Add RNasin to the reaction mix only after addition of reducing agent (DTT). RNasin® requires sulfhydryl reagents (1 mM DTT) for activity and is irreversibly denatured in the absence of these agents (Sambrook et al., 1989). Some preparations of ribonuclease inhibitor include inactivated RNases that copurify with the inhibitor. Treatments that denature the inhibitor (repeated freeze-thawing, heat, absence of DTT, etc.) can release these RNases, which can degrade RNA in subsequent procedures.*

2. Mix the solution thoroughly by gently vortexing; then place on ice.
3. Transfer 20-μl aliquots of the master mix into sterile 0.5-ml microfuge tubes on ice.
4. Add 5 μl test sample. (Typical levels of analyte assayed contain 100–10,000 copies. The final reaction composition is 50 mM Tris-HCl, pH 8.3, 175 mM potassium glutamate, 6% PEG 8000, 6 mM MgCl$_2$, 10 mM DTT, 0.01% Triton® ×-100, 0.25 mM each dNTP, 1 mM each NTP, 0.5 $\mu$$M$ cDNA primer, 0.15–0.2 $\mu$$M$ T7 promoter oligonucleotide, 16 U/μl T7 RNA polymerase, 4 U/μl Superscript™ II reverse transcriptase. 0.04 U/μl T4 DNA ligase, 0.002 U/μl *E. coli* RNase H, 0.2 U/μl RNase inhibitor, 100–10,000 copies target sequence.)
5. Mix and incubate at 37°C for 3 hr.
6. Add 25 μl 50 mM EDTA to terminate the reaction.

D. Detection of Amplification Products

Amplification products can be detected by a variety of nucleic acid hybridization techniques. We have used a modification of the dot-bolt filter hybridization method of Thomas (1983) using a ^{32}P-labeled oligonucleotide probe that is specific for either the RNA or the cDNA synthesized during the reaction. Additionally, we have developed a chemiluminescent immunocapture assay for the direct detection of RNA amplification products, employing a monoclonal antibody specific for RNA : DNA heteroduplex molecules and an alkaline phosphatase-labeled oligonucleotide probe.

Comparison of relative signal intensities between reaction products and dilutions of the original analyte (hybridization standards) provides a measure of the level of amplification attained from the LAT process.

1. Membrane Hybridizations

[32]P-Labeling of Oligonucleotide Probe (Kinase Forward Reaction)

1. To a 0.5-ml microfuge tube on ice, add:

   2.0 μl 5× Forward buffer [300 mM Tris-HCl, pH 7.8, 75 mM 2-mercaptoethanol (BME), 50 mM MgCl$_2$, 1.65 $\mu$$M$ ATP]

2.0 μl oligonucleotide probe (5 pmol/μl)
5.0 μl [γ-^{32}P]ATP (10 mCi/ml; 3000 Ci/mmol)
0.5 μl T4 polynucleotide kinase (10 units/μl)

for a final volume of 10.0 μl.

2. Incubate at 37°C for 10 min.
3. Terminate reactions by adding EDTA to 20 mM final concentration.
4. Remove unincorporated ATP from radiolabeled probe using any appropriate methodology (gel exclusion chromatography, ethanol precipitation, ion-exchange chromatography, etc.).

Glyoxal Dot-Blot Procedure

Materials

Glyoxal solution: 1 M glyoxal, 50% dimethylsulfoxide (DMSO), 50 mM NaPO$_4$, pH 7.0

1 volume deionized glyoxal (40%)
2 volumes DMSO
1 volume 200 mM sodium phosphate, pH 6.8–7.0

Hybridization solution: 1 M NaCl, 30% formamide, 5% SDS, 50 mM NaPo$_4$, pH 7.4, 1 mM EDTA, 0.1% gelatin, 50 μg/ml yeast tRNA

Protocol

1. Combine 5 μl LAT amplification reaction with 25 μl glyoxal solution.
2. Incubate at 50°C for 1 hr, then place on ice.

NOTE: *In addition to LAT reaction products, dilutions of the RNA or cDNA analyte can be treated with glyoxal and used as hybridization standards.*

3. Add 150 μl 20× SSC (3 M NaCl, 0.3 M sodium citrate, pH 7.0) to each glyoxal reaction. Mix and apply the entire volume to a suitable hybridization membrane with a microfiltration manifold according to standard protocols. We recommend using Biodyne® B nylon membrane (GIBCO/BRL).
4. Bake the membrane at 80°C for a minimum of 30 min to remove glyoxal adducts from the target and fix the target to the membrane. Because of the small size (~100 bases) of the LAT amplification products, membranes should also be fixed by UV treatment. Place the membrane on a 302-nm UV transilluminator (target side down) and expose for 2 min.

5. Place the filter in a heat-sealable plastic bag and prehybridize for 30 min. Use approximately 1 ml hybridization solution per 20 cm² membrane.

6. Replace the prehybridization solution with an equivalent volume of fresh hybridization solution containing 1×10^7 dpm/ml ^{32}P-labeled oligonucleotide probe (specific activity = 1×10^8 dpm/μg). Reseal the bag and hybridize overnight (16 hr).

7. Wash the membrane 5 times for 5 min each with 0.1× SSC, 0.1% SDS at 42°C.

8. Detect hybridized probe by phosphoimaging or autoradiography according to standard protocols (Sambrook *et al.*, 1989).

For detection of HPV16 sequences, prehybridization/hybridization was performed at 42°C when using the RNA probe oligo (5′-GTATATTGTGGA-GACCCTGGAACTATAGG-3′) and at 37°C when using the cDNA probe oligo (5′-TCCTATAGTTCCAGGGTCTCCACAATATAC-3′). Hybridization and wash conditions should be optimized for each individual probe.

2. Immunocapture Assay

a. *Preparation of anti-RNA : DNA-Coated Paramagnetic Beads* Mouse monoclonal IgG specific for RNA : DNA heteroduplex (anti-RNA : DNA) were prepared essentially as described by Boguslawski *et al.* 1986). A high affinity IgG2a isolate (unpublished results) was used for covalent coupling to magnetic polystyrene beads (Dynabeads® M-450; Dynal Inc., Great Neck, NY), essentially as recommended by the manufacturer. Uncoated M-450 magnetic beads require activation with *p*-toluenesulfonyl chloride before use. Optionally, tosyl-activated magnetic beads may be obtained from Dynal. Activated beads are used for direct covalent coupling of the primary monoclonal antibody.

Materials

Tosyl-activated magnetic beads (Dynabeads®)
anti-RNA : DNA, purified mouse monoclonal antibody (mg/ml in 50% glycerol and water)
Borate buffer: 0.2 M Na$_2$B$_4$O$_7$ · 10 H$_2$O (Borax), pH 9.2
1.0 M ethanolamine-HCl, pH 9.5
Tween20®
0.1 M phosphate buffered saline (PBS), pH 7.2

Tris-buffered saline (TBS): 0.05 M Tris-HCl, pH 7.5, 0.1 M NaCl, 0.1% BSA, 0.01% thimerosal

TBS with Tween 20®: 0.05 M Tris-HCl, pH 7.5, 0.1 M NaCl, 0.1% BSA, 0.01% thimerosal, 0.1% Tween20®

Protocol

1. Rinse tosyl-activated M-450 beads once with sterile distilled water and resuspend at a final concentration of 30 mg/ml.
2. Dilute anti-RNA : DNA in 0.2 M borate solution, pH 9.5, to a final concentration of 150 μg/ml.
3. Add equal volumes of activated beads and antibody solution to a 15-ml screw-cap polypropylene tube. Cap and seal the tube with parafilm.
4. Couple antibody to beads for 24 hr at room temperature with continuous mixing (end-over-end rotation).
5. Collect the beads with a magnetic concentrator (Dynal MPC® or equivalent) and discard the supernatant.
6. Wash antibody-coupled beads with 5 ml 0.1 M PBS for 10 min using end-over-end rotation. Collect the beads with magnetic concentration and discard the supernatant.
7. Add 5 ml 1 M ethanolamine-HCl, pH 9.5, 0.1% Tween20®. Incubate for 2 hr at room temperature with continuous mixing to block free tosyl groups. *Note:* Add Tween20® to the ethanolamine-HCl solution immediately before use.
8. Wash with 5 ml TBS with Tween20® for 12 hr at room temperature (end-over-end rotation).
9. Wash with 5 ml TBS (without Tween20®) for 2 hr at room temperature.
10. Collect the antibody-coated beads with a magnetic concentrator. Discard the supernatant and resuspend in TBS at a final concentration of 30 mg/ml. Store at 4°C.

NOTE: *Antibody-coated beads are stable for 6 mo at 4°C. Before each use, wash the beads with TBS, 0.1% Tween20® for 5 min at room temperature. Resuspend in TBS with Tween20® for a final concentration of 30 mg/ml.*

   b. Preparation of Alkaline Phosphatase-Coupled Oligonucleotide Probe   A 5' Aminoalkyl-oligonucleotide specific for RNA products from LAT amplification of HPV16 (5'-NH$_2$-GTATATTGTGGAGAC-CCTGGAACTATAGG-3') was synthesized by standard protocols using Aminolink-2 phosphoramidite (Applied Biosystems, Inc., Foster City, CA). Direct enzyme-labeled probe was prepared using the ACES Labeling System (GIBCO/BRL) according to the manufacturer's recommendations. Briefly, 20 nmol 5'-aminoalkyl oligonucleotide was activated by thiolation and purified by gel filtration. Maleimido-derivatized alkaline phosphatase was reacted with an excess of thiolated oligonucleotide to form a covalent adduct. Alkaline phosphatase-labeled oligonucleotide was purified from unconjugated oligonucleotide by gel filtration. Enzyme-linked probe is stable for up to 6 mo when stored at 4°C in 50 mM Tris-HCl, pH 7.5, 0.1% BSA, 0.2% sodium azide. For prolonged storage, probe should be aliquoted and stored at −20°C. Repeated freeze-thawing should be avoided since it may impair enzyme activity.

c. Solution Hybridization and Immunocapture

Materials

45% polyacrylic acid (sodium salt) (Polysciences, Inc., Warrington, PA)
RNase A (Sigma)
2× Hybridization solution: 6× SSC, 0.5% Tween20®, 4% sodium polyacrylate, 200 μg/ml sheared denatured herring sperm DNA
TNT buffer: 0.1 M Tris-HCl, pH 7.5, 0.6 M NaCl, 0.25% Tween20®
TE buffer: 10 mM Tris-HCl, pH 7.5, 1 mM EDTA
Enzyme-linked oligonucleotide probe: ~1.5 μM Tris-HCl, pH 7.5, 0.1% BSA, 0.2% sodium azide
Lumi-phos™ 530 reagent (GIBCO/BRL)
Control RNA analyte (prepared by *in vitro* transcription of cloned target sequence)

Protocol

1. Dilute alkaline phosphatase-lnked oligonucleotide probe in TE buffer to a final concentration of 25nM.
2. Prepare a master mix hybridization cocktail by combining 25 μl 2× hybridization solution and 20 μl 25 nM oligonucleotide probe/TE buffer solution for each hybridization. Prepare sufficient master mix to accommodate each LAT reaction as well as dilutions of an RNA control analyte (used to generate a standard curve).
3. Dispense 45-μl aliquots into 1.5-ml microfuge tubes.
4. Add 5 μl RNA sample (LAT reaction aliquot or control RNA) to each hybridization.

5. Incubate 30 min at 42°C.
6. Add 100 μl TNT containing 20 μg/ml RNase A and 5 μl antibody-coated beads to each hybridization reaction.
7. Mix; then incubate 30 min at room temperature.
8. Wash beads 6 times with 0.5 ml TNT. (a) Place tube in magnetic rack. (b) Allow beads to migrate to tube walls. (c) With tube in magnetic rack, remove supernatant by aspiration or decanting.

V. CHARACTERITICS OF THE LAT AMPLIFICATION SYSTEM

As described in Section II, the LAT amplification is an isothermal system designed for exponential amplification of nucleic acids. The system is capable of amplifying RNA or DNA. LAT amplification produces both RNA and single-stranded DNA, but RNA products are produced in larger quantities than DNA products. An overall amplification of 10^7- to 10^8-fold is usually achieved within 3 hr.

A model system based on the sequences of HPV16 has been used to characterize LAT amplification and its products (Fig. 4). The model system consisted of a 660-base *in vitro* transcript produced from a cloned fragment

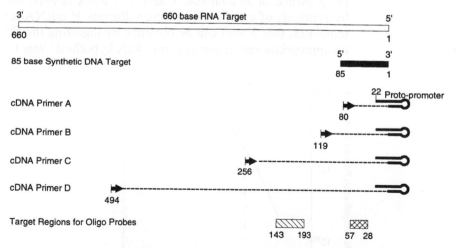

Figure 4 Schematic of HPV16 model system used for LAT amplification. A model analyte sequence was constructed by subcloning the *Alu–Bam*HI fragment of HPV16 (nucleotide positions 5505–6151) into a T7 transcription plasmid. The resulting 660-base *in vitro* transcript was used as the target for characterization of the LAT amplification process. Annealing positions, relative to the 660-base RNA, of the various primers and oligonucleotide probes used are indicated below each construct.

of the HPV16 genome. The system was designed so the 5' end region of this RNA template would be amplified. A series of different cDNA primers was used to generate cDNAs with different lengths that were subsequently amplified. The 3' ends of these cDNAs were complementary to the proto-promoter, which allowed amplification of all cDNAs using the same reagents.

A typical amplification reaction results in production of both RNA and DNA products. Quantitative hybridization of probes has shown that 5- to 10-fold more RNA is produced than DNA. The full-length transcript is the major RNA product, RNAse H degradation products make up the smaller species. The DNA products are of two species: (1) the full-length cDNA corresponding to the reverse transcript of the RNA produced in the reaction and (2) a slightly larger species produced as a result of ligation between cDNA and the proto-promoter.

We studied the kinetics of LAT amplification using a 119-base region of the *in vitro* RNA transcript just described. Amplification was shown to be biphasic. An exponential amplification occurred within 2–3 hr that produced an accumulation of approximately 10^{11} molecules of RNA. The subsequent decline in reaction rate is probably due to exhaustion of the reaction components. Amplification kinetics for an LAT reaction containing 1000 copies of RNA analyte are shown in Fig. 5. In other experiments, we have shown that the kinetics of amplification vary with the length of the product to be amplified; shorter fragments are amplified more rapidly.

As shown in Fig. 1, amplification of the DNA templates by LAT requires the presence of an available 3' end. For RNA targets, this is accomplished by synthesis of a cDNA exploiting a discrete 5' mRNA terminus. For DNA templates, this 3' end can be obtained by digesting the DNA template with an appropriate restriction enzyme. This hypothesis was tested using a clone

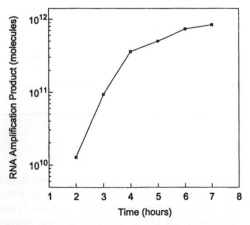

Figure 5 Time course of LAT amplification.

containing the genome of HPV16. The plasmid was digested with *Alu*1, which generated DNA restriction fragments with a 3' end region complementary to the proto-promoter used in LAT amplification. This allowed amplification of the DNA templates using the same reagents used for amplification of RNA. The results of one such amplification are shown in Table 1; different amounts of HPV16 DNA were used as the target. Amplification of DNA templates was demonstrated to be equally efficient to that of RNA templates. We also demonstrated that DNA amplification was not only dependent on the generation of the 3' end, but also on denaturation of DNA prior to starting the LAT reaction; double-stranded DNA templates fail to amplify. This feature provides a very selective procedure for specific amplification of RNA, even in the presence of DNA. Amplification of RNA may be important in some diagnostic procedures where expression of certain genes is more important than their mere presence. This may prove to be the case with HPV16, since the expression of transcripts derived from the E6 region has been implicated in tumorigenicity (Johnson *et al.*, 1990; Falcinelli *et al.*, 1993; Griep *et al.*, 1993; Lambert *et al.*, 1993).

VI. CHEMILUMINESCENCE DETECTION OF AMPLIFIED PRODUCTS

A nonradioactive chemiluminescence assay was developed for rapid detection of the amplified RNA from LAT. The assay used a synthetic oligonucleotide probe labeled with alkaline phosphatase. This probe is complementary to the RNA product of amplification and is used to detect and quantify the

TABLE 1
LAT Amplification of HPV16 DNA

Input target DNA prior to amplification[a]	Number of molecules	Hybridization with ^{32}P-labeled probe (counts)
No target control	0	230
0.1	10	712
1	100	1188
10	1,000	5782
100	10,000	35035
1000	100,000	61780

[a] Amplification was performed using HPV16 DNA digested with *Alu*I as target DNA. The RNA product of the amplification was immobilized on membranes and hybridized to a strand-specific internal oligonucleotide probe labeled with ^{32}P. The hybridization signal was quantified using a Beta-Gen solid phase counter.

levels of RNA generated. The products of amplification reactions were mixed and hybridized in solution to the oligonucleotide probe. The RNA : DNA hybrids formed as a result of hybridization were then captured using magnetic beads. The activity of the alkaline phosphatase associated with the hybrids was detected using the chemiluminescent substrate Lumiphos 530 [4-methoxy-4-(3-phosphate-phenol)spiro(1,2-dioxetane-3,2'-adamantane] (Schaap *et al.*, 1989). Use of a known amount of RNA target and the subsequent generation of a standard curve allowed the quantification of the RNA products generated by amplification. Figure 6 shows a standard curve and the results of a typical LAT amplification reaction detected by this method. A minimum of $1 \times : 10^8$ copies of target molecules can be detected using this hybridization procedure and a single oligonucleotide probe. The assay is linear up to 5×10^{10} molecules. These limits provide a useful range for the quantification of LAT amplification products. The results of an LAT amplification using various amounts of target RNA are shown in Table 2. LAT amplification was capable of amplifying as few as 10 copies of RNA target. The extend of amplification

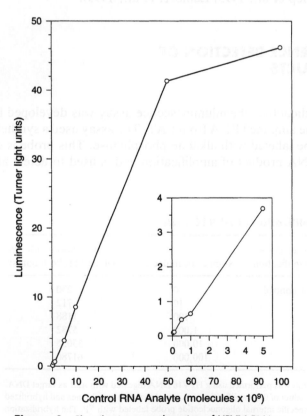

Figure 6   Chemiluminescent detection of HPV16 RNA.

TABLE 2
LAT Amplification of RNA and Nonradioactive Detection of Amplified Products Using Immunocapture Method

Input target RNA (molecules)	Chemiluminescent signal (arbitrary light units)	Amplification products (molecules)	Amplification factor
0	0.061	0	0
10	0.195	2×10^8	2×10^7
100	0.566	9×10^8	9×10^8
1,000	1.095	2×10^9	2×10^6
10,000	6.173	1×10^{10}	1×10^6
10,000 (no ligase control)	0.061	0	0

was relatively linear and the amount of amplified product correlated with the amount of input target.

In other experiments, we compared the accuracy of RNA quantification with dot-blot hybridization using radioactivity labeled oligonucleotide probes. These studies have shown equal sensitivity for both systems, as well as excellent agreement between results.

VII. AMPLIFICATION AND DETECTION OF HPV IN CLINICAL SPECIMENS

LAT amplification was applied to detection of HPV16 RNA in clinical specimens. The RNA molecule targeted for these assays was the major E6/E7 spliced transcript, E6* 1 (Smotkin and Wettstein, 1986; Smotkin et al., 1989; Doorbar et al., 1990; Nasseri et al., 1991), which is believed to be the primary molecular mechanism for expression of the oncogenic protein E7 in HPV16.

LAT amplification and RT-PCR were performed using total RNA isolated from 10 human cervical carcinoma biopsy samples; SiHa (HTB 35) cells, an HPV16-transformed cell line known to contain E6* mRNA (Smotkin and Wettstein, 1986, Smotkin et al., 1989), and C-33A (HTB 31) cells, a cell line derived from cervical cancer biopsies and devoid of HPV, as a negative control (Smotkin and Wettstein, 1986). Constructs used for LAT and RT-PCR amplification of HPV16 E6* mRNA are depicted schematically in Fig. 7. The 5′ terminus of the E6/E7 mRNAs initiated at the P97 promoter was targeted for amplification by both methods using oligonucleotide constructs that targeted identical sequence regions. An oligonucleotide probe designed to span the E6* I splice donor/acceptor sites was used for specific detection of the E6* I amplification product.

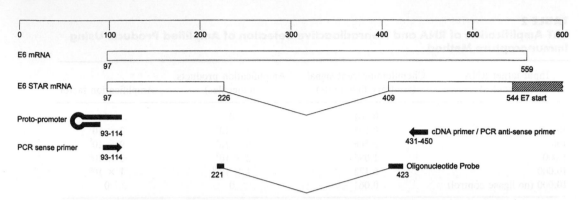

Figure 7 Schematic diagram of constructs used for amplification of the major E6/E7 spliced transcript, E6* I, from HPV16. Location of the E6 and E7 open reading frames (Seedorf *et al.,* 1985) and schematic of the E6–E7 collinear transcript, E6* I (Smotkin and Wettstein, 1986; Smotkin *et al.,* 1989; Doorbar *et al.,* 1990; Nasseri *et al.,* 1991) are depicted below the relative nucleotide positions for the HPV16 genome, indicated in base pairs on the top scale. Relative annealing positions for oligonucleotide constructs used for LAT and RT-PCR amplification of RNA transcripts derived from the HPV16 P97 promoter are shown below the schematic for the E6* I transcript.

First strand synthesis was primed using primer B (TGCTTTTCTTCAG-GACACAG), designed to be complementary to a region immediately 3' to the HPV16 E6/E7 major splice acceptor site (nucleotides 431–450). Approximately one-fourth of each cDNA synthesis reaction was used for PCR amplification (100-μl volumes) in conjunction with primer A (GAGAACTGCAAT-GTTTCAGG), which was specific for the 5' terminus of E6/E7 transcripts (homologous to the region from nucleotide 94 to 113 of the HPV16 genome).

TABLE 3
Detection of HPV16 E6* mRNA in Cervical Biopsy Samples by RT-PCR and LAT Amplification

Cervical biopsy sample	LAT results	RT-PCR results
1	−	−
2	−	−
3	+	++
4	++	++
5	+++	+++
6	+	+
7	−	−
8	−	−
9	+++	++
10	−	−

PCR amplification was carried through 25 temperature cycles according to standard protocols (Kawasaki, 1990). The 176-bp E6* I amplification product was detected by Southern blot using a ^{32}P-labeled oligonucleotide probe (TGACAGTTAATACACCTCACG) that spanned the splice donor and acceptor sites of the E6* I transcript.

LAT amplification reactions were performed with 5 μl total RNA sample in a final volume of 25 μl, as described in this chapter, using primer B as the cDNA primer and a proto-promoter containing a 3' single-stranded overhang homologous to the 5' terminus of HPV16 E6/E7 transcripts (nucleotides 94 to 113). Reactions included 20 U RNasin® to suppress endogenous RNase activities that may have persisted in the RNA isolation procedure. Reactions were incubated for 4 hr at 37°C, and E6* I amplification product was detected by glyoxal dot blot using conditions identical to those for PCR product Southern blot. The results of this comparison are shown in Table 3. Although relative signal intensities between LAT and PCR amplification product for a given sample sometimes differ, an excellent correlation is observed between the two methods for detection of E6* I amplification products.

VIII. SUMMARY

This chapter details the properties of a new method for amplification of nucleic acids. The method is based on the concerted action of four enzymes—T7 RNA polymerase, T4 DNA ligase, reverse transcriptase, and RNase H—in a homogeneous reaction. The LAT reaction is performed under isothermal conditions without thermocycling. The reaction is capable of using RNA or DNA as the starting template and produces both single-stranded RNA and DNA amplification products. The method has been demonstrated to amplify as few as 10 molecules of template RNA or DNA, and results in 10^7- to 18^8-fold amplification. The products of amplification are readily detectable by hybridization.

We have also described a solution hybridization method for detection of the amplified products. This solution-based immunocapture method utilizes DNA probes directly labeled with the enzyme alkaline phosphatase. The probe is hybridized in solution to the RNA produced in amplification; the resulting RNA : DNA hybrids are specifically captured using an immobilized antibody that is specific for RNA : DNA hybrids. The enzyme activity of the immobilized probe is then measured using a chemiluminescent substrate for alkaline phosphatase. This hybridization method provides a rapid and simple approach for nonradioactive detection of RNA samples.

The LAT amplification and immunocapture solution hybridization methodologies were combined for rapid amplification and detection of nucleic acid analytes. The results are quantitative for both amplification and nonradioactive detection. Application of LAT amplification to detection of HPV was demonstrated in a model system and in clinical specimens. The results have a strong correlation to those obtained by RT-PCR and demonstrate the validity of LAT amplification for the detection of biologically relevant nucleic acid sequences in clinical specimens.

ACKNOWLEDGMENTS

We wish to recognize and thank George Buchman and Charles Thornton for their many contributions to the development of the LAT method, Atilla Lorincz and Mindy Goldsborough for providing the clinical biopsy material and sharing their RT-PCR data on detection of E6* mRNA, and Gary Temple for his support and thoughtful discussions throughout this project.

REFERENCES

Barany, F. (1991a). Genetic disease detection and DNA amplification using cloned thermostable ligase. *Proc. Natl. Acad. Sci. USA* **88,** 189–193.

Barany, F. (1991b). The ligase chain reaction in a PCR world. *PCR Meth. Appl.* **1,** 5–16.

Barringer, K. J., Orgel, L., Wahl, G., and Gingeras, T. R. (1990). Blunt end and single-strand ligations by *Escherchia coli* ligase: Influence on an in vitro amplification scheme. *Gene* **89,** 117–122.

Boguslawski, S. J., Smith, D. E., Michalak, M. A., Mickelson, K. E., Yehle, C. O., Patterson, W. L., and Carrico, R. J. (1986). Characterization of monoclonal antibody to DNA.RNA and its application to immunodetection of hybrids. *J. Immunol. Meth.* **89,** 123–130.

Doorbar, J., Parton, A., Hartley, K., Banks, L., Crook, T., Stanley, M., and Crawford, L. (1990). Detection of novel splicing patterns in a HPV16-containing keratinocyte cell line. *Virology* **178,** 254–262.

Falcinelli, C., Van Belkum, A., Schrauen, L., Seldenrijk, K., and Quint, W. G. V. (1993). Absence of human papillomavirus type 16 E6 transcripts in HPV16-infected, cytologically normal cervical scrapings. *J. Med. Virol.* **40,** 261–265.

Griep, A. E., Herber, R., Jeon, S., Lohse, J. K., Dubielzig, R. R., and Lambert, P. F. (1993). Tumorigenicity by human papillomavirus type 16 E6 and E7 in transgenic mice correlates with alterations in epithelial cell growth and differentiation. *J. Virol.* **67,** 1373–1384.

Guatelli, J. C., Whitfield, K. M., Kwoh, D. Y., Barringer, K. J., Richman, D. D., and Gingeras, T. R. (1990). Isothermal, in vitro amplification of nucleic acids by a multienzyme reaction modeled after retroviral replication. *Proc. Natl. Acad. Sci. USA* **87,** 1874–1878.

Johnson, M. A., Blomfield, P. I., Bevan, I. S., Woodman, C. B., and Young, L. S. (1990). Analysis of human papillomavirus type 16 E6-E7 transcription in cervical carcinomas and normal cervical epithelium using the polymerase chain reaction. *J. Gen. Virol.* **71,** 1473–1479.

Kawasaki, E. S. (1990) Amplification of RNA. *In* "PCR Protocols: A Guide to Methods and Applications" (M. A. Innis, D. H. Gelfand, J. J. Sninsky, and T. J. White, eds.). Academic Press, San Diego.

Kwoh, D. Y., Davis, G. R., Whitfield, K. M., Chappell, H. L., DiMichele, L. J., and Gingeras, T. R. (1989). Transcription-based amplification system and detection of amplified human immunodeficiency virus type 1 with a bead-based sandwich hybridization format. *Proc. Natl. Acad. Sci. USA* **86**, 1173–1177.

Lambert, P. F., Pan, H., Pitot, H.C., Liem, A., Jackson, M., and Griep, A. E. (1993). Epidermal cancer associated with expression of human papillomavirus type 16 E6 and E7 oncogenes in the skin of transgenic mice. *Proc. Natl. Acad. Sci. USA* **90**, 5583–5587.

Lizardi, P. M., Guerra, C. E., Lomeli, H., Tussie-Luna, I., and Kramer, F. R. (1988). Exponential amplification of recombinant RNA hybridization probes. *Biotechnology* **6**, 1197–1202.

Mackey, J., Guan, N., and Rashtchian, A. (1992). Direct coupling of alkaline phosphatase to oligonucleotide probes for increased sensitivity and simplified non-radioactive detection. *Focus* **14**, 112–116.

Mulis, K. B., and Faloona, F. A. (1987). Specific synthesis of DNA in vitro via polymerase-catalyzed chain reaction. *Meth. Enzymol.* **155**, 335–350.

Nasseri, M., Gage, J. R., Lorincz, A., and Wettstein, F. O. (1991). Human papillomavirus type 16 immortalized cervical keratinocytes contain transcripts encoding E6, E7, and E2 initiated at the P97 promoter and express high levels of E7. *Virology* **184**, 131–140.

Rashtchian A., Eldrege, J., Ottaviani, M., Abbott, M., Mock, G., Lovem, D., Klinger, J., and Parsons, G. (1987). Immunological capture of nucleic acid hybrids and application to non-radioactive DNA probe assays. *Clin. Chem.* **33**, 1526–1530.

Rashtchian, A., Schuster, D., Buchman, G., Berninger, M., and Temple, G. F. (1990). A non-radioactive sandwich hybridization assay using paramagnetic microbeads and unlabeled RNA probes. Sixth International Congress on Rapid Methods and Automation in Microbiology and Immunology. Helsinki, Finland.

Sambrook, J., Fritsch, E. F., and Maniatis, T. (1989). "Molecular Cloning: A Laboratory Manual," 2d Ed. Cold Spring Harbor Laboratory Press, Cold Spring Harbor, NY.

Saiki, R. K., Scharf, S., Faloona, F. A., Mullis, K. B., Horn, G. T., Erlich, H. A., and Arnheim, N. (1985). Enzymatic amplification of beta- globin genomic sequences and restriction site analysis for diagnosis of sickle cell anemia. *Science* **230**, 1350–1354.

Schaap, A. P., and Akhaven-Tafti, H. (1990). New chemiluminescent 1,2-dioxetane derivatives. WO Patent 9007511.

Schaap, A. P., Akhaven, H., and Romano, L. J. (1989). Chemiluminescent substrates for alkaline phosphatase: Applications to ultra-sensitive enzyme-linked immunoassays and DNA probes. *Clin. Chem.* **35**, 1863–1864.

Seedorf, K., Kraemmer, G., Duerst, M., Suhai, S., and Roewekamp, W. G. (1985). Human papillomavirus type 16 DNA sequence. *Virology* **65**, 181–185.

Smotkin, D., and Wettstein, F. O. (1986). Transcription of human papillomavirus type 16 early genes in a cervical cancer and a cancer-derived cell line and identification of the E7 protein. *Proc. Natl. Acad. Sci. USA* **83**, 4680–4684.

Smotkin, D., Prokoph, H., and Wettstein, F. O. (1989). Oncogenic and nononcogenic human genital papillomaviruses generate the E7 mRNA by different mechanisms. *J. Virol.* **63**, 1441–1447.

Stollar, B. D., and Rashtchian, A. (1987). Immunochemical approaches to gene probe assays. *Anal. Biochem.* **161**, 387–394.

Thomas, P. (1983). Hybridization of denatured RNA transferred or dotted to nitrocellulose paper. *Meth. Enzymol.* **100**, 255–266.

Urdea, M. S., Running, J. A., Horn, T., Clyne, J., Ku, L., and Warner, B. D. (1987). A novel method for the rapid detection of specific nucleotide sequences in crude biologic samples without blotting or radioactivity: Application to the analysis of hepatitis B virus in human serum. *Gene* **61**, 253–264.

Urdea, M. S., Warner, B., Running, J. A., Kolberg, J. A., Clyne, J. M., Sanchez-Pescador, R., and Horn, T. (1990). Nucleic acid multimer for hybridization assays. WO Patent 8903891.

Index

Printed and bound by CPI Group (UK) Ltd, Croydon, CR0 4YY
03/10/2024
01040325-0012